22

TESI

THESES

tesi di perfezionamento in Matematica sostenuta il 9 settembre 2015

COMMISSIONE GIUDICATRICE
Fulvio Ricci, Presidente
Luigi Ambrosio
Stefano Bianchini
Alessio Figalli
Fabrizio Lillo

Maria Colombo
Institute for Theoretical Studies
ETH Zürich
Clausiusstrasse 47
CH-8092 Zürich
Switzerland

Flows of Non-smooth Vector Fields and Degenerate Elliptic Equations

Maria Colombo

Flows of Non-smooth Vector Fields and Degenerate Elliptic Equations

with Applications to the Vlasov-Poisson and Semigeostrophic Systems

EDIZIONI
DELLA
NORMALE

ISBN: 978-88-7642-606-3
e-ISBN: 978-88-7642-607-0

To my father

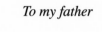

Contents

Introduction

In the last centuries, partial differential equations have been used to model many physical problems: the Navier-Stokes and Euler equations in fluid dynamics, the Boltzmann and Vlasov equations in statistical mechanics, the Schrodinger equation in quantum physics, and many other PDEs concerning, for instance, material science or meteorology. The richness of mathematical structure in these equations is always reason of surprise.

As a motivating example, we introduce the Vlasov-Poisson system. It describes the evolution of particles under their self-consistent electric or gravitational field. It is the continuous counterpart of the N-body problem, which describes the motion of N mass points under the influence of their mutual attraction governed by Newton's law of gravity. The N-body problem has applications in astronomy and plasma physics; for instance, it describes the solar system or the motion of galaxies. In the gravitational models, each element of unit mass with position x and velocity v obeys the equation

$$\begin{cases} \dot{x} = v \\ \dot{v} = -\partial_x V_t(x), \end{cases}$$

where $V_t(x)$ is the gravitational potential depending on time t and position x. Collisions between different masses are considered as an extremely unlikely event and are therefore neglected. Since the number of involved elements in a galaxy can be of order 10^{10}–10^{12}, the galaxy is described in the Vlasov-Poisson system in a statistical way rather than keeping track of each mass point. For this reason, we introduce the quantity $f_t(x, v)$, which describes the distribution of particles with given position x and velocity v at time t. The density f_t solves a first order conservation law on phase space

$$\partial_t f_t + v \cdot \nabla_x f_t - \nabla_x V_t \cdot \nabla_v f_t = 0 \qquad \text{in } (0, \infty) \times \mathbb{R}^d \times \mathbb{R}^d, \tag{1}$$

whose characteristics are the equations of motion of a single test particle.

In turn, the gravitational potential V_t is obtained from the physical density

$$\rho_t(x) = \int_{\mathbb{R}^d} f_t(x, v)\, dv \qquad \text{in } (0, \infty) \times \mathbb{R}^d \tag{2}$$

by solving the Poisson equation

$$-\Delta V_t = \sigma \rho_t \qquad \text{in } \mathbb{R}^d, \qquad \lim_{|x|\to\infty} V_t(x) = 0. \tag{3}$$

Here, $\sigma \in \{\pm 1\}$ distinguishes the gravitational (attractive) and the electrostatic (repulsive) problem.

The nonlinear system of partial differential equations (1), (2), and (3) has a transport structure: indeed it can be rewritten as

$$\partial_t f_t + \boldsymbol{b}_t \cdot \nabla_{x,v} f_t = 0, \tag{4}$$

where the vector field $\boldsymbol{b}_t(x, v) = (v, E_t(x)) : \mathbb{R}^{2d} \to \mathbb{R}^{2d}$ is coupled to f_t via the relation $E_t = \sigma c_d \rho_t * (x/|x|^d)$ and c_d is a dimensional constant. Indeed, the force field E_t is obtained as $-\nabla_x V_t$ and V_t can be written as the convolution of ρ_t with a singular kernel by solving (3). Since the vector field is divergence free, it can be also rewritten as a continuity equation

$$\partial_t f_t + \nabla_{x,v} \cdot \big(\boldsymbol{b}_t f_t\big) = 0. \tag{5}$$

Solutions of (5), when considering a fixed vector field $\boldsymbol{b}$, turn out to be obtained by flowing the initial datum f_0 along the characteristics of the vector field $\boldsymbol{b}$. The deep connection between the transport/continuity equation (Eulerian point of view) and the notion of flow (Lagrangian point of view) is one of the most fascinating aspects of this theory. It is the basis of many results regarding the continuity equation and the flows even in a non-smooth setting, starting from the fundamental papers of DiPerna and Lions [78] and Ambrosio [5].

Many questions regarding the Vlasov-Poisson equation are nowadays little understood and some of them are deeply related to the dual, Lagrangian and Eulerian, nature of the equation. One of the main open problems in statistical mechanics is, for instance, the rigorous derivation of the equation. It amounts in proving that, when a sequence of configurations with finitely many particles approximates a continuous initial distribution of particles, the solutions of the approximate systems converge to the solution of the Vlasov-Poisson equation. As well as the Boltzmann equation, the Vlasov equation has been rigorously derived only under restrictive smallness assumptions on the time of observation, the total mass of matter, or the distance of the distribution function to equilibrium. Moreover, all derivations of the Vlasov equation assume that the interaction at small scales is either smooth or not too singular.

As we saw above, the Vlasov-Poisson equation can be seen as a transport equation in the phase space, coupled with a PDE which determines the gravitational field in terms of the distribution of particles. The main scope of our thesis is a further step in understanding some aspects of the interaction between transport equations and PDEs. More precisely, we consider the following problems, which regard the DiPerna-Lions theory and the regularity of degenerate elliptic equations, together with the analysis of the interaction between these points of view in models coming from mathematical physics.

- The Di Perna-Lions and Ambrosio theory for flows of non-smooth vector fields: We develop a local version of the DiPerna-Lions theories for ODE's, providing a complete analogy with the Cauchy-Lipschitz theory. More precisely, we prove existence and uniqueness of a maximal regular flow for non-smooth vector fields using only local regularity and summability assumptions on the vector field, in analogy with the classical theory, which uses only local regularity assumptions.
- The quantitative estimates for the ODE: They constitute a different approach to the DiPerna-Lions theory, this time relying on a priori estimates on solutions of the ODE rather than on the connection between Lagrangian and Eulerian structure. We apply these estimates in the Eulerian setting to obtain renormalized solutions of the continuity equation with a linear source term; this equation is not easily covered by the methods of DiPerna and Lions.
- The regularity of very degenerate elliptic equations: This problem comes from a model in traffic dynamic and it is a variant of the optimal transport problem, which takes into account congestion effects in the transportation. It leads to different equivalent formulations; they employ in one case some concepts related to flows of vector fields, in another case the minimization of a variational integral, where the convexity of the integrand degenerates on a full convex set. We are interested in the regularity of solutions.
- The Vlasov-Poisson system: This equation, introduced above, couples the transport structure in the phase space (namely, the space of positions and velocities of particles) with the Laplace equation, which describes the force field. The existence of classical solutions is limited to dimensions $d \leq 3$ under strong assumptions on the initial data, while weak solutions are known to exist under milder conditions. However, in the setting of weak solutions it is unclear whether the Eulerian description provided by the equation physically corresponds to a Lagrangian evolution of the particles. Through general tools concerning the Lagrangian structure of transport equations with non-smooth vector

fields, we show that weak solutions of Vlasov-Poisson are Lagrangian and we obtain global existence of weak solutions under minimal assumptions on the initial data.

- The semigeostrophic system: It was introduced in meteorology to describe atmospheric/ocean flows. After a suitable change of variable, it has a dual version which couples a transport equation with a nonlinear elliptic PDE, namely the Monge-Ampère equation. We study the problem of existence of distributional solutions to the original system.

In the following, we give a quick overview on all these problems and an outline of the thesis' content, postponing a more detailed mathematical and bibliographical description of the single problems to the beginning of each chapter. The results in this thesis are the final outcome of several collaborations developed during the PhD studies and have been presented in a series of papers, already published or submitted.

Flows of non-smooth vector fields. Given a vector field $\boldsymbol{b} : (0, T) \times \mathbb{R}^d \to \mathbb{R}^d$ we consider the ordinary differential equation

$$\begin{cases} \partial_t \boldsymbol{X}(t, x) = \boldsymbol{b}(t, \boldsymbol{X}(t, x)) & \forall t \in (0, T) \\ \boldsymbol{X}(0, x) = x, \end{cases} \tag{6}$$

which is strictly related (via the method of characteristics) to the continuity equation

$$\begin{cases} \partial_t u + \nabla \cdot (\boldsymbol{b} u) = 0 & \text{in } (0, T) \times \mathbb{R}^d \\ u_0 = \bar{u} \quad \text{given}, \end{cases} \tag{7}$$

where $u : (0, T) \times \mathbb{R}^d \to \mathbb{R}$. If the vector field $\boldsymbol{b}$ is Lipschitz with respect to space uniformly in time, the Cauchy-Lipschitz theory and classical PDE arguments provide existence and uniqueness of a solution to (6) and (7). In their fundamental papers, exploiting the connection between (6) and (7), Di Perna and Lions [81] and Ambrosio [5] proved existence and uniqueness of a so called *regular lagrangian flow*, namely a certain solution to (6), even in the case of Sobolev and BV vector fields. However, the Cauchy-Lipschitz theory is not only pointwise but also purely local, meaning that existence and uniqueness for small intervals of time depend *only* on local regularity properties of the vector fields $\boldsymbol{b}_t(x)$. On the other hand, not only the DiPerna-Lions theory is an almost everywhere theory (and this really seems to be unavoidable) but also the existence results for the flow depend on *global* in space growth estimates on $|\boldsymbol{b}|$, the most typical one being

$$\frac{|\boldsymbol{b}_t(x)|}{1 + |x|} \in L^1\big((0, T); L^1(\mathbb{R}^d)\big) + L^1\big((0, T); L^\infty(\mathbb{R}^d)\big),$$

which prevent the trajectories of the flow from blowing up in finite time. In Chapter 2, based on a joint work [10] with Ambrosio and Figalli, under purely local and natural assumptions on the vector field, we prove existence of a unique *maximal regular flow* $\boldsymbol{X}(t,x)$, defined up to a maximal time $T_{\boldsymbol{X}}(x)$ which is positive $\mathscr{L}^d$-a.e. in $\mathbb{R}^d$, with

$$\limsup_{t\to T_{\boldsymbol{X}}(x)} |\boldsymbol{X}(t,x)| = \infty \qquad \text{for } \mathscr{L}^d\text{-a.e. } x \in \mathbb{R}^d \text{ such that } T_{\boldsymbol{X}}(x) < T.$$

We then study, in Chapter 3, the natural semigroup and stability properties of this object; finally we analyze the blow-up of the maximal regular flow $\boldsymbol{X}(\cdot,x)$ at the maximal time $T_{\boldsymbol{X}}(x)$. Surprisingly enough, indeed, the proper blow up of trajectories, namely

$$\lim_{t\to T_{\boldsymbol{X}}(x)} |\boldsymbol{X}(t,x)| = \infty \qquad \text{for } \mathscr{L}^d\text{-a.e. } x \in \mathbb{R}^d \text{ such that } T_{\boldsymbol{X}}(x) < T$$

happens only under a global bound on the divergence of $\boldsymbol{b}$, whereas there are counterexamples if only local bounds are assumed.

Quantitative estimates for the continuity equation. Another aspect of the theory of regular lagrangian flows are the so called "quantitative estimates", developed in the Lagrangian case (namely, for solutions of (6)) by Ambrosio, Lecumberry, and Maniglia [22], Crippa and De Lellis [67]. This theory allows to prove uniqueness and stability of flows, in an independent way with respect to the analysis of the solutions to the continuity equation. More precisely, the fundamental a-priori estimate is the following: given a small parameter $\delta > 0$, if $\boldsymbol{X}_1$ and $\boldsymbol{X}_2$ are the flows of two vector fields $\boldsymbol{b}_1$ and $\boldsymbol{b}_2$ we consider the functional

$$\Phi_\delta(t) := \int_{\mathbb{R}^d} \log\left(1 + \frac{|\boldsymbol{X}_1(t,x) - \boldsymbol{X}_2(t,x)|^2}{\delta}\right) dx,$$

whose time derivative is bounded independently on δ under suitable assumptions on the vector fields. A similar functional can be employed also in the Eulerian setting to estimate the distance of two solutions of the continuity equation (7). This approach is followed in joint works with Crippa and Spirito [56, 57], presented in Chapter 5, where we consider (7) with a non-smooth vector field and a linear source term, called *damping term* (although its sign may be either positive or negative), namely a right-hand side of the form cu with $c : (0,T) \times \mathbb{R}^d \to \mathbb{R}$. In their fundamental paper [81], DiPerna and Lions proved that, when c is bounded in space and time, the equation is well posed in the class of distributional solutions and the solution is transported by suitable characteristics of the vector field. Thanks to the quantitative estimates for the solution of the continuity equation, existence and uniqueness of solutions holds under more general assumptions on the data, for instance, assuming only integrability of the damping term.

Regularity of degenerate elliptic PDEs. In Chapter 6 and 7 we study the gradient regularity of local minimizers of the functional

$$\int_\Omega \mathcal{F}(\nabla u) + fu, \tag{8}$$

where we are given a bounded open subset Ω of $\mathbb{R}^d$, a convex function $\mathcal{F} : \mathbb{R}^d \to \mathbb{R}$ which exhibits a large degeneracy set, and an integrable function $f : \Omega \to \mathbb{R}$. Our model function is

$$\mathcal{F}(v) = \frac{1}{p}(|v| - 1)_+^p \qquad \forall\, v \in \mathbb{R}^d, \tag{9}$$

so that the degeneracy set is the entire unit ball. This problem comes from a model by Beckmann [30], where, given an urban area where people move from home to work, the optimal traffic flow σ solves the minimum problem

$$\min\left\{\int_\Omega \mathcal{F}^*(\sigma) : \sigma \in L^{p'}(\Omega),\ \nabla\cdot\sigma = f,\ \sigma\cdot\nu_{\partial\Omega} = 0\right\}. \tag{10}$$

Here, $\mathcal{F}^*$ denotes the convex conjugate of the function $\mathcal{F}$; by the choice of $\mathcal{F}$ in (9), we have that

$$\mathcal{F}^*(\sigma) = |\sigma| + \frac{1}{p'}|\sigma|^{p'} \qquad \forall \sigma \in \mathbb{R}^d$$

where p' satisfies $1/p + 1/p' = 1$. The function $\mathcal{F}$ is chosen so that its convex conjugate $\mathcal{F}^*$ has more than linear growth at infinity (so to avoid "congestion") and satisfies $\liminf_{w\to 0} |\nabla\mathcal{F}^*(w)| > 0$ (which means that moving in an empty street has a nonzero cost).

Problem 10 is equivalent to the problem of minimizing the energy (8) with the particular choice of $\mathcal{F}$ given by (9). The unique optimal minimizer $\bar\sigma$ in problem (10) turns out to be exactly $\nabla\mathcal{F}(\nabla u)$, where $\mathcal{F}$ is defined by (9). The continuity of $\bar\sigma$ is meaningful in terms of traffic models, as shown in [49]. Indeed, one can consider measures on the space of possible paths and select an optimal measure which satisfies a Wardrop equilibrium principle: no traveler wants to change his path, provided all the other ones keep the same strategy. According to this optimal measure, every path is a geodesic with respect to a metric on Ω of the form

$g(|\bar{\sigma}(x)|)Id$ (where $g(t) = 1 + t^{p-1}$ is the so-called "congestion function"), which is defined in terms of the optimal traffic distribution itself. The continuity of $\bar{\sigma}$ and, therefore, of the metric allows to set and study the geodesic problem in the usual sense.

In order to understand the regularity of minimizers of functionals as in (8), we first recall that, when $\nabla^2\mathcal{F}$ is uniformly elliptic, namely there exist $\lambda, \Lambda > 0$ such that $\lambda Id \leq \nabla^2\mathcal{F} \leq \Lambda Id$, the regularity results of u rely on De Giorgi theorem and Schauder estimates. If the ellipticity of $\mathcal{F}$ degenerates at only one point, then several results are still available. For instance, in the model case of the p-Laplace equation, that is when $\mathcal{F}(v) = |v|^p$ and $f = 0$, the $C^{1,\alpha}$ regularity of u has been proved by Uraltseva for $p \geq 2$, initiating a wide literature.

With the choice of $\mathcal{F}$ in (9), the Lipschitz regularity of a local minimizer u follows by standard techniques [87], since the equation is the classical p-Laplace equation when the gradient is large. In general no more regularity than L^∞ can be expected on ∇u. Indeed, when $\mathcal{F}$ is given by (6.5) and f is identically 0, every 1-Lipschitz function is a global minimizer of (6.4). However, in Chapters 6 and 7, based on joint works with Figalli [55, 59] we prove the continuity of $\nabla\mathcal{F}(\nabla u)$, extending a previous result of Santambrogio and Vespri [114] which holds only in dimension 2.

The Vlasov-Poisson system. The structure of transport equation hidden in the nonlinear Vlasov-Poisson system, presented at the beginning of this Introduction, has been exploited in a huge literature, in order to obtain existence and uniqueness of classical solutions, namely, solutions where all the relevant derivatives exist. The first existence results were obtained in dimension 1 by Iordanskii [99], in dimension 2 by Ukai and Okabe [120], in dimension 3 for small data by Bardos and Degond [26], and for symmetric initial data in [29, 123, 95, 116]. Finally, in 1989 Pfaffelmöser [111] and Lions and Perthame [105] were able to prove global existence of classical solutions starting from general data. Moreover, the uniqueness problem has been addressed under more restrictive assumptions on the initial datum in [105] and [108], and both proofs employ the Lagrangian flow associated to the solution, which is regular enough under a global bound on the space density.

In recent years, an interesting direction of research in the context of the Vlasov-Poisson system is given by the analysis of existence, uniqueness and properties of weak solutions. In particular, when one drops the assumption of boundedness of the initial density (this assumption is preserved along solutions thanks to the transport structure of the equation) and assumes only that $f_t \in L^1(\mathbb{R}^{2d})$, the term $E_t f_t$ appearing in the equation is not even locally integrable. For this reason, Di Perna and Lions [78] introduced the concept of *renormalized solution*, which is equivalent to the notion of weak (distributional) solution under suitable integrability assumptions on f_t. In this context, DiPerna and Lions announced global existence of solutions when the total energy is finite and $f_0 \log(1 + f_0) \in L^1(\mathbb{R}^{2d})$.

In the setting of weak solutions, due to the low regularity of the density and of the vector field, it is unclear whether the *Eulerian description* provided by the equation physically corresponds to a *Lagrangian evolution* of the particles. In Chapter 8 (based on a joint work with Ambrosio and Figalli [11]), we investigate this problem and we apply the general tools developed in Chapter 4 to prove that the Lagrangian structure holds even in the context of weak/renormalized solutions. We obtain also global existence of weak solutions under minimal assumptions on the initial data and improve the result in [78], dropping the hypothesis $f_0 \log(1 + f_0) \in L^1(\mathbb{R}^{2d})$ and assuming only the finiteness of energy.

The semigeostrophic system. The semigeostrophic system models athmosperic/ocean flows on large scales. The problem can be described in the case of periodic solutions in $\mathbb{R}^2$, namely on the 2-dimensional torus $\mathbb{T}^2$

$$\begin{cases} \partial_t \nabla P_t(x) + \big(u_t(x) \cdot \nabla\big) \nabla P_t(x) = J(\nabla P_t(x) - x) & (x,t) \in \mathbb{T}^2 \times (0,\infty) \\ \nabla \cdot u_t(x) = 0 & (x,t) \in \mathbb{T}^2 \times [0,\infty) \\ P_0(x) = P^0(x) & x \in \mathbb{T}^2. \end{cases} \tag{11}$$

where P^0 is the initial datum, $J \in \mathbb{R}^{2\times 2}$ is a rotation matrix, u_t represents the velocity, and ∇P_t is related to the pressure of the fluid.

Energetic considerations show that it is natural to assume the convexity of the function $P_t(x)$. The system (11) has a *dual formulation* obtained with a *change of variable*

$$\begin{cases} \partial_t \rho_t + \nabla \cdot (U_t \rho_t) = 0 \\ U_t(x) = J(x - \nabla P_t^*(x)) \\ \rho_t = (\nabla P_t)_\sharp \mathscr{L}_{\mathbb{T}^2} \\ P_0(x) = p^0(x) + |x|^2/2, \end{cases}$$

where P_t^* is the convex conjugate of P_t. The existence of dual solutions was proved in 1998 by Benamou and Brenier [31], and, starting from the lagrangian solutions of the dual equation, in [69] the authors managed to build a very weak solution of (11) of lagrangian type, by reversing the change of variables. The formal expression for the velocity u_t of the original system, given a solution (P_t, ρ_t) of the dual system, is given by

$$u_t(x) := [\partial_t \nabla P_t^*](\nabla P_t(x)) + [\nabla^2 P_t^*](\nabla P_t(x)) J(\nabla P_t(x) - x). \quad (12)$$

However, the existence of distributional solutions to (11) stayed as an open problem due to the low regularity of the change of variable, since a priori $\nabla^2 P_t^*$ is only a matrix-valued measure and one needs also differentiability in time of ∇P_t^* to give a meaning to (12). The existence of Eulerian solutions is shown in joint works with Ambrosio, De Philippis, and Figalli [7, 8], thanks to the recent regularity results on solutions of the Monge-Ampère equation [73], and it is the content of Chapter 9.

In the final part of this introduction, we outline other works developed during the PhD that present some common underlying ideas and techniques with the ones outlined above in this introduction.

Regularity of double phase variational problems. Degenerate elliptic problems arise also to model strongly anisotropic materials. Given $\Omega \subset \mathbb{R}^d$, $d \geq 2$, we are here interested in the regularity of local minimizers $u \colon \Omega \to \mathbb{R}$ of a class of variational integrals whose model is given by the functional

$$\mathcal{P}(w) := \int_\Omega (|Dw|^p + a(x)|Dw|^q)\, dx\,, \quad (13)$$

which is naturally defined on $W^{1,1}(\Omega)$, where

$$1 < p < q\,, \qquad 0 \leq a \in C^{0,\alpha}(\Omega)\,, \qquad \alpha \in (0, 1]\,.$$

The functional $\mathcal{P}$ belongs to the class of functionals with non-standard growth conditions, which have been widely studied in recent years. These are integral functionals of the type

$$w \mapsto \int_\Omega f(x, Dw)\, dx\,,$$

where the integrand $f \colon \Omega \times \mathbb{R}^n \to \mathbb{R}$ satisfies unbalanced polynomial growth conditions of the type

$$|z|^p \lesssim f(x, z) \lesssim |z|^q + 1 \qquad \text{for every } z \in \mathbb{R}^d.$$

In (13), the coefficient $a(x)$ describes the geometry of a composite, made of two different materials, with power hardening of rate p and q, respectively. From the mathematical viewpoint, the integrand of (13) switches between two different types (phases) of elliptic behaviors according to the coefficient $a(\cdot)$. Since a interacts directly with the ellipticity of the problem, the presence of x is not any longer a perturbation, and this has direct consequences on the regularity of minimizers. More precisely, the regularity of the minimizer holds if the gap between the exponents p and q is controlled in terms of the regularity of a by

$$q \leq p + \alpha. \tag{14}$$

This condition is sharp, as shown in the counterexample in [83]. In [62], Mingione and I proved that bounded local minimizers of (13) under the assumption (14) have Hölder continuous gradients, namely $\nabla u \in C^{0,\beta}$ for some $\beta > 0$. Boundedness is a rather common feature since it for instance follows by maximum principle when considering solutions of Dirichlet problems involving a bounded boundary datum $u_0 \in L^\infty(\Omega) \cap W^{1,p}(\Omega)$. In a companion paper [61] we prove that the same regularity holds also in the case of unbounded local minimizers, but this time we assume a different relation between the exponents p, q and the regularity of a:

$$q < p + \frac{\alpha p}{n}.$$

The proofs in [61, 62] rely on many different technical tools, going from the p-harmonic approximation lemma to a fractional Caccioppoli inequality. A common underlying idea is to consider, *at each scale*, namely on every ball $B_R \subset \Omega$, an alternative according to the fact that

$$\sup_{x \in B_R} \frac{a(x)}{R^\alpha} \leq M$$

holds or not, for a threshold M to be chosen. If it holds, then *at this fixed scale* we are in the *p-phase* and we compare our minimizer to a solution of the p-Laplace equation in the same ball. Otherwise, we are in the *(p, q)-phase* and the solution is compared to the solution of a functional like (13) with frozen coefficient $a(\cdot) = a_0$. The regularity for the frozen problem has been studied in [104].

Many questions arise from the results presented above. For instance, in collaboration with Baroni and Mingione [27, 28], we see that Harnack inequalities, in analogy with the results of [76], hold also for minimizers of double phase integrals and that the regularity theory developed in [61] can be generalized to different ellipticity types. In particular, we consider a functional of the type

$$\mathcal{P}_{ln}(w) := \int_\Omega \left[|Dw|^p + a(x)|Dw|^p \ln(1 + |Dw|) \right] dx$$

and correspondingly, the coefficient a is allowed to have a logarithmic modulus of continuity in order to obtain the Hölder continuity of the minimizer.

Optimal transport with Coulomb cost. In some recent papers, Buttazzo, De Pascale and Gori-Giorgi [40] and Cotar, Friesecke and Klüppelberg [66] consider a mathematical model for the strong interaction limit of the density functional theory (DFT). In particular, the model for the minimal interaction of N electrons is formulated in terms of a multimarginal Monge transport problem. Let $c : (\mathbb{R}^d)^N \to \mathbb{R}$ be the Coulomb cost function

$$c(x_1, \dots, x_N) = \sum_{1 \le i < j \le N} \frac{1}{|x_i - x_j|} \qquad \forall (x_1, \dots, x_N) \in (\mathbb{R}^d)^N, \quad (15)$$

$\rho \in \mathcal{P}(\mathbb{R}^d)$ be a given probability measure on $\mathbb{R}^d$, and $\mathcal{T}(\rho)$ be the set of transport maps $\mathcal{T}(\rho) = \{T : \mathbb{R}^d \to \mathbb{R}^d \text{ Borel} : T_\sharp \rho = \rho\}$, where $T_\sharp \rho$ represents the pushforward measure of the measure ρ through the Borel map T. We consider the Monge multimarginal problem

$$(M) = \inf \left\{ \int_{\mathbb{R}^d} c(x, T_2(x), \dots, T_N(x))\, d\mu(x) : T_2, \dots, T_N \in \mathcal{T}(\rho) \right\}$$

and its cyclical version

$$(M_{cycl}) = \inf \left\{ \int_{\mathbb{R}^d} c(x, T(x), \dots, T^{(N-1)}(x)) d\mu(x) : T \in \mathcal{T}(\rho), T^{(N)} = Id \right\},$$

which is meaningful since the cost function is symmetric. Following the standard theory of optimal transport, we introduce the set of transport plans

$$\Pi(\rho) = \left\{ \gamma \in \mathcal{P}(\mathbb{R}^{dN}) \ : \ \pi^i_\sharp \gamma = \rho, \ i = 1, \dots, N \right\},$$

where $\pi^i : (\mathbb{R}^d)^N \to \mathbb{R}^d$ are the projections on the i-th component for $i = 1, \dots, N$, and the Kantorovich multimarginal problem

$$(K) = \min \left\{ \int_{(\mathbb{R}^d)^N} c(x_1, \dots, x_N) d\gamma(x_1, \dots, x_N) : \gamma \in \Pi(\rho) \right\},$$

where, in contrast with (M), we allow the splitting of mass. To every $(N-1)$-uple of transport maps $T_2, \dots, T_N \in \mathcal{T}(\rho)$ we associate the transport plan

$$\gamma = (Id, T_2, \dots, T_N)_\sharp \rho \in \Pi(\rho).$$

We remark that the existence of an optimal transport plan, namely a minimizer of (K), follows from the lower semicontinuity of the cost, from the linearity of the cost of a plan γ with respect to γ and from the fact that the admissible plans form a tight subset of the set of measures on $(\mathbb{R}^d)^N$. In a joint paper with Di Marino [53], under the sharp assumption that ρ is non-atomic, we prove that $(K) = (M) = (M_{cycl})$. In particular, if an optimal transport map exists, it has the cyclical structure that appears in (M_{cycl}). This result reduces the optimization problem (K) over measures on $\mathbb{R}^{Nd}$ to the problem (M_{cycl}) over functions on $\mathbb{R}^N$ and is useful in deriving numerical methods to compute the value of (K). In a companion paper [54], joint work with Di Marino and De Pascale, we address the problem of existence of optimal transport maps in dimension $d = 1$, providing an explicit construction of the optimal map. For $N = 2$, in any dimension, existence follows from the standard optimal transport theory (see [124]) since the so called "twist condition" is formally satisfied by the Coulomb cost (15). In the multimarginal case $N \geq 3$, there is no general theory for the existence of optimal maps and the construction in [54] heavily relies on the assumption $d = 1$. The generalization of this result to higher dimensions is open. Finally, in a paper with Stra [64] we begin the analysis of the case of spherically symmetric data, which model for instance Litium and Berillium atoms. We disprove a conjecture on the structure of the optimal transport, showing that some special maps, introduced by Seidl, Gori Giorgi and Savin, are not always optimal in the corresponding transport problem. We also provide examples of maps satisfying optimality conditions for special classes of data.

Geometric characterizations of rigidity in symmetrization inequalities and nonlocal perimeters. Symmetrization inequalities are among the most basic tools of the Calculus of Variations. They include the Polya-Szego inequality for the Dirichlet energy, the Steiner symmetrization and its analogous in the Gaussian setting, named Ehrhard symmetrization, which is a well-known tool in Probability Theory, arising in the study of geometric variational problems in Gauss space.

The study of their equality cases plays a fundamental role in the explicit characterization of minimizers, thus in the computation of optimal constants in geometric and functional inequalities. Although it is usually easy to derive useful necessary conditions for equality cases, the analysis of *rigidity of equality cases* (that is, the situation when every set realizing equality in the given symmetrization inequality turns out to be symmetric) is a much subtler issue. Sufficient conditions for rigidity have been known, and largely used, in the case of the Polya-Szego inequality for the Dirichlet energy [39], and of Steiner inequality for perimeter [50]. However, these sufficient conditions fail to be also necessary: for example, the one proposed in [50] fails to characterize rigidity even in the class of polyhedra in $\mathbb{R}^3$. A preliminary analysis of some examples indicates that, in order to formulate geometric conditions which could possibly be suitable for characterizing rigidity, one needs a measure-theoretic notion which describes when a Borel set "disconnects" another Borel set. This notion, called *essential connectedness*, was first introduced in a joint paper with Cagnetti, De Philippis, and Maggi [47] and it is inspired by the notion of indecomposable current adopted in Geometric Measure Theory (see [86, 4.2.25]). It allows to formulate in its terms a simple geometric condition that characterizes rigidity in Ehrhard inequality for Gaussian perimeter. The same notion can be employed, together with a fine analysis of the differentiability properties of the barycenter function of a set of finite perimeter whose sections are segments, to provide various characterizations of rigidity in Steiner inequality for Euclidean perimeter. This was done in collaboration with Cagnetti, De Philippis, and Maggi [48].

Chapter 1
An overview on flows of vector fields and on optimal transport

The aim of this Chapter is twofold. On one side, we give an overview on the classical results regarding flows of vector fields, the regularity of degenerate elliptic PDEs and, in particular, the Monge-Ampère equation. These results and ideas will be fundamental for the development of all the subsequent chapters. On the other side, we present the classical theory according to a point of view that will be useful in the rest of this thesis, showing refinements of the known theorems that suit the subsequent discussions.

1.1. Classical and nonsmooth theory

Given a vector field $\boldsymbol{b} : (0, T) \times \mathbb{R}^d \to \mathbb{R}^d$ we consider the ordinary differential equation

$$\begin{cases} \partial_t \boldsymbol{X}(t, x) = \boldsymbol{b}_t(\boldsymbol{X}(t, x)) & \forall t \in (0, T) \\ \boldsymbol{X}(0, x) = x, \end{cases} \tag{1.1}$$

In the smooth setting, namely when $\boldsymbol{b}$ is locally Lipschitz with respect to the space variable, existence and uniqueness of a solution to (1.1) is guaranteed by the Cauchy-Lipschitz theorem.

Theorem 1.1 (Cauchy-Lipschitz). *Let $T > 0$, $\boldsymbol{b} \in L^1((0, T);$ $\mathrm{Lip}_{\mathrm{loc}}(\mathbb{R}^d; \mathbb{R}^d))$. Then for every $x \in \mathbb{R}^d$ there exists a unique maximal solution $\boldsymbol{X}(\cdot, x)$ of* (1.1) *defined in a nonempty maximal existence time $[0, T_{\boldsymbol{X}}(x))$. Moreover, the map $T_{\boldsymbol{X}}$ is lower semicontinuous, for every $x \in \mathbb{R}^d$ such that $T_{\boldsymbol{X}}(x) < T$ the trajectory $\boldsymbol{X}(\cdot, x)$ blows up properly, namely*

$$\lim_{t \to T_{\boldsymbol{X}}(x)} |\boldsymbol{X}(t, x)| = \infty,$$

and the map $\boldsymbol{X}(t, \cdot)$ is locally Lipschitz in space on its finiteness domain.

The ODE (1.1) is strictly related (via the method of characteristics) to the transport equation

$$\begin{cases} \partial_t u + \boldsymbol{b} \cdot \nabla u = 0 & \text{in } (0,T) \times \mathbb{R}^d \\ u_0 = \bar{u} \quad \text{given.} \end{cases} \tag{1.2}$$

Indeed, if u is a smooth solution of (1.2) and $x \in \mathbb{R}^d$, we have

$$\begin{aligned} \frac{d}{dt} u_t(\boldsymbol{X}(t,x)) &= \partial_t u_t(\boldsymbol{X}(t,x)) + \partial_t \boldsymbol{X}(t,x) \cdot \nabla u_t(\boldsymbol{X}(t,x)) \\ &= \partial_t u_t(\boldsymbol{X}(t,x)) + \boldsymbol{b}_t(\boldsymbol{X}(t,x)) \cdot \nabla u_t(\boldsymbol{X}(t,x)) = 0, \end{aligned}$$

so that u is constant along the characteristics of $\boldsymbol{b}$. Hence, given an initial datum $u_0 = \bar{u}$, we expect

$$u_t(x) = \bar{u}(\boldsymbol{X}(t,\cdot)^{-1}(x))$$

to be a solution of the transport equation, and this can be easily checked by direct computation. In the last thirty years, a huge effort has been made in order to develop a theory of flows of vector fields in the non-smooth setting, in view of applications to physical systems. In the following, we precise the meaning of the ODE (1.1) and of the continuity and transport equation in a non-smooth setting. The continuity equation is

$$\begin{cases} \partial_t u + \nabla \cdot (\boldsymbol{b} u) = 0 & \text{in } (0,T) \times \mathbb{R}^d \\ u_0 = \bar{u} \text{ given,} \end{cases} \tag{1.3}$$

where $u : (0,T) \times \mathbb{R}^d \to \mathbb{R}$; in the case of a divergence-free vector field, it is equivalent to the transport equation (1.2). We mostly use standard notation, denoting by $\mathscr{L}^d$ the Lebesgue measure in $\mathbb{R}^d$, and by $f_\# \mu$ the push-forward of a Borel nonnegative measure μ under the action of a Borel map f, namely $f_\# \mu(B) = \mu(f^{-1}(B))$ for any Borel set B in the target space. We denote by $\mathcal{B}(\mathbb{R}^d)$ the family of all Borel sets in $\mathbb{R}^d$. In the family of positive finite measures in an open set Ω, we will consider both the weak topology induced by the duality with $C_b(\Omega)$ that we will call *narrow* topology, and the *weak* topology induced by $C_c(\Omega)$. Also, $\mathscr{M}_+(\mathbb{R}^d)$ will denote the space of finite Borel measures on $\mathbb{R}^d$, while $\mathscr{P}(\mathbb{R}^d)$ denotes the space of probability measures.

In the non-smooth setting, given a Borel vector field $\boldsymbol{b} : (0,T) \times \mathbb{R}^d \to \mathbb{R}^d$, an integral curve $\gamma : [0,T] \to \mathbb{R}^d$ of the equation $\partial_t \gamma = \boldsymbol{b}_t(\gamma)$ (see (1.1)) is an absolutely continuous curve in $AC([0,T];\mathbb{R}^d)$ which satisfies the previous ODE for almost every $t \in [0,T]$. The continuity equation is intended in distributional sense, according to the following definition.

Definition 1.2 (Distributional solutions). A family $\{\mu_t\}_{t\in[0,T]}$ of locally finite signed measures on $\mathbb{R}^d$ such that $\boldsymbol{b}_t\mu_t$ is a locally finite measure is a solution of the continuity equation if it solves

$$\partial_t \mu_t + \nabla \cdot (\boldsymbol{b}_t \mu_t) = 0$$

in the sense of distributions, namely for every $\phi \in C_c^\infty((0,T)\times\mathbb{R}^d)$

$$\int_0^T \int_{\mathbb{R}^d} \left[\partial_t \phi_t(x) + \nabla_x \phi_t(x) \cdot \boldsymbol{b}_t(x)\right] d\mu_t(x)\, dt = 0.$$

The family $\{\mu_t\}_{t\in[0,T]}$ is a solution of the continuity equation with initial datum μ_0 if for every $\phi \in C_c^\infty([0,T)\times\mathbb{R}^d)$

$$\int_{\mathbb{R}^d} \phi_0(x)\mu_0(x) + \int_0^T \int_{\mathbb{R}^d} \left[\partial_t \phi_t(x) + \nabla_x \phi_t(x) \cdot \boldsymbol{b}_t(x)\right] d\mu_t(x)\, dt = 0.$$

When we consider possibly singular measures μ_t, the vector field $\boldsymbol{b}_t$ has to be defined pointwise and not only $\mathscr{L}^d$-a.e., since the product $\boldsymbol{b}_t\mu_t$ is sensitive to modifications of $\boldsymbol{b}_t$ in $\mathscr{L}^d$-negligible sets. In the following, in particular with Sobolev or BV vector fields, we will often consider only measures μ_t which are absolutely continuous with respect to $\mathscr{L}^d$, so everything is well posed and does depend only on the equivalence class of $\boldsymbol{b}$ in $L^1_{\mathrm{loc}}((0,T)\times\mathbb{R}^d)$.

If we consider a function $\beta \in C^1(\mathbb{R})$ and we multiply the transport equation (1.2) by $\beta'(u)$, we see that, if u is a smooth solution of the transport equation, so is $\beta(u)$. The previous observation is encoded in the following definition.

Definition 1.3 (Renormalized solutions). Let $\boldsymbol{b}\in L^1_{\mathrm{loc}}((0,T); L^1_{\mathrm{loc}}(\mathbb{R}^d;\mathbb{R}^d))$ be a vector field with div $\boldsymbol{b}\in L^1_{\mathrm{loc}}((0,T); L^1_{\mathrm{loc}}(\mathbb{R}^d;\mathbb{R}^d))$. Let $u\in L^\infty_{\mathrm{loc}}((0,T); L^\infty_{\mathrm{loc}}(\mathbb{R}^d))$ and assume that, in the sense of distributions, there holds

$$c := \partial_t u + \boldsymbol{b}\cdot\nabla u \in L^1_{\mathrm{loc}}((0,T); L^1_{\mathrm{loc}}(\mathbb{R}^d;\mathbb{R}^d)). \tag{1.4}$$

Then, u is a renormalized solution of (1.4) if for every $\beta \in C^1(\mathbb{R}) \cap L^\infty(\mathbb{R})$

$$\partial_t \beta(u) + \boldsymbol{b}\cdot\nabla\beta(u) = c\beta'(u).$$

in the sense of distributions. Analogously, we say that u is a renormalized solution starting from a Borel function $u_0 : \mathbb{R}^d \to \mathbb{R}$ if

$$\int_{\mathbb{R}^d} \phi_0(x)\beta(u_0(x))dx + \int_0^T \int_{\mathbb{R}^d} [\partial_t\phi_t(x) + \nabla\phi_t(x)\cdot\boldsymbol{b}_t(x)]\beta(u_t(x))dxdt = 0$$

for all $\phi \in C_c^\infty([0,T)\times\mathbb{R}^d)$ and all $\beta \in C^1 \cap L^\infty(\mathbb{R})$.

The renormalization property describes a property of solutions of a wide class of PDEs related to the transport equation (1.2); for this reason, we will introduce in the following Chapters a few definitions of renormalized solutions that capture better the features of each single problem. The renormalization property can be also used to give a meaning to equation (1.3) when the boundedness (or even the integrability) of u is not any more assumed as an assumption. Indeed, although the product $\boldsymbol{b}_t u_t$ may not even be locally integrable if $\boldsymbol{b}_t \in L^1_{\rm loc}((0,T) \times \mathbb{R}^d)$ and $u_t \in L^1_{\rm loc}((0,T) \times \mathbb{R}^d)$, the term $\boldsymbol{b}_t \beta(u_t)$ appearing in (5.11) is always locally integrable. This will be used in Chapter 8 to give a general notion of solution to the Vlasov-Poisson equation and in Chapter 5 for the continuity equation with an integrable damping term (see Definition 8.1 and 5.3 respectively).

If the vector field $\boldsymbol{b}$ is not assumed to be smooth, namely locally Lipschitz in space, but only Sobolev or BV, easy one dimensional examples show that the uniqueness of trajectories of the ODE 1.1 fails. For instance, if we consider the autonomous vector field $\boldsymbol{b}(x) = \sqrt{|x|}$, $x \in \mathbb{R}$, then we have many solutions of the ODE, which start from $x_0 = -c^2 < 0$, reach the origin in time $2c$, stay at the origin for any time $T \geq 0$, and continue as $(t - T - 2c)^2$.

However, one can still associate to the vector field $\boldsymbol{b}$ a notion of flow, made of a selection of trajectories of the ODE. Among all possible selections, we prefer the ones that do not allow for concentration, as presented in the following definition.

Definition 1.4. Let $T > 0$ and $\boldsymbol{b} : (0,T) \times \mathbb{R}^d \to \mathbb{R}^d$ a Borel, locally integrable vector field. We say that the Borel map $\boldsymbol{X} : \mathbb{R}^d \times [0,T] \to \mathbb{R}^d$ is a regular Lagrangian flow of $\boldsymbol{b}$ if the following two properties hold:

(i) for $\mathscr{L}^d$-a.e. $x \in \mathbb{R}^d$, $\boldsymbol{X}(\cdot, x) \in AC([0,T];\mathbb{R}^d)$ and solves the ODE $\dot{x}(t) = \boldsymbol{b}_t(x(t))$ $\mathscr{L}^1$-a.e. in $(0,T)$, with the initial condition $\boldsymbol{X}(0,x) = x$;

(ii) there exists a constant $C = C(\boldsymbol{X})$ satisfying $\boldsymbol{X}(t,\cdot)_\# \mathscr{L}^d \leq C \mathscr{L}^d$ for every $t \in [0,T]$.

It can be easily checked that the definition of regular Lagrangian flow depends on the equivalence class of $\boldsymbol{b}$ in $L^1_{\rm loc}((0,T) \times \mathbb{R}^d)$ rather then on the pointwise values of $\boldsymbol{b}$.

The well-celebrated papers of DiPerna and Lions [81] and Ambrosio [5] provide existence and uniqueness of the regular Lagrangian flow assuming local Sobolev or BV regularity of $\boldsymbol{b}$, boundedness of the distributional divergence $\operatorname{div} \boldsymbol{b}$, and some growth conditions on $\boldsymbol{b}$.

Theorem 1.5. *Let $\boldsymbol{b} \in L^1((0,T); BV_{\mathrm{loc}}(\mathbb{R}^d;\mathbb{R}^d))$ be a vector field that satisfies the bound on the divergence* $(\operatorname{div} \boldsymbol{b})_- \in L^1((0,T); L^\infty(\mathbb{R}^d))$ *and the growth condition*

$$\frac{|\boldsymbol{b}_t(x)|}{1+|x|} \in L^1((0,T); L^1(\mathbb{R}^d)) + L^1((0,T); L^\infty(\mathbb{R}^d)).$$

Then there exists a unique regular Lagrangian flow $\boldsymbol{X}$ of $\boldsymbol{b}$.

The previous theorem has been extended to different classes of vector fields; some of them are listed in Remark 1.9 below. Thanks to the existence and uniqueness of a regular Lagrangian flow, it is possible to define the notion of Lagrangian solution for the continuity and transport equation. These are solutions obtained by flowing the initial datum according to the regular Lagrangian flow of $\boldsymbol{b}$.

The proof of the previous theorem is based on the interaction between the PDE point of view on the continuity equation and the Lagrangian techniques. In the following two sections, we present two key ideas behind Theorem 1.5, which in turn will be fundamental in order to develop a local version of Theorem 1.5.

1.2. A bridge between Lagrangian and Eulerian solutions: the superposition principle

This section is devoted to the so called "superposition principle", which encodes the connection between the Eulerian and the Lagrangian formulation of the continuity equation, namely between nonnegative distributional solutions of the PDE and solutions transported by a set of (possibly branching) curves. The aim of Section 1.3 is, then, to show that, under more restrictive assumptions on the vector field, this set of curves is given exactly by the flow of $\boldsymbol{b}$.

Let us fix $T \in (0,\infty)$ and consider a weakly continuous family $\mu_t \in \mathscr{M}_+(\mathbb{R}^d)$, $t \in [0,T]$, solving in the sense of distributions the continuity equation

$$\frac{d}{dt}\mu_t + \nabla \cdot (\boldsymbol{b}_t \mu_t) = 0 \qquad \text{in } (0,T) \times \mathbb{R}^d$$

for a Borel vector field $\boldsymbol{b} : (0,T) \times \mathbb{R}^d \to \mathbb{R}^d$, locally integrable with respect to the space-time measure $\mu_t dt$. When we restrict ourselves to probability measures μ_t, then weak and narrow continuity with respect to. t are equivalent; analogously, we may equivalently consider compactly supported test functions $\varphi(t,x)$ in the weak formulation of the continuity equation, or functions with bounded C^1 norm whose support is contained in $I \times \mathbb{R}^d$ with $I \Subset (0,T)$. If $J \subset \mathbb{R}$ is an interval and $t \in J$, we denote by $e_t : C(J;\mathbb{R}^d) \to \mathbb{R}^d$ the evaluation map at time t, namely $e_t(\eta) := \eta(t)$ for any continuous curve $\eta : J \to \mathbb{R}^d$.

We now recall the so-called superposition principle. We prove it under the general assumption that μ_t may a priori vanish for some $t \in [0,T]$, but satisfies (1.5); we see in Remark 1.7 that this assumption implies that there is no mass loss, namely $\mu_t(\mathbb{R}^d) = \mu_0(\mathbb{R}^d)$ for every $t \in [0,T]$. Remark 1.7 allows the reduction of the superposition principle, as stated below, to [12, Theorem 12], which presents the same result assuming that the family μ_t is made of probability measures. We mention also [19, Theorem 8.2.1], where a proof is presented in the even more special case of L^p integrability on $\boldsymbol{b}$ for some $p > 1$

$$\int_0^T \int_{\mathbb{R}^d} |\boldsymbol{b}_t(x)|^p \, d\mu_t(x)\, dt < \infty.$$

The superposition principle will play a role in the proof of the comparison principle stated in Proposition 1.11, in the blow-up criterion of Theorem 3.13 and in Theorem 4.9, where a completely local version of the superposition principle is presented.

Theorem 1.6 (Superposition principle and approximation). *Let $\boldsymbol{b} : (0,T)\times\mathbb{R}^d \to \mathbb{R}^d$ be a Borel vector field. Let $\mu_t \in \mathscr{M}_+(\mathbb{R}^d)$, $0 \le t \le T$, with μ_t weakly continuous in $[0,T]$ solution to the equation $\frac{d}{dt}\mu_t + \operatorname{div}(\boldsymbol{b}\mu_t) = 0$ in $(0,T)\times\mathbb{R}^d$, with*

$$\int_0^T \int_{\mathbb{R}^d} \frac{|\boldsymbol{b}_t(x)|}{1+|x|}\, d\mu_t(x)\, dt < \infty. \tag{1.5}$$

Then there exists $\boldsymbol{\eta} \in \mathscr{M}_+\big(C([0,T];\mathbb{R}^d)\big)$ satisfying:

(i) *$\boldsymbol{\eta}$ is concentrated on absolutely continuous curves η in $[0,T]$, solving the ODE $\dot\eta = \boldsymbol{b}_t(\eta)$ $\mathscr{L}^1$-a.e. in $(0,T)$;*
(ii) *$\mu_t = (e_t)_\#\boldsymbol{\eta}$ (so, in particular, $\mu_t(\mathbb{R}^d) = \mu_0(\mathbb{R}^d)$) for all $t \in [0,T]$.*

Moreover, there exists a family of measures $\mu_t^R \in \mathscr{M}_+\big(\mathbb{R}^d\big)$, narrowly continuous in $[0,T]$, solving the continuity equation and supported on $\overline{B}_R$, such that $\mu_t^R \uparrow \mu_t$ as $R \to \infty$ for all $t \in [0,T]$.

Remark 1.7. We show that, if μ_t and $\boldsymbol{b}_t$ are taken as in Theorem 1.6 then μ_t does not loose or gain mass, namely

$$\mu_t(\mathbb{R}^d) = \mu_0(\mathbb{R}^d) \qquad \forall t \in [0,T]. \tag{1.6}$$

Indeed, let $R \ge 1$ and $\chi_R \in C_c^\infty(B_{3R})$ be a cut-off function with $0 \le \chi_R \le 1$, $\chi_R \equiv 1$ on a neighborhood of B_R and $|\nabla\chi_R| \le \chi_{B_{3R}\setminus B_R}$. Since

μ_t solves the continuity equation and since $1/R \le 4/(1+|x|)$ for $|x| \in B_{3R} \setminus B_R$, we have

$$\begin{aligned}\left|\int_{\mathbb{R}^d} \chi_R \, d\mu_0 - \int_{\mathbb{R}^d} \chi_R \, d\mu_t\right| &\le \int_0^T \left|\frac{d}{dt}\int_{\mathbb{R}^d} \chi_R \, d\mu_t\right| dt \\ &= \int_0^T \left|\int_{B_{3R}\setminus B_R} \boldsymbol{b}_t \cdot \nabla \chi_R \, d\mu_t\right| dt \\ &\le \frac{1}{R}\int_0^T \int_{B_{3R}\setminus B_R} |\boldsymbol{b}_t| \, d\mu_t \, dt \\ &\le 4\int_0^T \int_{B_{3R}\setminus B_R} \frac{|\boldsymbol{b}_t(x)|}{1+|x|} \, d\mu_t(x) \, dt.\end{aligned}$$

Hence we deduce that

$$\begin{aligned}\mu_0(B_R) - \mu_t(B_{3R}) &\le \int_{\mathbb{R}^d} \chi_R \, d\mu_0 - \int_{\mathbb{R}^d} \chi_R \, d\mu_t \\ &\le 4\int_0^T \int_{B_{3R}\setminus B_R} \frac{|\boldsymbol{b}_t(x)|}{1+|x|} \, d\mu_t(x) \, dt\end{aligned} \tag{1.7}$$

and

$$\begin{aligned}\mu_t(B_R) - \mu_0(B_{3R}) &\le \int_{\mathbb{R}^d} \chi_R \, d\mu_t - \int_{\mathbb{R}^d} \chi_R \, d\mu_0 \\ &\le 4\int_0^T \int_{B_{3R}\setminus B_R} \frac{|\boldsymbol{b}_t(x)|}{1+|x|} \, d\mu_t(x) \, dt.\end{aligned} \tag{1.8}$$

Letting $R \to \infty$ in (1.7) and (1.8), the right-hand sides converge to 0 by (1.5) and we find (1.6).

The proof of the superposition principle, as stated in Theorem 1.6, can be found in [12, Theorem 12], once Remark 1.7 is taken into account. The proof is based on a clever regularization argument: we consider a family of convolution kernels $\{\rho_\varepsilon\}_{\varepsilon\in(0,1)}$, having integral 1 and supported on the whole $\mathbb{R}^d$, and we define

$$\mu_t^\varepsilon := \mu_t * \rho^\varepsilon, \qquad \boldsymbol{b}^\varepsilon := \frac{(\boldsymbol{b}\mu_t) * \rho^\varepsilon}{\mu_t * \rho^\varepsilon}.$$

We call $\boldsymbol{X}^\varepsilon$ the flow of the vector field $\boldsymbol{b}^\varepsilon$, so that μ^ε solves the continuity equation and it is transported by $\boldsymbol{X}^\varepsilon$, since $\boldsymbol{b}^\varepsilon$ satisfies some local Lipschitz bounds, uniformly in time. Then, we define $\boldsymbol{\eta}^\varepsilon \in \mathscr{M}(AC([0,T];\mathbb{R}^d))$ as the law under μ_0^ε of the map $x \mapsto \boldsymbol{X}^\varepsilon(\cdot,x)$, namely $\boldsymbol{\eta}^\varepsilon := \boldsymbol{X}^\varepsilon(\cdot,x)_\# \mu_0^\varepsilon$. Assumption (1.5) (which holds uniformly also for $\boldsymbol{b}^\varepsilon$ and μ^ε) allows to conclude that the sequence $\boldsymbol{\eta}^\varepsilon$ is tight and hence it converges to some $\boldsymbol{\eta}$ (up to subsequences). Finally, one can show that

$$\int \frac{\left|\eta(t) - x - \int_0^t \boldsymbol{b}_s(\eta(s))\,ds\right|}{1 + \max_{[0,T]}|\eta|}\,d\boldsymbol{\eta}(\eta) = 0 \qquad \text{for every } t \in [0,T],$$

which proves that $\boldsymbol{\eta}$ is concentrated on integral curves of $\boldsymbol{b}$.

The last statement in Theorem 1.6 can simply be obtained by restricting $\boldsymbol{\eta}$ to the class of curves contained in $\overline{B}_R$ for all $t \in [0,T]$ to obtain positive finite measures $\boldsymbol{\eta}^R \le \boldsymbol{\eta}$ which satisfy $\boldsymbol{\eta}^R \uparrow \boldsymbol{\eta}$, and then defining $\mu_t^R := (e_t)_\# \boldsymbol{\eta}^R$.

1.3. Uniqueness of bounded solutions of the continuity equation

In Section 1.2 we saw that, under very general assumptions, nonnegative distributional solutions of the continuity equation are transported by a set of curves. The aim of this section is to exploit the connection between the well posedness of the continuity equation and the fact that solutions of the continuity equation are Lagrangian, namely, are transported by the flow of $\boldsymbol{b}$. In particular, we show in Theorem 1.12 that, if the vector field $\boldsymbol{b}$ satisfies a local uniqueness property of solutions of the continuity equation, then the disintegration of every representation of a bounded distributional solution with respect to the evaluation at time 0 gives a family of deltas, which in turn represent the regular Lagrangian flow.

Given a closed interval $I \subset \mathbb{R}$ and an open set $\Omega \subset \mathbb{R}^d$, let us define the class $\mathcal{L}_{I,\Omega}$ of all nonnegative functions which are essentially bounded, nonnegative, and compactly supported in Ω:

$$\mathcal{L}_{I,\Omega} := L^\infty\big(I; L^\infty_+(\Omega)\big) \cap \big\{w : \operatorname{supp} w \text{ is a compact subset of } I \times \Omega\big\}. \quad (1.9)$$

We say that $\rho \in \mathcal{L}_{I,\Omega}$ is weakly* continuous if there is a representative ρ_t with $t \mapsto \rho_t$ continuous in I with respect to. the weak* topology of $L^\infty(\Omega)$. Notice that, in the class $\mathcal{L}_{I,\Omega}$, weak* continuity of ρ is equivalent to the narrow continuity of the corresponding measures $\mu_t := \rho_t \mathscr{L}^d \in \mathscr{M}_+(\mathbb{R}^d)$.

For $T \in (0,\infty)$ we are given a Borel vector field $\boldsymbol{b} : (0,T) \times \Omega \to \mathbb{R}^d$ satisfying:

(a-Ω) $\int_0^T \int_{\Omega'} |\boldsymbol{b}(t,x)|\,dxdt < \infty$ for any $\Omega' \Subset \Omega$;
(b-Ω) for any nonnegative $\bar\rho \in L^\infty_+(\Omega)$ with compact support in Ω and any closed interval $I = [a,b] \subset [0,T]$, the continuity equation

$$\frac{d}{dt}\rho_t + \operatorname{div}(\boldsymbol{b}\rho_t) = 0 \qquad \text{in } (a,b) \times \Omega$$

has at most one weakly* continuous solution $I \ni t \mapsto \rho_t \in \mathcal{L}_{I,\Omega}$ with $\rho_a = \bar{\rho}$.

Remark 1.8. Theorem 1.5 holds also if the local regularity of $\boldsymbol{b}$, namely the hypothesis $\boldsymbol{b} \in L^1((0,T); BV_{\text{loc}}(\mathbb{R}^d; \mathbb{R}^d))$, is substituted by assumptions (a-$\mathbb{R}^d$) and (b-$\mathbb{R}^d$). This can be seen from the proof of Theorem 1.5 and will be clear after the discussion in Chapter 2.

Remark 1.9. Assumption (b-Ω) is known to be true in many cases. The following list does not pretend to be exhaustive:

– Sobolev vector fields [81], BV vector fields whose divergence is a locally integrable function in space [34, 5, 51, 52], some classes of vector fields of bounded deformation [14];

– vector fields $\boldsymbol{B}(x, y) = (\boldsymbol{b}_1(x, y), \boldsymbol{b}_2(x, y))$ with different regularity with respect to. x and y [101, 102];

– two-dimensional Hamiltonian vector fields [4] (within this class, property (b-Ω) has been characterized in terms of the so-called weak Sard property);

– vector fields arising from the convolution of L^1 functions with singular integrals [35, 36]. In this case, the authors proved uniqueness of the regular Lagrangian flow associated to $\boldsymbol{b}$; we outline in the next remark how to obtain the eulerian uniqueness property (b-Ω) following their argument.

– vector fields with a particular structure, one of whose components is obtained from the convolution of a finite measure with a singular kernel [32] (see also Section 1.4).

Remark 1.10. Under the assumptions on the vector field $\boldsymbol{b}$ considered in [36], the authors proved in [36, Theorem 6.2] the uniqueness of the Lagrangian flow. In their key estimate, the authors take two regular Lagrangian flows $\boldsymbol{X}$ and $\boldsymbol{Y}$, provide an upper and lower bound for the quantity

$$\Phi_\delta(t) := \int \log\left(1 + \frac{|\boldsymbol{X}(t,x) - \boldsymbol{Y}(t,x)|}{\delta}\right) dx \qquad t \in [0, T] \quad (1.10)$$

in terms of a parameter $\delta > 0$, and eventually let $\delta \to 0$. To show that property (b-Ω) holds, we consider two nonnegative bounded solutions of the continuity equation with the same initial datum which are compactly supported in $[a, b] \times \Omega$. By Theorem 1.6 there exist $\boldsymbol{\eta}^1, \boldsymbol{\eta}^2 \in \mathscr{P}\big(C([a, b]; \mathbb{R}^d)\big)$ which are concentrated on absolutely continuous solutions $\eta \in AC([a, b]; \Omega)$ of the ODE $\dot{\eta} = \boldsymbol{b}(t, \eta)$ $\mathscr{L}^1$-a.e. in (a, b), and satisfy $(e_t)_\# \boldsymbol{\eta}^i \leq C \mathscr{L}^d$ for any $t \in [a, b]$, $i = 1, 2$. Moreover, we have that $(e_a)_\#\boldsymbol{\eta}^1 = (e_a)_\#\boldsymbol{\eta}^2$. Given $\delta > 0$, we consider the quantity

$$\Psi_\delta(t) := \int_\Omega \int \int \log\left(1 + \frac{|\gamma(t) - \eta(t)|}{\delta}\right) d\boldsymbol{\eta}^1_x(\gamma) d\boldsymbol{\eta}^2_x(\eta)\, d[(e_a)_\#\boldsymbol{\eta}^1](x) \qquad t \in [a, b], \tag{1.11}$$

where $\boldsymbol{\eta}^1_x$, $\boldsymbol{\eta}^2_x$ are the disintegrations of $\boldsymbol{\eta}^1$ and $\boldsymbol{\eta}^2$ with respect to the map e_a. Since $\boldsymbol{\eta}^1$ and $\boldsymbol{\eta}^2$ are concentrated on curves in $C([a, b]; \Omega)$, to show that $\boldsymbol{\eta}^1 = \boldsymbol{\eta}^2$ we can neglect the behavior of $\boldsymbol{b}$ outside Ω. Following the same computations of [36] with the functional (1.11) instead of (1.10), we show that $\boldsymbol{\eta}^1_x = \boldsymbol{\eta}^2_x$ for $(e_a)_\#\boldsymbol{\eta}^1$-a.e. $x \in \Omega$ and this implies the validity of property (b-Ω).

More recently, these well-posedness results have also been extended to vector fields in infinite-dimensional spaces (see [16] and the bibliography therein). It is interesting to observe that the uniqueness assumption in (b-Ω) actually implies the validity of a comparison principle.

Proposition 1.11 (Comparison principle). *If (a-Ω) and (b-Ω) are satisfied, then the following implication holds:*

$$\rho^1_0 \le \rho^2_0 \qquad \Longrightarrow \qquad \rho^1_t \le \rho^2_t \quad \forall t \in [0, T]$$

for all weakly continuous solutions of* (1.3) *in the class* $\mathcal{L}_{[0,T],\Omega}$.

Proof. Let $\boldsymbol{\eta}^i$ be representing $\mu^i_t := \rho^i_t \mathscr{L}^d$ according to Theorem 1.6, and let $\boldsymbol{\eta}^i_x$ be the conditional probability measures induced by e_0, that is

$$\int F(\eta)\, d\boldsymbol{\eta}^i = \int_{\mathbb{R}^d} \left(\int F(\eta)\, d\boldsymbol{\eta}^i_x \right) d\mu^i_0(x) \quad \forall\, F : C([0, T]; \mathbb{R}^d) \to \mathbb{R} \text{ bounded},$$

or (in a compact form) $\boldsymbol{\eta}^i(d\eta) = \int \boldsymbol{\eta}^i_x(d\eta)\, d\mu^i_0(x)$. Defining

$$\tilde{\boldsymbol{\eta}}(d\eta) := \int \boldsymbol{\eta}^2_x\, d\mu^1_0(x), \qquad \tilde{\mu}_t := (e_t)_\#\tilde{\boldsymbol{\eta}},$$

because $\mu^1_0 \le \mu^2_0$, we get $\tilde{\boldsymbol{\eta}} \le \boldsymbol{\eta}^2$. Moreover, the densities of measures $\tilde{\mu}_t$ and μ^1_t provide two elements in $\mathcal{L}_{[0,T],\Omega}$, solving the continuity equation with the same initial condition μ^1_0. Therefore assumption (b-Ω) gives $\tilde{\mu}_t = \mu^1_t$ for all $t \in [0, T]$, and $\mu^1_t = \tilde{\mu}_t = (e_t)_\#\tilde{\boldsymbol{\eta}} \le (e_t)_\#\boldsymbol{\eta}^2 = \mu^2_t$ for all $t \in [0, T]$, as desired. □

Theorem 1.12. *Assume that* $\boldsymbol{b}$ *satisfies (a-Ω) and (b-Ω), and let* $\boldsymbol{\lambda} \in \mathscr{P}\big(C([0, T]; \mathbb{R}^d)\big)$ *satisfy:*

(i) $\boldsymbol{\lambda}$ *is concentrated on*

$$\left\{\eta \in AC([0,T];\Omega) : \dot\eta(t) = \boldsymbol{b}_t(\eta(t)) \text{ for } \mathscr{L}^1\text{-a.e. } t \in (0,T)\right\};$$

(ii) *there exists* $C_0 \in (0,\infty)$ *such that*

$$(e_t)_\#\boldsymbol{\lambda} \le C_0 \mathscr{L}^d \qquad \forall t \in [0,T]. \tag{1.12}$$

Then the conditional probability measures $\boldsymbol{\lambda}_x$ *induced by the map* e_0 *are Dirac masses for* $(e_0)_\#\boldsymbol{\lambda}$*-a.e.* x*; equivalently, there exist curves* $\eta_x \in AC([0,T];\Omega)$ *solving the Cauchy problem* $\dot\eta = \boldsymbol{b}_t(\eta)$ *with the initial condition* $\eta(0) = x$*, satisfying*

$$\boldsymbol{\lambda} = \int \delta_{\eta_x}\, d[(e_0)_\#\boldsymbol{\lambda}](x).$$

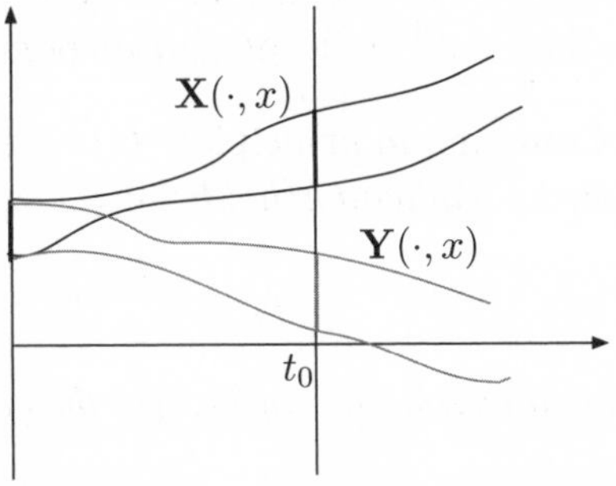

Figure 1.1. The trajectories $\boldsymbol{X}(\cdot,x)$ and $\boldsymbol{Y}(\cdot,x)$ on which a certain $\boldsymbol{\lambda}$ may be concentrated.

The simplest situation where the thesis of Theorem 1.12 does not hold is given by a measure $\boldsymbol{\lambda}$ such that, for a set of positive measure $A \subseteq \mathbb{R}^d$ of initial positions, there are two integral curves of $\boldsymbol{b}$, $\boldsymbol{X}(\cdot,x)$ and $\boldsymbol{Y}(\cdot,x)$ on which $\boldsymbol{\lambda}_x$ is concentrated, and the measure $\boldsymbol{\lambda}_x$ weights them equally. Up to reducing the set A, we may assume that all the trajectories $\boldsymbol{X}(\cdot,x)$ and $\boldsymbol{Y}(\cdot,x)$ starting from A live in a compact set and, after some time $t_0 > 0$, the two sets $\boldsymbol{X}(t_0,A)$ and $\boldsymbol{Y}(t_0,B)$ are disjoint (see Figure 1.1). Hence, the two solutions obtained by flowing the initial datum $(e_0)_\#\boldsymbol{\lambda} \llcorner A$ according to $\boldsymbol{X}$ and $\boldsymbol{Y}$ are concentrated on disjoint sets at time t_0. This contradicts the well-posedness of the continuity equation.

Proof of Theorem 1.12. Let $\{A_n\}_{n\in\mathbb{N}}$ be an increasing family of open subsets of Ω whose union is Ω, with $A_n \Subset A_{n+1} \Subset \Omega$ for every n. Possibly considering the restriction of $\boldsymbol{\lambda}$ to the sets

$$\left\{\eta \in C([0,T];\mathbb{R}^d) : \eta(t) \in \overline{A}_n \text{ for every } t \in [0,T]\right\}$$

it is not restrictive to assume that $\boldsymbol{\lambda}$ is concentrated on a family Γ of curves satisfying $\bigcup_{\eta\in\Gamma}\eta([0,T]) \Subset \Omega$. Then, using the uniqueness assumption for uniformly bounded and compactly supported solutions to the continuity equation, the result follows from the decomposition procedure of [12, Theorem 18] (notice that the latter slightly improves the original argument of [5, Theorem 5.4], where comparison principle for the continuity equation was assumed, see also Proposition 1.11 and its proof). For the sake of completeness, we describe briefly the idea of the argument. By contradiction, we assume that $\boldsymbol{\lambda}_x$ are not Dirac masses in a set of $(e_0)_\#\boldsymbol{\lambda}$ positive measure. Hence we can find $t_0 \in (0,T]$, two disjoint Borel sets $E, E' \subseteq \mathbb{R}^d$, and a Borel set C with $[(e_0)_\#\boldsymbol{\lambda}](C) > 0$, such that

$$\boldsymbol{\lambda}_x\big(\{\gamma : \gamma(t_0) \in E\}\big)\,\boldsymbol{\lambda}_x\big(\{\gamma : \gamma(t_0) \in E'\}\big) > 0 \qquad \forall x \in C$$

and more precisely

$$0 < \frac{\boldsymbol{\lambda}_x\big(\{\gamma : \gamma(t_0) \in E\}\big)}{\boldsymbol{\lambda}_x\big(\{\gamma : \gamma(t_0) \in E'\}\big)} \le M \qquad \forall x \in C \tag{1.13}$$

for some $M > 0$ (see [12, Lemma 16]). Setting $f(x)$ the ratio in (1.13), we introduce

$$\begin{aligned}
\boldsymbol{\lambda}^1 &:= \boldsymbol{\lambda} \llcorner \{\gamma : \gamma(0) \in C,\ \gamma(t_0) \in E\},\\
\boldsymbol{\lambda}^2 &:= f(\gamma(0))\boldsymbol{\lambda} \llcorner \big\{\gamma : \gamma(0) \in C,\ \gamma(t_0) \in E'\big\}.
\end{aligned}$$

The measures $(e_t)_\#\boldsymbol{\lambda}^1$ and $(e_t)_\#\boldsymbol{\lambda}^2$ are the two bounded distributional solutions of the continuity equation with the same initial datum (by the definition of f), hence they should coincide by our assumption on the vector field. On the other hand, $(e_{t_0})_\#\boldsymbol{\lambda}^1$ and $(e_{t_0})_\#\boldsymbol{\lambda}^2$ are orthogonal, and this gives a contradiction. □

Remark 1.13. The assumption (b-Ω) is purely local, as it is proved in Section 2.3. Moreover, it could be reformulated in terms of a local uniqueness property of regular Lagrangian flows: for any $t_0 \ge 0$, $x_0 \in \Omega$ there exists $\varepsilon := \varepsilon(t_0, x_0) > 0$ such that for any Borel set $B \subset B_\varepsilon(x_0) \subset \Omega$ and any closed interval $I = [a,b] \subset [t_0-\varepsilon, t_0+\varepsilon] \cap [0,T]$, there exists at most one regular Lagrangian flow in $B \times [a,b]$ with values in $B_\varepsilon(x_0)$ (see Definition 2.1).

Indeed, (b-Ω) implies the local uniqueness of regular Lagrangian flows by Theorem 1.12 applied to $\boldsymbol{\lambda} = \frac{1}{2}\int_B (\delta_{\boldsymbol{X}(\cdot,x)} + \delta_{\boldsymbol{Y}(\cdot,x)})\, d\mathscr{L}^d(x)$, where $\boldsymbol{X}$ and $\boldsymbol{Y}$ are regular Lagrangian flows in $B \times [a,b]$; on the other hand, we obtain the converse implication through the superposition principle. This approach has the advantage to state the assumptions and the results of Chapter 2, 3, and 4, and only in terms of the Lagrangian point of view on the continuity equation. On the other hand, in concrete examples it is usually easier to verify assumption (b-Ω) than the corresponding Lagrangian formulation.

1.4. Uniqueness for the continuity equation and singular integrals

In this section we deal with uniqueness of solutions to the continuity equation when the gradient of the vector field is given by the singular integral of a time dependent family of measures. This kind of vector fields appear when considering weak solutions of the Vlasov-Poisson system, as in Chapter 8. The theorem is a minor variant of a result by Bohun, Bouchut, and Crippa [32] (see also [36], where the uniqueness is proved for vector fields whose gradient is the singular integral of an L^1 function). We give the proof of the theorem under the precise assumptions that we need later on, since [32] deals with globally defined regular flows (hence the authors need to assume global growth conditions on the vector field), whereas here we present a local version of such result.

Theorem 1.14. *Let* $\boldsymbol{b} : (0,T) \times \mathbb{R}^{2d} \to \mathbb{R}^{2d}$ *be given by* $\boldsymbol{b}_t(x,v) = (\boldsymbol{b}_{1t}(v), \boldsymbol{b}_{2t}(x))$, *where*

$$\boldsymbol{b}_1 \in L^\infty((0,T); W^{1,\infty}_{\text{loc}}(\mathbb{R}^d;\mathbb{R}^d)), \qquad \boldsymbol{b}_{2t} = K * \rho_t$$

with $\rho \in L^\infty((0,T); \mathscr{M}_+(\mathbb{R}^d))$ *and* $K(x) = x/|x|^d$.
Then $\boldsymbol{b}$ *satisfies (b) of Section 3.5, namely the uniqueness of bounded, compactly supported, nonnegative, distributional solutions of the continuity equation.*

Proof. To simplify the notation we give the proof in the case of autonomous vector fields, but the same computations work for the general statement.

It is enough to show that, given $B_R \subset \mathbb{R}^d$ and $\boldsymbol{\eta} \in \mathscr{P}\big(C([0,T]; B_R \times B_R)\big)$ concentrated on integral curves of $\boldsymbol{b}$ and such that $(e_t)_\#\boldsymbol{\eta} \le C_0\mathscr{L}^d$ for all $t \in [0,T]$, the disintegration $\boldsymbol{\eta}_x$ of $\boldsymbol{\eta}$ with respect to the map e_0 is a Dirac delta for $e_{0\#}\boldsymbol{\eta}$-a.e. x. Indeed, any two nonnegative, bounded, compactly supported, distributional solutions with the same initial datum $\bar\rho$ can be represented through the superposition principle (see Theorem 1.6 or [12, Theorem 12]) by $\boldsymbol{\eta}_1, \boldsymbol{\eta}_2 \in \mathscr{P}\big(C([0,T]; B_R \times B_R)\big)$. Hence, setting $\boldsymbol{\eta} = (\boldsymbol{\eta}_1 + \boldsymbol{\eta}_2)/2$, if we can prove that $\boldsymbol{\eta}_x$ is a Dirac delta for $\bar\rho$-a.e. x we deduce that $(\boldsymbol{\eta}_1)_x = (\boldsymbol{\eta}_2)_x = \boldsymbol{\eta}_x$ for $\bar\rho$-a.e. x, thus $\boldsymbol{\eta}_1 = \boldsymbol{\eta}_2$.

To show that $\boldsymbol{\eta}_x$ is a Dirac delta for $e_{0\#}\boldsymbol{\eta}$-a.e. x, let us consider the function

$$\Phi_{\delta,\zeta}(t) := \iiint \log\left(1 + \frac{|\gamma^1(t) - \eta^1(t)|}{\zeta\delta} + \frac{|\gamma^2(t) - \eta^2(t)|}{\delta}\right) d\boldsymbol{\eta}_x(\gamma) d\boldsymbol{\eta}_x(\eta)\, d\bar{\rho}(x),$$

where δ, $\zeta \in (0, 1)$ are small parameters to be chosen later, $t \in [0, T]$, $\bar{\rho} := (e_0)_\#\boldsymbol{\eta}$, and we use the notation $\gamma(t) = (\gamma^1(t), \gamma^2(t)) \in \mathbb{R}^d \times \mathbb{R}^d$. It is clear that $\Phi_{\delta,\zeta}(0) = 0$.

Let us define the probability measure $\mu \in \mathscr{P}\left(\mathbb{R}^d \times C([0, T]; \mathbb{R}^d)^2\right)$ by $d\mu(x, \eta, \gamma) := d\boldsymbol{\eta}_x(\eta) d\boldsymbol{\eta}_x(\gamma) d\bar{\rho}(x)$, and assume by contradiction that $\boldsymbol{\eta}_x$ is not a Dirac delta for $\bar{\rho}$-a.e. x. This means that there exists a constant $a > 0$ such that

$$\iiint \left(\int_0^T \min\{|\gamma(t) - \eta(t)|, 1\}\, dt\right) d\mu(x, \eta, \gamma) \geq a.$$

By Fubini's Theorem this implies that there exists a time $t_0 \in (0, T]$ such that

$$\iiint \min\{|\gamma(t_0) - \eta(t_0)|, 1\}\, d\mu(x, \eta, \gamma) \geq \frac{a}{T}.$$

Since the integrand is bounded by 1 and the measure μ has mass 1, this means that the set

$$A := \left\{(x, \eta, \gamma) : \min\{|\gamma(t_0) - \eta(t_0)|, 1\} \geq \frac{a}{2T}\right\}$$

has μ-measure at least $a/(2T)$. Then, assuming without loss of generality that $a \leq 2T$, this implies that $|\gamma(t_0) - \eta(t_0)| \geq a/(2T)$ for all $(x, \eta, \gamma) \in A$, hence

$$\begin{aligned} \Phi_{\delta,\zeta}(t_0) &\geq \iiint_A \log\left(1 + \frac{|\gamma^1(t_0) - \eta^1(t_0)|}{\zeta\delta} + \frac{|\gamma^2(t_0) - \eta^2(t_0)|}{\delta}\right) d\mu(x, \eta, \gamma) \\ &\geq \frac{a}{2T} \log\left(1 + \frac{a}{2\delta T}\right). \end{aligned} \tag{1.14}$$

We now want to show that this is impossible.

Computing the time derivative of $\Phi_{\delta,\zeta}$ we see that

$$\begin{aligned}\frac{d\Phi_{\delta,\zeta}}{dt}(t) \le \int_{\mathbb{R}^d}\int\int\Bigg(&\frac{|\boldsymbol{b}_1(\gamma^2(t)) - \boldsymbol{b}_1(\eta^2(t))|}{\zeta\left(\delta + |\gamma^2(t) - \eta^2(t)|\right)} \\ &+ \frac{\zeta|\boldsymbol{b}_2(\gamma^1(t)) - \boldsymbol{b}_2(\eta^1(t))|}{\zeta\delta + |\gamma^1(t) - \eta^1(t)|}\Bigg)\, d\mu(x,\eta,\gamma).\end{aligned} \tag{1.15}$$

By our assumption on $\boldsymbol{b}_1$, the first summand is easily estimated using the Lipschitz regularity of $\boldsymbol{b}_1$ in B_R:

$$\int_{\mathbb{R}^d}\int\int \frac{|\boldsymbol{b}_1(\gamma^2(t)) - \boldsymbol{b}_1(\eta^2(t))|}{\zeta(\delta + |\gamma^2(s) - \eta^2(s)|)}\, d\mu(x,\eta,\gamma) \le \frac{\|\nabla \boldsymbol{b}_1\|_{L^\infty(B_R)}}{\zeta}. \tag{1.16}$$

To estimate the second integral we show that for some constant C, which depends only on d, $|\rho|(\mathbb{R}^d)$ and R, one has

$$\begin{aligned}\iiint &\frac{\zeta|K*\rho(\gamma^1(t)) - K*\rho(\eta^1(t))|}{\zeta\delta + |\gamma^1(t) - \eta^1(t)|}\, d\mu(x,\eta,\gamma) \\ &\le C\zeta\Big(1 + \log\Big(\frac{C}{\zeta\delta}\Big)\Big).\end{aligned} \tag{1.17}$$

To this end, we first recall the definition of weak L^p norm of a μ-measurable function $f : X \to \mathbb{R}$ in a measure space (X, μ):

$$|||f|||_{M^p(X,\mu)} := \sup\{\lambda\, \mu(\{|f| > \lambda\})^{1/p} : \lambda > 0\}.$$

By [36, Proposition 4.2 and Theorem 3.3(ii)], there exists a modified maximal operator $\tilde{M}$, which associates to every function of the form $DK * \sigma$, $\sigma \in \mathscr{M}_+(\mathbb{R}^d)$, the function $\tilde{M}(DK*\sigma) \in L^1(\mathbb{R}^d)$ with the following properties: there exists a set L with $\mathscr{L}^d(L) = 0$ such that

$$\begin{aligned}|K*\sigma(x) - K*\sigma(y)| \le C\big[\tilde{M}(DK*\sigma)(x) + \tilde{M}(DK*\sigma)(y)\big]\,|x-y| \\ \forall x,\ y \in \mathbb{R}^d \setminus L,\end{aligned} \tag{1.18}$$

and the weak-L^1 estimate

$$|||\tilde{M}(DK*\rho)|||_{M^1(B_R)} \le C|\rho|(\mathbb{R}^d) \tag{1.19}$$

holds with a constant C which depends only on d and R. Applying (1.18), we see that

$$\iiint \frac{|K*\rho(\gamma^1(t)) - K*\rho(\eta^1(t))|}{\zeta\delta + |\gamma^1(t) - \eta^1(t)|}\, d\mu \le \int g_t(x,\eta,\gamma)\, d\mu, \tag{1.20}$$

where

$$g_t(x,\eta,\gamma) := \min\left\{C\tilde{M}(DK*\rho)(\gamma^1(t)) + C\tilde{M}(DK*\rho)(\eta^1(t)),\right.$$
$$\left.\frac{|K*\rho|(\gamma^1(t)) + |K*\rho|(\eta^1(t))}{\zeta\delta}\right\}.$$

Let us fix $p := \frac{d}{d-1/2} \in \left(1, \frac{d}{d-1}\right)$, so that $|K| \in L^p_{\mathrm{loc}}(\mathbb{R}^d)$. The last term in (1.20) can be estimated thanks to the following interpolation inequality (see [36, Lemma 2.2])

$$\|g_t\|_{L^1(\mu)} \le \frac{p}{p-1}|||g_t|||_{M^1(\mu)}\left(1+\log\left(\frac{|||g_t|||_{M^p(\mu)}}{|||g_t|||_{M^1(\mu)}}\right)\right).$$

Then, the first term in the right-hand side above can be estimated using our assumption $(e_t)_\#\boldsymbol{\eta} \le C_0\mathscr{L}^d$ and (1.19):

$$\begin{aligned}
|||g_t|||_{M^1(\mu)} &\le 2|||\tilde{M}(DK*\rho)(\eta^1(t))|||_{M^1(\mu)} \\
&= 2|||\tilde{M}(DK*\rho)(\eta^1(t))|||_{M^1(\boldsymbol{\eta})} \\
&= 2|||\tilde{M}(DK*\rho)(x)|||_{M^1(B_R\times B_R, e_{t\#}\boldsymbol{\eta})} \\
&\le 2C_0|||\tilde{M}(DK*\rho)(x)|||_{M^1(B_R\times B_R,\mathscr{L}^{2d})} \\
&\le 2C_0\mathscr{L}^d(B_R)|||\tilde{M}(DK*\rho)(x)|||_{M^1(B_R,\mathscr{L}^d)} \\
&\le 2C_0C\mathscr{L}^d(B_R)|\rho|(\mathbb{R}^d).
\end{aligned}$$

Similarly, the second term in the right hand side can be estimated using $(e_t)_\#\boldsymbol{\eta} \le C_0\mathscr{L}^d$ and Young's inequality:

$$\begin{aligned}
|||g_t|||_{M^p(\mu)} &\le 2(\zeta\delta)^{-1}\|(K*\rho)(\eta^1(t))\|_{L^p(\mu)} \\
&= 2(\zeta\delta)^{-1}\|(K*\rho)(\eta^1(t))\|_{L^p(\boldsymbol{\eta})} \\
&\le 2C_0(\zeta\delta)^{-1}\|(K*\rho)(x)\|_{L^p(B_R\times B_R)} \\
&\le 2C_0(\zeta\delta)^{-1}\mathscr{L}^d(B_R)\|(K*\rho)\|_{L^p(B_R)} \\
&\le 2C_0(\zeta\delta)^{-1}\mathscr{L}^d(B_R)\,\|K\|_{L^p(B_R)}|\rho|(\mathbb{R}^d) \\
&\le C(\zeta\delta)^{-1},
\end{aligned}$$

where C depends on d, R, and $|\rho|(\mathbb{R}^d)$. Combining these last estimates with (1.20), we obtain (1.17).

Then, using (1.15), (1.16), and (1.17), we deduce that

$$\frac{d\Phi_{\delta,\zeta}}{dt}(t) \le \frac{C}{\zeta} + C\zeta + C\zeta\log\left(\frac{C}{\zeta\delta}\right)$$

for some constant C depending only on $d, R, |\rho|(\mathbb{R}^d)$, and $\|\nabla b_1\|_{L^\infty(\mathbb{R}^d)}$. Integrating with respect to time in $[0, t_0]$, we find that

$$\Phi_{\delta,\zeta}(t_0) \le C t_0 \left(\frac{1}{\zeta} + \zeta + \zeta \log\left(\frac{C}{\zeta}\right) + \zeta \log\left(\frac{1}{\delta}\right) \right).$$

Choosing first $\zeta > 0$ small enough in order to have $Ct_0\zeta < a/(2T)$ and then letting $\delta \to 0$, we find a contradiction with (1.14), which concludes the proof. □

1.5. Optimal transport

In this section we present the basic optimal transport tools that are needed in order to build physical solutions of the semigeostrophic system, as we will do in Chapter 9. We refer to [124, 18] for a presentation of the topic and to [94] for the regularity theory of the related Monge-Ampère equation.

Given two probability measures $\mu, \nu \in \mathscr{P}(\mathbb{R}^d)$ we consider all *transport maps* T that "move μ onto ν", namely that satisfy the relation $T_{\#}\mu = \nu$. Among these maps, we look for the minimizers for the Monge problem

$$\inf\left\{ \int_{\mathbb{R}^d} |x - S(x)|^2 \, d\mu(x) : S_\sharp \mu = \nu \right\}. \tag{1.21}$$

A natural relaxation of this problem is the one where we move μ onto ν and we allow the splitting of mass. In other words, we consider as *transport plan* every $\gamma \in \mathscr{P}(\mathbb{R}^d \times \mathbb{R}^d)$ with $(\pi_1)_\sharp\gamma = \mu$ and $(\pi_2)_\sharp\gamma = \nu$ (here π_1 and π_2 are, respectively, the projections on the first and second factor).

The relaxed minimization problem, due to Kantorovich, is

$$\min\left\{ \int_{\mathbb{R}^{2d}} |x - y|^2 d\gamma(x, y) : \gamma \in \mathscr{P}(\mathbb{R}^{2d}),\ (\pi_1)_\sharp\gamma = \mu,\ (\pi_2)_\sharp\gamma = \nu \right\}, \tag{1.22}$$

A fundamental result of Brenier says that, under mild assumptions on the initial and final measures, the unique optimizer for problem (1.22) is given by a map, which can be represented as the gradient of a convex function.

Theorem 1.15 (Brenier). *Let $\mu, \nu \in \mathscr{P}(\mathbb{R}^d)$ with $\mu \ll \mathscr{L}^d$ and*

$$\int_{\mathbb{R}^d} |x|^2 d\mu(x) + \int_{\mathbb{R}^d} |y|^2 d\nu(y) < \infty. \tag{1.23}$$

Then there exists a unique minimizer γ in (1.22). *Moreover the plan γ is induced by the gradient of a convex function u, that is $\gamma = (\mathrm{Id} \times \nabla u)_\sharp \mu$ and thus ∇u is also a solution to* (1.21).

If we assume the function ∇u of the previous theorem to be a smooth diffeomorphism between two smooth densities $\rho_1\,dx$ and $\rho_2\,dx$, by the change of variable formula we see that for every test function $\varphi \in C_c^\infty(\mathbb{R}^d)$

$$\int_{\mathbb{R}^d} \varphi(\nabla u(x))\rho_1(x)\,dx = \int_{\mathbb{R}^d} \varphi(y)\rho_2(y)\,dy$$
$$= \int_{\mathbb{R}^d} \varphi(\nabla u(x))\rho_2(\nabla u(x))\det \nabla^2 u(x)\,dx.$$

Hence, u solves the *Monge-Ampère equation*

$$\det \nabla^2 u = \frac{\rho_1}{\rho_2 \circ \nabla u} \qquad \text{in } \mathbb{R}^d.$$

When the function u is simply convex, without any smoothness assumption, one may consider different notions of solution of the Monge-Ampère equation. We give here the definition of Aleksandrov solution, which is a key concept in order to study the regularity of optimal maps.

In order to introduce this notion, we consider a convex domain $\Omega \subseteq \mathbb{R}^d$ and a convex function $u : \Omega \to \mathbb{R}$; we recall that the subdifferential of u is given by

$$\partial u(x) = \{p \in \mathbb{R}^n : u(y) \geq u(x) + p \cdot (y - x)\ \forall\, y \in \Omega\}$$

and we define the *Monge-Ampère measure* of u as

$$\mu_u(E) = \mathscr{L}^d(\partial u(E)) = \mathscr{L}^d\Big(\bigcup_{x\in E} \partial u(x)\Big) \quad \text{for every set } E \subset \Omega, \tag{1.24}$$

where $\mathscr{L}^d$ denotes in the previous formula the Lebesgue *outer* measure. The main properties of the Monge-Ampère measure are the following:

- if $u \in C^2(\Omega)$, the Area Formula implies $\mu_u = \det \nabla^2 u \mathscr{L}^d$;
- the restriction of μ_u to the Borel σ-algebra is a measure;
- the absolutely continuous part of μ_u with respect to the Lebesgue measure is given by $\frac{d\mu_u}{d\mathscr{L}^d} = \det \nabla^2 u$ [1].

Thanks to these properties, we can give the following definition.

Definition 1.16. Given an open convex set Ω and a Borel measure μ on Ω, a convex and continuous function $u : \Omega \to \mathbb{R}$ is said an *Aleksandrov solution* to the Monge-Ampère equation

$$\det \nabla^2 u = \mu,$$

if $\mu = \mu_u$ as Borel measures.

[1] Since u is a convex function, its Hessian is a matrix-valued locally finite measure. In this case, we denote by $\nabla^2 u$ the density of the absolutely continuous part of the Hessian.

We finally mention that an important connection between the optimal transport problem and the theory of flows of vector fields is given by the Benamou-Brenier formula. Under the assumptions of Theorem 1.15, we can consider the optimal map ∇u; the optimal cost between μ and ν can be found by looking at all vector fields which move μ onto ν and then minimizing the kinetic energy of this "dynamical transport plan"

$$\begin{aligned}&\int_{\mathbb{R}^d} |x - \nabla u(x)|^2 \, d\mu(x)\\ &= \min \left\{ \int_0^1 \int_{\mathbb{R}^d} |\boldsymbol{v}_t|^2 \, d\rho_t : \partial_t \rho_t + \nabla \cdot (\boldsymbol{v}_t \rho_t) = 0, \ \rho_0 = \mu, \ \rho_1 = \nu \right\}.\end{aligned}$$

In the next two subsections we give more precise statements on the 2-dimensional torus $\mathbb{T}^2$, which are suitable for the application in Chapter 9.

1.5.1. Existence and uniqueness of optimal transport maps on the torus

The following theorem can be found in [65].

Theorem 1.17 (Existence of optimal maps on $\mathbb{T}^2$). *Let μ and ν be $\mathbb{Z}^2$-periodic Radon measures on $\mathbb{R}^2$ such that $\mu([0,1)^2) = \nu([0,1)^2) = 1$ and $\mu = \rho\mathscr{L}^2$ with $\rho > 0$ almost everywhere. Then there exists a unique (up to an additive constant) convex function $P : \mathbb{R}^2 \to \mathbb{R}$ such that $(\nabla P)_\sharp \mu = \nu$ and $P - |x|^2/2$ is $\mathbb{Z}^2$-periodic. Moreover*

$$\nabla P(x+h) = \nabla P(x) + h \qquad \text{for a.e. } x \in \mathbb{R}^2, \ \forall h \in \mathbb{Z}^2, \tag{1.25}$$

$$|\nabla P(x) - x| \le \operatorname{diam}(\mathbb{T}^2) = \frac{\sqrt{2}}{2} \qquad \text{for a.e. } x \in \mathbb{R}^2. \tag{1.26}$$

In addition, if $\mu = \rho\mathscr{L}^2$, $\nu = \sigma\mathscr{L}^2$, and there exist constants $0 < \lambda \le \Lambda < \infty$ such that $\lambda \le \rho, \sigma \le \Lambda$, then P is a strictly convex Alexandrov solution of

$$\det \nabla^2 P(x) = f(x), \qquad \textit{with } f(x) = \frac{\rho(x)}{\sigma(\nabla P(x))}.$$

Proof. Existence of P follows from [65]. To prove uniqueness we observe that, under our assumption, also the convex conjugate $p^*(y) := P^*(y) - |y|^2/2$ is $\mathbb{Z}^2$-periodic. Hence, since

$$P(x) = \sup_{y \in \mathbb{R}^2} x \cdot y - P^*(y),$$

we get that the function $p(x) := P(x) - |x|^2/2$ satisfies

$$\begin{aligned} p(x) &= \sup_{y\in\mathbb{R}^2} \Big(- \frac{|y-x|^2}{2} - P^*(y) + \frac{|y|^2}{2} \Big) \\ &= \sup_{y\in[0,1]^2} \sup_{h\in\mathbb{Z}^2} \Big(- \frac{|y+h-x|^2}{2} - p^*(y+h) \Big) \\ &= \sup_{y\in\mathbb{T}^2} \Big(- \frac{d^2_{\mathbb{T}^2}(x,y)}{2} - p^*(y) \Big), \end{aligned}$$

where $d_{\mathbb{T}^2}$ is the quotient distance on the torus, and we used that $p^*(y)$ is $\mathbb{Z}^2$-periodic. This means that the function p is $d^2_{\mathbb{T}^2}$-convex, and that p^* is its $d^2_{\mathbb{T}^2}$-transform (compare with [124, Chapter 5]). Hence $\nabla P = Id + \nabla p : \mathbb{T}^2 \to \mathbb{T}^2$ is the unique (μ-a.e.) optimal transport map sending μ onto ν ([109, Theorem 9]), and since $\rho > 0$ almost everywhere this uniquely characterizes P up to an additive constant. Finally, all the other properties of P follow from [65]. □

1.5.2. Regularity of optimal transport maps on the torus

Theorem 1.17 can be combined with the regularity results for strictly convex Alexandrov solutions of the Monge-Ampère equation (see [41, 42, 43, 65, 73, 91]), which are completely local and therefore work in $\mathbb{R}^d$ as well as on the torus. The main regularity results are summarized in the next theorem.

Theorem 1.18 (Space regularity of optimal maps on $\mathbb{T}^2$). *Let $\mu = \rho\mathscr{L}^2$, $\nu = \sigma\mathscr{L}^2$ be $\mathbb{Z}^2$-periodic Radon measures on $\mathbb{R}^2$ such that $\mu([0,1)^2) = \nu([0,1)^2) = 1$, let $0 < \lambda \le \Lambda < \infty$ such that $\lambda \le \rho, \sigma \le \Lambda$, and let P be as in Theorem 1.17 with $\int_{\mathbb{T}^2} P\,dx = 0$. Then:*

(i) *$P \in C^{1,\beta}(\mathbb{T}^2)$ for some $\beta = \beta(\lambda,\Lambda) \in (0,1)$, and there exists a constant $C = C(\lambda,\Lambda)$ such that*

$$\|P\|_{C^{1,\beta}} \le C.$$

(ii) *$P \in W^{2,1}(\mathbb{T}^2)$, more precisely for any $k \in \mathbb{N}$ there exists a constant $C = C(\lambda,\Lambda,k)$ such that*

$$\int_{\mathbb{T}^2} |\nabla^2 P| \log^k_+ |\nabla^2 P|\,dx \le C$$

and there exist a constant $C = C(\lambda)$ and an exponent $\gamma_0 = \gamma_0(\lambda) > 1$ such that

$$\int_{\mathbb{T}^2} |\nabla^2 P|^{\gamma_0}\,dx \le C. \tag{1.27}$$

(iii) *If $\rho, \sigma \in C^{k,\alpha}(\mathbb{T}^2)$ for some $k \in \mathbb{N}$ and $\alpha \in (0,1)$, then $P \in C^{k+2,\alpha}(\mathbb{T}^2)$ and there exists a constant $C = C(\lambda, \Lambda, \|\rho\|_{C^{k,\alpha}}, \|\sigma\|_{C^{k,\alpha}})$ such that*

$$\|P\|_{C^{k+2,\alpha}} \leq C.$$

Moreover, there exist two positive constants c_1 and c_2, depending only on λ, Λ, $\|\rho\|_{C^{0,\alpha}}$, and $\|\sigma\|_{C^{0,\alpha}}$, such that

$$c_1 Id \leq \nabla^2 P(x) \leq c_2 Id \qquad \forall x \in \mathbb{T}^2.$$

1.6. A few glimpses of classical regularity theory for elliptic equations

The aim of this section is to give the basic setting and the fundamental tools for the development of Chapters 6 and 7, which deal with the regularity theory of local minimizers of certain variational integrals. To this end, we first provide a general introduction to some aspects of the classical regularity theory. We also focus on some useful ideas and lemmas, that will be employed in Chapters 6 and 7.

As in the setting of Hilbert's XIX problem, given an open set $\Omega \subseteq \mathbb{R}^d$, a convex function $\mathcal{F} : \mathbb{R}^d \to \mathbb{R}$, and an integrable function $f : \Omega \to \mathbb{R}$, we consider local minimizers $u : \Omega \to \mathbb{R}$ of the functional

$$\int_\Omega \mathcal{F}(\nabla u) + fu \tag{1.28}$$

whose Euler-Lagrange equation can be written as

$$\nabla \cdot (\nabla \mathcal{F}(\nabla u)) = f \qquad \text{in } \Omega \tag{1.29}$$

or equivalently $\partial_i(\partial_i \mathcal{F}(\nabla u)) = f$ (here and in the following we use the Einstein's summation convention, omitting the summation sign).

Given bounded, measurable coefficients a_{ij} and an integrable function $g : \Omega \to \mathbb{R}$, we also consider solutions of the equation

$$\partial_i(a_{ij}\partial_j v) = \partial_i g^i \qquad \text{in } \Omega \tag{1.30}$$

(notice that the partial derivatives $\partial_e u$ of the solution of (1.29) formally solve this kind of equation, where the coefficients a_{ij} are taken to be $\partial_{ij}\mathcal{F}$, as it can be easily seen differentiating the Euler-Lagrange equation with respect to ∂_e).

The fundamental result of De Giorgi says that any solution to a uniformly elliptic operator is locally Hölder continuous.

Theorem 1.19 (De Giorgi-Nash-Moser). *Let $\Lambda,\ \lambda > 0$, $q > d$, $(a_{ij})_{i,j=1,\dots,d} : B_1 \to \mathbb{R}^{d\times d}$ be measurable coefficients with*

$$\lambda I \le \big(a_{ij}(x)\big) \le \Lambda I \qquad \text{for } \mathscr{L}^d\text{-a.e. } x \in B_1.$$

Let $g \in L^q(B_1;\mathbb{R}^d)$ and let $v \in W^{1,2}(B_1)$ be a distributional solution of (1.30).

Then there exist constants $\alpha := \alpha(d,\lambda,\Lambda) \in (0,1)$ and $C := C(d,\lambda,\Lambda) > 0$ such that

$$\|v\|_{C^\alpha(B_{1/2})} \le C\|v\|_{L^2(B_1)}.$$

For the sake of completeness, we also mention the regularity result of Schauder, which assumes some regularity of the coefficients and of the right-hand side.

Theorem 1.20 (Schauder). *Let $\lambda > 0$, $k \in \mathbb{N}\cup\{0\}$, $\alpha \in (0,1)$, $(a_{ij})_{i,j=1,\dots,d} \in C^{k,\alpha}(B_1;\mathbb{R}^{d\times d})$ be such that*

$$\lambda I \le \big(a_{ij}(x)\big) \qquad \text{for every } x \in B_1.$$

Let $g \in C^{k,\alpha}(B_1;\mathbb{R}^d)$ and let $v \in W^{1,2}(B_1)$ be a distributional solution of (1.30). *Then $v \in C^{k+1,\alpha}(B_{1/2})$.*

By the previous results, it follows that any minimizer of (1.28) is of class $C^\infty_{\rm loc}(\Omega)$ as soon as we assume $\mathcal{F}$, $f \in C^\infty(\Omega)$ and

$$\lambda I \le \nabla^2\mathcal{F}(x) \le \Lambda I \qquad \text{for any } x \in \Omega,$$

for some $0 < \lambda < \Lambda < \infty$.

To conclude this introductory section, we recall the validity of weak Harnack inequalities for supersolutions of elliptic equations (see [91, Theorem 8.18]) that will play a crucial role in the proof of Lemma 6.10. A function $v \in W^{1,2}_{\rm loc}(\Omega)$ is said to be a distributional supersolution of (1.30) if for every smooth, nonnegative, compactly supported test function $\varphi \in C^\infty_c(\Omega)$, we have

$$\int_\Omega a_{ij}(x)\partial_j v(x)\partial_i\varphi(x)\,dx \le \int_\Omega g^i(x)\partial_i\varphi(x)\,dx.$$

Theorem 1.21. *Let $\Lambda, \lambda > 0$, $q > d$, $r > 0$, let a_{ij} be measurable coefficients with*

$$\lambda I \le \big(a_{ij}(x)\big) \le \Lambda I \qquad \text{for any } x \in B_r,$$

and let $g \in L^q(B_r; \mathbb{R}^d)$. Let $v \in W^{1,2}(B_r)$ be a nonnegative supersolution of

$$\partial_i(a_{ij}(x)\partial_j v) \leq \partial_i g^i \qquad \text{in } B_r. \tag{1.31}$$

Then there exists a constant $c_0 := c_0(d, \lambda, \Lambda) > 0$ such that

$$\inf\{v(x) : x \in B_{r/4}\} \geq c_0 \fint_{B_{r/2}} v(x)\, dx - r^{1-d/q}\|g\|_{L^q(B_r)}.$$

1.6.1. The p-Laplacian

When the ellipticity condition on the hessian of $\mathcal{F}$ fails at one point, several regularity results for local minimizers of (1.28) are still available. For instance, we consider the model case is given by the p-Dirichlet energy, that is

$$\int_\Omega |\nabla u|^p + fu, \tag{1.32}$$

whose Euler-Lagrange equation reads as

$$p\,\nabla \cdot (|\nabla u|^{p-2}\nabla u) = f \qquad \text{in } \Omega. \tag{1.33}$$

The $C^{1,\alpha}$ regularity of any local minimizer u (even in the vectorial case) has been proved in a series of papers by Uraltseva [121], Uhlenbeck [119], and Evans [85] for $p \geq 2$, and by Lewis [103] and Tolksdorff [118] for $p > 1$ (see also [75, 122]). Notice that in this case the equation is uniformly elliptic outside the origin.

Theorem 1.22. *Let $p \in (1, \infty)$, $q > d$, $g \in L^q(B_1)$ and let $u \in W^{1,p}(\Omega)$ be a local minimizer of (1.28). Then we have that $u \in C^{1,\alpha}(B_{1/2})$ for some $\alpha > 0$.*

One cannot expect, in general, more than Hölder continuity of the gradient of local minimizers of (1.32). Indeed, the function $u(x) = |x|^{1+\alpha}$, whose gradient is $\nabla u(x) = |x|^\alpha x$, satisfies

$$p\,\nabla\cdot(|\nabla u|^{p-2}\nabla u) = p\,\nabla\cdot(|x|^{(p-1)\alpha-1}x) = ((p-1)\alpha-1+n)|x|^{(p-1)\alpha-1}$$

Hence, if we choose $\alpha = (p-1)^{-1}$, u solves the Euler-Lagrange equation (1.33) with $f = n$, hence it is a local minimizer of (1.32), but it is not twice differentiable at the origin for $p > 2$.

1.6.2. The non-variational problem

The following lemma concerns elliptic equations in non-divergence form

$$a_{ij}\partial_{ij}u = f \qquad \text{in } B_1. \tag{1.34}$$

Although it could be stated for non-smooth viscosity supersolutions of the equation above, for simplicity we state it as an a priori estimate on smooth solutions.

Theorem 1.23 (Alexandroff-Bakelman-Pucci estimate). *Let $\Lambda, \lambda > 0$, $(a_{ij})_{i,j=1,\dots,d} \in C(B_1; \mathbb{R}^{d\times d})$ be such that*

$$\lambda I \le \big(a_{ij}(x)\big) \le \Lambda I \qquad \textit{for every } x \in B_1.$$

Let $g \in C(B_1)$ and let $u \in C^2(B_1)$ be a supersolution of (1.34) such that $u \ge 0$ on $\partial B_{1/2}$.

Then, denoting by Γ_u the convex envelope of u, namely the largest non-positive convex function in B_1 that lies below u in $B_{1/2}$, there exists a constant $C := C(d, \Lambda, \lambda)$ such that

$$\sup_{B_{1/2}}(u_-)^d \le C \int_{\{x\in B_1: u=\Gamma_u\}} (f_+)^d \, dx.$$

1.6.3. A few basic lemmas

In this section we present two classical criteria to prove the Hölder regularity of a function (or of its gradient).

Given a locally integrable function $u : \mathbb{R}^d \to \mathbb{R}$, we denote its average on $B_r(x)$ by

$$(u)_{B_r(x)} = \fint_{B_r(x)} u(y)\, dy = \frac{1}{|B_r(x)|} \int_{B_r(x)} u(y)\, dy$$

and its mean oscillation in $B_r(x)$ by

$$\fint_{B_r(x)} |u(y) - (u)_{B_r(x)}|\, dy.$$

The first result exploits the classical equivalence between Hölder spaces and Campanato spaces; the proof can be found in [89, Theorem 1.3, Section III].

Lemma 1.24 (Campanato's description of Hölder continuity). *Let $M > 0$, $\alpha \in (0, 1)$, and $u \in L^2(B_1)$ be a function. Let us assume that for every $x \in B_{1/2}$ and $r \in [0, 1/2)$ we have*

$$\left(\fint_{B_r(x)} |u - (u)_{B_r(x)}|^2\right)^{1/2} \le Mr^\alpha.$$

Then $u \in C^{0,\alpha}(B_{1/2})$ and $[u]_{C^{0,\alpha}(B_{1/2})} \le cM$ for some constant $c := c(d)$.

The following lemma is a classical description of $C^{1,\alpha}$ regularity of a function and was used, for instance, in the context of the regularity theory of minimal surfaces. Its proof can be found in [45, Section 1] or [73, Lemma 3.1] (under the assumption that u is also convex).

Lemma 1.25. *Let $\alpha \in (0,1]$, $M > 0$, $\rho \in (0,1/2)$. Let $u : B_1 \to \mathbb{R}$ be a Lipschitz function such that for every $x \in B_{1/2}$ there exists $A_x \in \mathbb{R}^d$ such that the plane passing through x and of slope A_x well approximates u*

$$|u(y) - u(x) - A_x \cdot (y - x)| \leq M|y - x|^{1+\alpha} \qquad \forall\, y \in B_\rho(x).$$

Then $u \in C^{1,\alpha}(B_{1/2})$ and $[\nabla u]_{C^{0,\alpha}(B_{1/2})} \leq cM$ for some constant $c := c(d)$.

Chapter 2
Maximal regular flows for non-smooth vector fields

Given a vector field $\boldsymbol{b}_t(x)$ in $\mathbb{R}^d$, the theory of DiPerna-Lions, introduced in the seminal paper [81], provides existence and uniqueness of the flow (in the almost everywhere sense, with respect to Lebesgue measure $\mathscr{L}^d$) under weak regularity assumptions on $\boldsymbol{b}$, for instance when $\boldsymbol{b}_t(\cdot)$ is Sobolev [81] or BV [5] and satisfies global bounds on the divergence. In this respect, this theory could be considered as a weak Cauchy-Lipschitz theory for ODE's. This analogy is confirmed by many global existence results, by a kind of Lusin type approximation of DiPerna-Lions flows by Lipschitz flows [14, 67], and even by differentiability properties of the flow [101]. However, this analogy is presently not perfect, and the main aim of this Chapter is to fill this gap.

Indeed, the Cauchy-Lipschitz theory is not only pointwise but also purely local, whereas the DiPerna-Lions theory is an almost everywhere theory and relies on *global* in space growth estimates on $|\boldsymbol{b}|$, like

$$\frac{|\boldsymbol{b}_t(x)|}{1+|x|} \in L^1\big((0,T); L^1(\mathbb{R}^d)\big) + L^1\big((0,T); L^\infty(\mathbb{R}^d)\big). \tag{2.1}$$

This is in contrast with the fact that the so-called "renormalization property", which plays a key role in the theory, seems to depend only on local properties of $\boldsymbol{b}$, because it deals with distributional solutions to a continuity/transport equation with a source term: as a matter of fact, it is proved using only local regularity properties of $\boldsymbol{b}$.

Given an open set $\Omega \subset \mathbb{R}^d$, in this Chapter we consider vector fields $\boldsymbol{b} : (0,T) \times \Omega \to \mathbb{R}^d$ satisfying only the local integrability property $\int_0^T \int_{\Omega'} |\boldsymbol{b}| dx dt < \infty$ for all $\Omega' \Subset \Omega$, a local one-sided bound on the distributional divergence, and the property that the continuity equation with velocity $\boldsymbol{b}$ is well-posed in the class of nonnegative bounded and compactly supported functions in Ω. Some of these assumptions have already been introduced in Section 1.3 and, as illustrated in Remark 1.9, the last assumption is fulfilled in many cases of interest and it is known to be deeply linked to the uniqueness of the flow; in addition, building on the superposition principle (Theorem 1.6), it is proved in Section 2.3 that even this assumption is purely local, as well as the other two ones concerning integrability and bounds on divergence.

Under these three assumptions we prove existence of a unique *maximal regular flow* $\boldsymbol{X}(t,x)$ in Ω, defined up to a *maximal time* $T_{\Omega,\boldsymbol{X}}(x)$ which is positive $\mathscr{L}^d$-a.e. in Ω, with

$$\limsup_{t\uparrow T_{\Omega,\boldsymbol{X}}(x)} V_\Omega(\boldsymbol{X}(t,x)) = \infty \qquad \text{for } \mathscr{L}^d\text{-a.e. } x \in \{T_{\Omega,\boldsymbol{X}} < T\}. \tag{2.2}$$

Here $V_\Omega : \Omega \to [0,\infty)$ is a given continuous "confining potential", namely with $V(x) \to \infty$ as $x \to \partial\Omega$; hence, (2.2) is a synthetic way to state that, for any $\Omega' \Subset \Omega$, $\boldsymbol{X}(t,x)$ is not contained in Ω' for t close to $T_{\Omega,\boldsymbol{X}}(x)$.

In our axiomatization, which parallels the one of [5] and slightly differs from the one of the DiPerna-Lions theory (being only based on one-sided bounds on divergence and independent of the semigroup property), "maximal" refers to (2.2), while "regular" means the existence of constants $C(\Omega',\boldsymbol{X})$ such that

$$\int_{\Omega'\cap\{\mathsf{h}_{\Omega'}>t\}} \phi(\boldsymbol{X}(t,x))\,dx \le C(\Omega',\boldsymbol{X}) \int_{\mathbb{R}^d} \phi(y)\,dy \qquad \text{for all } \phi \in C_c(\mathbb{R}^d) \text{ nonnegative} \tag{2.3}$$

for all $t \in [0,T]$, $\Omega' \Subset \Omega$, where $\mathsf{h}_{\Omega'}(x) \in [0, T_{\Omega,\boldsymbol{X}}(x)]$ is the first time that $\boldsymbol{X}(\cdot,x)$ hits $\mathbb{R}^d \setminus \Omega'$. Under global bounds on the divergence, (2.3) can be improved to

$$\int_{\Omega\cap\{T_{\Omega,\boldsymbol{X}}>t\}} \phi(\boldsymbol{X}(t,x))\,dx \le C_* \int_{\mathbb{R}^d} \phi(y)\,dy \qquad \text{for all } \phi \in C_c(\mathbb{R}^d) \text{ nonnegative} \tag{2.4}$$

for all $t \in [0,T]$, but many structural properties can be proved with (2.3) only.

Uniqueness of the maximal regular flow follows basically from the "probabilistic" techniques developed in [5], which allow one to transfer uniqueness results at the level of the PDE (the continuity equation), here axiomatized, into uniqueness results at the level of the ODE. Existence follows by analogous techniques; the main new difficulty here is that even if we truncate $\boldsymbol{b}$ by multiplying it by a $C^\infty_c(\Omega)$ cut-off function, the resulting vector field has not divergence in L^∞ (just L^1, actually, when $|\boldsymbol{b}_t| \notin L^\infty_{\mathrm{loc}}(\Omega)$), hence the standard theory is not applicable. Hence, several new ideas and techniques need to be introduced to handle this new situation. These results are achieved in Section 2.2.

Besides existence and uniqueness, we discuss in the next chapter the natural semigroup and stability properties of maximal regular flows, as well as the proper blow up of trajectories. The concepts introduced in this Chapter, together with their properties described in Chapter 3 and 4, will be applied in Chapter 8 to describe the lagrangian structure of weak solutions of the Vlasov-Poisson equation and to prove existence of weak solutions with L^1 summability of the initial datum.

2.1. Regular flow, hitting time, maximal flow

Definition 2.1 (Local regular flow). Let $B \in \mathcal{B}(\mathbb{R}^d)$, $\tau > 0$, and $\boldsymbol{b} : (0, \tau) \times \mathbb{R}^d \to \mathbb{R}^d$ Borel. We say that $\boldsymbol{X} : [0, \tau] \times B \to \mathbb{R}^d$ is a *local regular flow* starting from B (relative to $\boldsymbol{b}$) up to τ if the following two properties hold:

(i) for $\mathscr{L}^d$-a.e. $x \in B$, $\boldsymbol{X}(\cdot, x) \in AC([0, \tau]; \mathbb{R}^d)$ and solves the ODE $\dot{x}(t) = \boldsymbol{b}_t(x(t))$ $\mathscr{L}^1$-a.e. in $(0, \tau)$, with the initial condition $\boldsymbol{X}(0, x) = x$;

(ii) there exists a constant $C = C(\boldsymbol{X})$ satisfying $\boldsymbol{X}(t, \cdot)_\#(\mathscr{L}^d \llcorner B) \leq C\mathscr{L}^d$.

In the previous definition, as long as the image of $[0, \tau] \times B$ through $\boldsymbol{X}$ is contained in an open set Ω, it is not necessary to specify the vector field $\boldsymbol{b}$ outside Ω. By Theorem 1.12 we obtain a consistency result of the local regular flows with values in Ω in the intersection of their domains.

Lemma 2.2 (Consistency of local regular flows). *Assume that $\boldsymbol{b}$ satisfies (a-Ω) and (b-Ω). Let $\boldsymbol{X}_i$ be local regular flows starting from B_i up to τ_i, $i = 1, 2$, with $\boldsymbol{X}_i([0, \tau_i] \times B_i) \subset \Omega$. Then*

$$\boldsymbol{X}_1(\cdot, x) \equiv \boldsymbol{X}_2(\cdot, x) \quad \text{in } [0, \tau_1 \wedge \tau_2], \text{for } \mathscr{L}^d\text{-a.e. } x \in B_1 \cap B_2. \tag{2.5}$$

Proof. Take $B \subset B_1 \cap B_2$ Borel with $\mathscr{L}^d(B)$ finite, and apply Theorem 1.12 with $T = \tau_1 \wedge \tau_2, m = d$, and

$$\boldsymbol{\lambda} := \frac{1}{2} \int \left(\delta_{\boldsymbol{X}_1(\cdot,x)} + \delta_{\boldsymbol{X}_2(\cdot,x)}\right) d\mathscr{L}^d_B(x),$$

where $\mathscr{L}^d_B$ is the normalized Lebesgue measure on B. □

If we consider a smooth vector field $\boldsymbol{b}$ in a domain Ω, a maximal flow of $\boldsymbol{b}$ in Ω would be given by the trajectories of $\boldsymbol{b}$ until they hit the boundary of Ω. In order to deal at the same time with bounded and unbounded domains (including the case $\Omega = \mathbb{R}^d$) we introduce a continuous potential function $V_\Omega : \Omega \to [0, \infty)$ satisfying

$$\lim_{x \to \partial\Omega} V_\Omega(x) = \infty, \tag{2.6}$$

meaning that for any $M > 0$ there exists $K \Subset \Omega$ with $V_\Omega > M$ on $\Omega \setminus K$ (in particular, when $\Omega = \mathbb{R}^d$, $V_\Omega(x) \to \infty$ as $|x| \to \infty$). For instance, an admissible potential is given by $V_\Omega(x) = \max\{[\mathrm{dist}(x, \mathbb{R}^d \setminus \Omega)]^{-1}, |x|\}$.

Definition 2.3 (Hitting time in Ω). Let $\tau > 0$, $\Omega \subset \mathbb{R}^d$ open and $\eta : [0, \tau) \to \mathbb{R}^d$ continuous. We define the *hitting time* of η in Ω as

$$\mathsf{h}_\Omega(\eta) := \sup\{t \in [0, \tau) : \max_{[0,t]} V_\Omega(\eta) < \infty\},$$

with the convention $\mathsf{h}_\Omega(\eta) = 0$ if $\eta(0) \notin \Omega$.

It is easily seen that this definition is independent of the choice of V_Ω, that $\mathsf{h}_\Omega(\eta) > 0$ whenever $\eta(0) \in \Omega$, and that

$$\mathsf{h}_\Omega(\eta) < \tau \quad \Longrightarrow \quad \limsup_{t \uparrow \mathsf{h}_\Omega(\eta)} V_\Omega(\eta(t)) = \infty. \tag{2.7}$$

Using V_Ω we can also define the concept of maximal regular flow, where "regular" refers to the local bounded compression condition (2.8).

Definition 2.4 (Maximal regular flow in an open set Ω). Let $\boldsymbol{b}:(0,T)\times \Omega \to \mathbb{R}^d$ be a Borel vector field. We say that a Borel map $\boldsymbol{X}$ is a *maximal regular flow* relative to $\boldsymbol{b}$ in Ω if there exists a Borel map $T_{\Omega,\boldsymbol{X}} : \Omega \to (0, T]$ such that $\boldsymbol{X}(t, x)$ is defined in the set $\{(t, x) : t < T_{\Omega,\boldsymbol{X}}(x)\}$ and the following properties hold:

(i) for $\mathscr{L}^d$-a.e. $x \in \Omega$, $\boldsymbol{X}(\cdot, x) \in AC_{\mathrm{loc}}([0, T_{\Omega,\boldsymbol{X}}(x)); \mathbb{R}^d)$, and solves the ODE $\dot{x}(t) = \boldsymbol{b}_t(x(t))$ $\mathscr{L}^1$-a.e. in $(0, T_{\Omega,\boldsymbol{X}}(x))$, with the initial condition $\boldsymbol{X}(0, x) = x$;

(ii) for any $\Omega' \Subset \Omega$ there exists a constant $C(\Omega', \boldsymbol{X})$ such that

$$\boldsymbol{X}(t, \cdot)_\#(\mathscr{L}^d \llcorner \{T_{\Omega'} > t\}) \le C(\Omega', \boldsymbol{X}) \mathscr{L}^d \llcorner \Omega' \quad \forall t \in [0, T], \tag{2.8}$$

where

$$T_{\Omega'}(x) := \begin{cases} \mathsf{h}_{\Omega'}(\boldsymbol{X}(\cdot, x)) & \text{for } x \in \Omega', \\ 0 & \text{otherwise;} \end{cases} \tag{2.9}$$

(iii) $\limsup_{t \uparrow T_{\Omega,\boldsymbol{X}}(x)} V_\Omega(\boldsymbol{X}(t,x)) = \infty$ for $\mathscr{L}^d$-a.e. $x \in \Omega$ such that $T_{\Omega,\boldsymbol{X}}(x) < T$.

Notice that (2.8) could be equivalently written as

$$X(t,\cdot)_\#(\mathscr{L}^d \llcorner \{T_{\Omega'} > t\}) \le C(\Omega', X)\mathscr{L}^d \qquad \text{for all } t \in [0,T],$$

because the push-forward measure is concentrated on Ω'; so the real meaning of this requirement is that the push forward measure must have a bounded density w.r.t. $\mathscr{L}^d$. In turn, (2.8) is *not* equivalent to require that $(X(t,\cdot)_\#\mathscr{L}^d)\llcorner\Omega' \le C(\Omega', X)\mathscr{L}^d$, since trajectories may be compressed while they are outside Ω' and then enter Ω' again. Let us emphasize that our assumption (3.41) is *not* equivalent to require that $(X(t,\cdot)_\#\mathscr{L}^d)\llcorner\Omega' \le C(\Omega', X)\mathscr{L}^d$. Indeed, with our assumption trajectories may be compressed when they are outside Ω' and then

Indeed in our case we are only assuming that the flow has bounded compression as long as the trajectories remain inside Ω', while the latter assumption In turn, (3.41) is *not* equivalent to require that $(X(t,\cdot)_\#\mathscr{L}^d)\llcorner\Omega' \le C(\Omega', X)\mathscr{L}^d$, since trajectories may be compressed while they are outside Ω' and then enter Ω' again.

Remark 2.5 (Maximal regular flows induce regular flows). Given any maximal regular flow X in Ω, $\tau \in (0,T)$, and a Borel set $B \subset \Omega$ such that $T_{\Omega,X} > \tau$ on B and

$$\big\{X(t,x) : x \in B,\ t \in [0,\tau]\big\} \Subset \Omega,$$

we have an induced local regular flow in the set B up to time τ.

Remark 2.6 (Invariance in the equivalence class of b). It is important and technically useful (see for instance [6]) to underline that the concepts of local regular flow and of maximal regular flow are invariant in the Lebesgue equivalent class, exactly as our constitutive assumptions (a-Ω), (b-Ω), and the global/local bounds on the divergence of b. Indeed, for local regular flows, Definition 2.1(ii) in conjunction with Fubini's theorem implies that for any $\mathscr{L}^{1+d}$-negligible set $N \subset (0,T)\times\mathbb{R}^d$ the set

$$\big\{x \in B : \mathscr{L}^1(\{t \in (0,\tau) : (t, X(t,x)) \in N\}) > 0\big\}$$

is $\mathscr{L}^d$-negligible. An analogous argument, based on (2.8), applies to maximal regular flows.

2.2. Existence and uniqueness of the maximal regular flow

In this section we consider a Borel vector field $b : (0,T)\times\Omega \to \mathbb{R}^d$ such that the spatial divergence $\operatorname{div} b_t(\cdot)$ in the sense of distributions satisfies

$$\forall\, \Omega' \Subset \Omega, \quad \operatorname{div} \boldsymbol{b}_t(\cdot) \geq m(t) \quad \text{in } \Omega', \text{with } L(\Omega', \boldsymbol{b}) := \int_0^T |m(t)|\, dt < \infty \tag{2.10}$$

and which satisfies the assumptions (a-Ω), (b-Ω) of Section 1.3, namely

(a-Ω) $\int_0^T \int_{\Omega'} |\boldsymbol{b}_t(x)|\, dxdt < \infty$ for any $\Omega' \Subset \Omega$;
(b-Ω) for any nonnegative $\bar\rho \in L^\infty_+(\Omega)$ with compact support in Ω and any closed interval $I = [a, b] \subset [0, T]$, the continuity equation

$$\frac{d}{dt}\rho_t + \operatorname{div}(\boldsymbol{b}_t \rho_t) = 0 \qquad \text{in } (a, b) \times \Omega$$

has at most one weakly* continuous solution $I \ni t \mapsto \rho_t \in \mathcal{L}_{I,\Omega}$ (defined in (1.9)) with $\rho_a = \bar\rho$.

Remark 2.7. Assumption (2.10) could be weakened to $m \in L^1(0, T_0)$ for all $T_0 \in (0, T)$, but we made it global in time to avoid time-dependent constants in our estimates (and, in any case, the maximal flow could be obtained in this latter case by a simple gluing procedure w.r.t. time).

In order to construct a maximal regular flow, we would like to approximate the vector field $\boldsymbol{b}$ by convolution and then consider a suitable weak limit of the approximated flows. However, due to the lack of global bounds on $\boldsymbol{b}$, we have to exclude the possibility that trajectories escape to infinity faster and faster as the convolution parameter vanishes, because in this case the existence time of the limit trajectory would be 0. The following a priori estimate excludes this phenomenon by showing that the blow-up time is strictly positive almost everywhere.

Remark 2.8 (An a priori estimate on the existence time). Let us consider a locally integrable vector field $\boldsymbol{b} : (0, T) \times \mathbb{R}^d \to \mathbb{R}^d$ and a maximal regular flow $\boldsymbol{X}$ defined in the set $\{(s, x) : s < T_{\boldsymbol{X}}(x)\}$. Then for every $t \in (0, T)$ and $r > 0$

$$|\{x \in B_{r/2} : T_{\boldsymbol{X}}(x) \leq t\}| \leq \frac{2}{r} C(B_r, \boldsymbol{X}) \int_0^t \int_{B_r} |\boldsymbol{b}_s(x)|\, dx\, ds, \tag{2.11}$$

where $C(B_r, \boldsymbol{X})$ is the constant appearing in (2.8). In particular, letting $t \to 0$ in (2.11) with r fixed, we find that for every $r > 0$ the existence time $T_{\boldsymbol{X}}$ is strictly positive $\mathscr{L}^d$-a.e. in $B_{r/2}$.

Indeed, let us consider the existence time $T_{B_r,\boldsymbol{X}}$ defined as in (2.9) and let $T'_{B_r,\boldsymbol{X}} = \min\{t, T_{B_r,\boldsymbol{X}}\}$. We notice that $r/2 \leq |\boldsymbol{X}(T'_{B_r,\boldsymbol{X}}(x), x) -$

$X(0,x)|$ on the set $|\{x \in B_{r/2} : T_{B_r,X}(x) \le t\}|$ and therefore, by the properties of the maximal regular flow, we have

$$\begin{aligned}
|\{x \in B_{r/2} : T_X(x) \le t\}| &\le |\{x \in B_{r/2} : T_{B_r,X}(x) \le t\}| \\
&\le \frac{2}{r}\int_{B_r} |X(T'_{B_r,X}(x),x) - X(0,x)|\,dx \\
&\le \frac{2}{r}\int_{B_r}\int_0^{T'_{B_r,X}(x)} |\dot{X}(s,x)|\,ds\,dx \\
&\le \frac{2}{r}\int_{B_r}\int_0^{T'_{B_r,X}(x)} |b_s(X(s,x))|\,ds\,dx \\
&\le \frac{2}{r}\int_0^t \int_{B_r\cap\{T_{B_r,X}(x)>s\}} |b_s(X(s,x))|\,dx\,ds \\
&\le \frac{2}{r}C(B_r,X)\int_0^t\int_{B_r} |b_s(x)|\,dx\,ds.
\end{aligned}$$

This proves (2.8).

Estimate (2.11) is enough to prove local existence of a regular flow, but once this object has been built it is not clear how to extend each trajectory up to the blow-up time. For this reason, we first build maximal regular flows in every bounded open set A, compactly contained in our given domain Ω, where $\boldsymbol{b}$ is integrable (in this context "maximal" refers to the fact that a trajectory may hit the boundary of the open set A). Then, we glue these flows together to obtain a global existence result.

We also mention that a different approach to construct maximal regular flows in $\mathbb{R}^d$, instead of building the local ones first, consists in employing the one-point compactification of $\mathbb{R}^d$ and a "damped" stereographic projection, with damping chosen in therms of the vector field $\boldsymbol{b}$, as it will be done in Section 4.3. However, this method of proof requires a global bound on the divergence of $\boldsymbol{b}$ that can be replaced with a local bound following the approach below.

The first step in the construction of the maximal regular flow will be the following local existence result.

Theorem 2.9 (Local existence). *Let $\boldsymbol{b} : (0,T) \times \Omega \to \mathbb{R}^d$ be a Borel vector field which satisfies (a-Ω), (b-Ω), (2.10), and let $A \Subset \Omega$ be open. Then there exist a Borel map $T_A : A \to (0,T]$ and a Borel map $X(t,x)$, defined for $x \in A$ and $t \in [0, T_A(x)]$, such that:*

(a) *for $\mathscr{L}^d$-a.e. $x \in A$, $X(\cdot,x) \in AC([0,T_A(x)];\mathbb{R}^d)$, $X(0,x) = x$, $X(t,x) \in A$ for all $t \in [0,T_A(x))$, and $X(T_A(x),x) \in \partial A$ when $T_A(x) < T$;*

(b) *for* $\mathscr{L}^d$*-a.e.* $x \in A$, $\boldsymbol{X}(\cdot, x)$ *solves the ODE* $\dot{\gamma} = \boldsymbol{b}_t(\gamma)$ *in* $(0, T_A(x))$*;*
(c) $\boldsymbol{X}(t, \cdot)_\#(\mathscr{L}^d \llcorner \{T_A > t\}) \leq e^{L(A,\boldsymbol{b})} \mathscr{L}^d \llcorner A$ *for all* $t \in [0, T]$*, where* $L(A, \boldsymbol{b})$ *is the constant in* (2.10).

Notice that since the statement of the theorem is local (see also Section 2.3, in connection with property (b-Ω)), we need only to prove it under the assumption $|\boldsymbol{b}| \in L^1((0, T) \times \Omega)$, which is stronger than (a-Ω).

We will obtain Theorem 2.9 via an approximation procedure which involves the concept of regular generalized flow in closed domains, where now "regular" refers to the fact that the bounded compression condition is imposed only in the interior of the domain.

Definition 2.10 (Regular generalized flow in $\overline{A}$). Let $A \subset \mathbb{R}^d$ be an open set and let $\boldsymbol{c} : (0, T) \times \overline{A} \to \mathbb{R}^d$ be a Borel vector field. A probability measure $\boldsymbol{\eta}$ in $C([0, T]; \mathbb{R}^d)$ is said to be a *regular generalized flow* on $\overline{A}$ if the following two conditions hold:

(i) $\boldsymbol{\eta}$ is concentrated on

$$\left\{\eta \in AC([0, T]; \overline{A}) : \dot{\eta}(t) = \boldsymbol{c}_t(\eta(t)) \text{ for } \mathscr{L}^1\text{-a.e. } t \in (0, T)\right\};$$

(ii) there exists $C := C(\boldsymbol{\eta}) \in (0, \infty)$ satisfying

$$((e_t)_\# \boldsymbol{\eta}) \llcorner A \leq C \mathscr{L}^d \qquad \forall t \in [0, T]. \tag{2.12}$$

Any constant C for which (2.12) holds is called a *compressibility constant* of $\boldsymbol{\eta}$.

The class of regular generalized flows enjoys good tightness and stability properties. We recall that a sequence $\boldsymbol{\eta}^n \in \mathscr{P}\big(C([0, T]; \overline{A})\big)$ is said *tight* if for every $\varepsilon > 0$ there exists a compact set $\Gamma_\varepsilon \subseteq C([0, T]; \overline{A})$ such that $\boldsymbol{\eta}^n(C([0, T]; \overline{A}) \setminus \Gamma_\varepsilon) \leq \varepsilon$ for every $n \in \mathbb{N}$. Equivalently, the sequence $\boldsymbol{\eta}^n$ is said to be tight if there exists a coercive, lower semicontinuous functional $\Sigma : C([0, T]; \overline{A}) \to [0, \infty]$ such that $\sup_{n \in \mathbb{N}} \int \Sigma \, d\boldsymbol{\eta}^n < \infty$. We state the tightness and stability properties in the case of interest for us, namely when the velocity vanishes at the boundary. Notice that, in the following theorem, assumption (2.13) requires the convergence of the vector fields in $L^1((0, T) \times A; \mathbb{R}^d)$ and not only in L^1_{loc}; this allows to apply Dunford-Pettis's theorem to prove the tightness of any sequence of generalized regular flows with bounded compressibility constants.

Theorem 2.11. (Tightness and stability of regular generalized flows in $\overline{A}$). *Let* $A \subset \mathbb{R}^d$ *be a bounded open set, let* $\boldsymbol{c}, \boldsymbol{c}^n : (0, T) \times \overline{A} \to \mathbb{R}^d$ *be Borel vector fields such that* $\boldsymbol{c} = \boldsymbol{c}^n = 0$ *on* $(0, T) \times (\mathbb{R}^d \setminus A)$ *and*

$$\lim_{n \to \infty} \boldsymbol{c}^n = \boldsymbol{c} \qquad \text{in } L^1((0, T) \times A; \mathbb{R}^d). \tag{2.13}$$

Let $\boldsymbol{\eta}^n \in \mathscr{P}\big(C([0,T];\overline{A})\big)$ *be regular generalized flows of* $\boldsymbol{c}^n$ *in* $\overline{A}$ *and let us assume that the best compressibility constants* C_n *of* $\boldsymbol{\eta}^n$ *satisfy* $\sup_n C_n < \infty$. *Then* $(\boldsymbol{\eta}^n)$ *is tight, any limit point* $\boldsymbol{\eta}$ *is a regular generalized flow of* $\boldsymbol{c}$ *in* $\overline{A}$, *and the following implication holds:*

$$\begin{aligned}&((e_t)_\#\boldsymbol{\eta}^n \llcorner \Gamma)\llcorner A' \leq c_n \mathscr{L}^d \\ &\text{for some } c_n > 0 \quad \Longrightarrow \quad ((e_t)_\#\boldsymbol{\eta} \llcorner \Gamma)\llcorner A' \leq \liminf_n c_n \mathscr{L}^d\end{aligned} \tag{2.14}$$

for any choice of open sets $\Gamma \subset C([0,T];\overline{A})$ *and* $A' \subset A$.

In the previous theorem the assumption that all the vector fields vanish on the boundary of A allows us to say the following: if an integral curve of $\boldsymbol{c}^n$ in A hits ∂A and stops there, then it is still an integral curve of $\boldsymbol{c}^n$ on the whole $\overline{A}$. We remark that the previous theorem is invariant if the vector fields $\boldsymbol{c}^n$ are modified on a set of Lebesgue measure zero in $(0,T)\times A$, thanks to the compressibility condition (2.12) required in A; on the contrary, the value of $\boldsymbol{c}^n$ on ∂A has to be understood in a pointwise sense.

Proof of Theorem 2.11. By Dunford-Pettis' theorem, since the family $\{\boldsymbol{c}^n\}$ is compact in $L^1(\overline{A};\mathbb{R}^d)$ (recall that $\boldsymbol{c}_n(t,\cdot)$ vanish outside of A), there exists a modulus of integrability for $\boldsymbol{c}^n$, namely an increasing, convex, superlinear function $F:[0,\infty)\to[0,\infty)$ such that $F(0)=0$ and

$$\sup_{n\in\mathbb{N}} \int_0^T \int_{\overline{A}} F(|\boldsymbol{c}^n(t,x)|)\,dx\,dt < \infty. \tag{2.15}$$

Let us introduce the functional $\Sigma : C([0,T];\mathbb{R}^d) \to [0,\infty]$ as follows

$$\Sigma(\eta) := \begin{cases} \int_0^T F(|\dot\eta(t)|)\,dt & \text{if } \eta \in AC([0,T];\overline{A}), \\ \infty & \text{if } \eta \in C([0,T];\mathbb{R}^d)\setminus AC([0,T];\overline{A}). \end{cases}$$

Using Ascoli-Arzelà theorem, the compactness of $\overline{A}$, and a well-known lower semicontinuity result due to Ioffe (see for instance [17, Theorem 5.8]), it turns out that Σ is lower semicontinuous and coercive, namely its sublevels $\{\Sigma \leq M\}$ are compact.

Since $\boldsymbol{\eta}^n$ is concentrated on $AC([0,T];\overline{A})$ we get

$$\begin{aligned}\int \Sigma\,d\boldsymbol{\eta}^n &= \int\int_0^T F(|\dot\eta|)\,dt\,d\boldsymbol{\eta}^n(\eta) = \int_0^T \int_{\overline{A}} F(|\boldsymbol{c}^n|)\,d[(e_t)_\#\boldsymbol{\eta}^n]\,dt \\ &\leq C_n \int_0^T \int_{\overline{A}} F(|\boldsymbol{c}^n|)\,dx\,dt,\end{aligned}$$

so that that $\int \Sigma\,d\boldsymbol{\eta}^n$ is uniformly bounded thanks to (2.15). Therefore Prokhorov compactness theorem provides the existence of limit points.

Since Σ is lower semicontinuous we obtain that any limit point $\boldsymbol{\eta}$ satisfies $\int \Sigma \, d\boldsymbol{\eta} < \infty$, therefore $\boldsymbol{\eta}$ is concentrated on $AC([0,T];\overline{A})$.

Let $C := \liminf_{n\in\mathbb{N}} C_n < \infty$. Since $(e_t)_\# \boldsymbol{\eta}^n$ narrowly converge to $(e_t)_\#\boldsymbol{\eta}$, we know that for any open set $A' \subset A$ there holds

$$(e_t)_\#\boldsymbol{\eta}(A') \le \liminf_{n\to\infty} (e_t)_\#\boldsymbol{\eta}^n(A') \le C\mathscr{L}^d(A') \qquad \forall t \in [0,T].$$

Since A' is arbitrary we deduce that $\boldsymbol{\eta}$ satisfies (2.12). A similar argument provides its localized version (2.14). To show that $\boldsymbol{\eta}$ is concentrated on integral curves of $\boldsymbol{c}$, it suffices to show that

$$\int \left| \eta(t) - \eta(0) - \int_0^t \boldsymbol{c}_s(\eta(s))ds \right| d\boldsymbol{\eta}(\eta) = 0 \tag{2.16}$$

for any $t \in [0,T]$. The technical difficulty is that this test function, due to the lack of regularity of $\boldsymbol{c}$, is not continuous with respect to η. To this aim, we prove that

$$\int \left| \eta(t) - \eta(0) - \int_0^t \boldsymbol{c}'_s(\eta(s))ds \right| d\boldsymbol{\eta}(\eta) \le C \int_{(0,T)\times A} |\boldsymbol{c} - \boldsymbol{c}'| \, dx \, dt \tag{2.17}$$

for any continuous vector field $\boldsymbol{c}' : [0,T] \times \overline{A} \to \mathbb{R}^d$ with $\boldsymbol{c}' = 0$ in $[0,T] \times \partial A$. Then, choosing a sequence $(\boldsymbol{c}'_n)$ of such vector fields converging to $\boldsymbol{c}$ in $L^1(\overline{A};\mathbb{R}^d)$ and noticing that

$$\begin{aligned}\int\int_0^T |\boldsymbol{c}_s(\eta(s)) - \boldsymbol{c}'_{ns}(\eta(s))| \, ds d\boldsymbol{\eta}(\eta) &= \int_0^T \int_A |\boldsymbol{c} - \boldsymbol{c}'_n| \, d(e_s)_\#\boldsymbol{\eta} \, ds \\ &\le C_n \int_{(0,T)\times A} |\boldsymbol{c} - \boldsymbol{c}'_n| \, dx \, dt,\end{aligned}$$

converges to 0 as n goes to ∞, we can take the limit in (2.17) with $\boldsymbol{c}' = \boldsymbol{c}'_n$ to obtain (2.16).

It remains to show (2.17). This is a limiting argument based on the fact that (2.16) holds for $\boldsymbol{c}^n$, $\boldsymbol{\eta}^n$:

$$\begin{aligned}&\int \left| \eta(t) - \eta(0) - \int_0^t \boldsymbol{c}'_s(\eta(s)) \, ds \right| d\boldsymbol{\eta}^n(\eta) \\ &= \int \left| \int_0^t \left(\boldsymbol{c}^n_s(\eta(s) - \boldsymbol{c}'_s(\eta(s)) \right) ds \right| d\boldsymbol{\eta}^n(\eta) \\ &\le \int \int_0^t |\boldsymbol{c}^n_s - \boldsymbol{c}'_s|(\eta(s)) \, ds \, d\boldsymbol{\eta}^n(\eta) \\ &= \int_0^t \int_A |\boldsymbol{c}^n_s - \boldsymbol{c}'_s| \, d[(e_s)_\#\boldsymbol{\eta}^n] \, ds \le C_n \int_0^t \int_A |\boldsymbol{c}^n - \boldsymbol{c}'| \, dx \, ds.\end{aligned}$$

Taking the limit in the chain of inequalities above we obtain (2.17). □

Now we show how Theorem 2.9 can be deduced from the existence of regular generalized flows in $\overline{A}$; indeed, assumption (b-Ω) allows to prove through Theorem 1.12 that generalized regular flows on $\overline{A}$ are induced by a proper regular flow defined on A. In the second part of the proposition, we show that flows associated to sufficiently smooth vector fields induce regular generalized flows (actually even classical ones, but we will need them in generalized form to take limits).

Proposition 2.12. (i) *Let* $\boldsymbol{b} : (0,T) \times \Omega \to \mathbb{R}^d$ *be a Borel vector field which satisfies (a-Ω) and (b-Ω), let* $A \Subset \Omega$ *be an open set, and let* $\boldsymbol{\eta}$ *be a regular generalized flow in* $\overline{A}$ *relative to* $\boldsymbol{c} = \chi_A \boldsymbol{b}$ *with compressibility constant* C *and that satisfies* $(e_0)_\# \boldsymbol{\eta} = \rho_0 \mathscr{L}^d$ *with* $\rho_0 > 0$ $\mathscr{L}^d$*-a.e. in* A*. Then there exist* $\boldsymbol{X}$ *and* T_A *as in Theorem 2.9(a)-(b) that satisfy*

$$\boldsymbol{X}(t,\cdot)_\#(\rho_0 \llcorner \{T_A > t\}) \le C \mathscr{L}^d \llcorner A \tag{2.18}$$

for all $t \in [0,T]$.

(ii) *Let* $\boldsymbol{b} \in C^\infty([0,T] \times \overline{A}; \mathbb{R}^d)$*. Then there exists a regular generalized flow* $\boldsymbol{\eta}$ *associated to* $\boldsymbol{b}\chi_A$*, with* $(e_0)_\# \boldsymbol{\eta}$ *equal to the normalized Lebesgue measure in* A *and satisfying*

$$((e_t)_\# \boldsymbol{\eta} \llcorner \{\mathsf{h}_{A'}(\cdot) > t\}) \llcorner A' \le \frac{e^{L(A',\boldsymbol{b})}}{\mathscr{L}^d(A)} \mathscr{L}^d \qquad \forall t \in [0,T] \tag{2.19}$$

for any open set $A' \Subset A$.

Proof. We first prove (i). Set $\mu_0 = \rho_0 \mathscr{L}^d$ and consider a family $\{\boldsymbol{\eta}_x\} \subset \mathscr{P}\big(C([0,T];\overline{A})\big)$ of conditional probability measures, concentrated on

$$\big\{\eta \in AC([0,T];\overline{A}) : \dot{\eta} = \boldsymbol{c}_t(\eta)\ \mathscr{L}^1\text{-a.e. in } (0,T), \eta(0) = x\big\}$$

and representing $\boldsymbol{\eta}$, *i.e.*, $\int \boldsymbol{\eta}_x \, d\mu_0(x) = \boldsymbol{\eta}$. We claim that μ_0-almost every $x \in A$:

(1) $\mathsf{h}_A(\eta)$ is equal to a positive constant for $\boldsymbol{\eta}_x$-a.e. η;
(2) if $T_A(x)$ is the constant in (1), $(e_t)_\# \boldsymbol{\eta}_x$ is a Dirac mass for all $t \in [0, T_A(x)]$.

By our assumption on μ_0, the properties stated in the claim hold $\mathscr{L}^d$-a.e. in A. Hence, given the claim, if we define

$$\boldsymbol{X}(t,x) := \int \eta(t) \, d\boldsymbol{\eta}_x(\eta)$$

then for $\mathscr{L}^d$-a.e. $x \in A$ the integrand $\eta(t)$ is independent of η as soon as $t < T_A(x)$, hence $\boldsymbol{X}(t,x)$ satisfies (a) and (b) in the statement of

Theorem 2.9. The compressibility property (2.18) follows immediately from (2.12).

Let us prove our claim. We notice that the hitting time is positive for μ_0-a.e. $x \in A$. For $q \in \mathbb{Q} \cap (0, T)$, we shall denote by Γ_q the set $\{\eta : \mathsf{h}_A(\eta) > q\}$ and by $\Sigma^q : \Gamma_q \to C([0, q]; A)$ the map induced by restriction to $[0, q]$, namely $\Sigma^q(\eta) = \eta|_{[0,q]}$.

In order to prove the claim it clearly suffices to show that, for all $q \in \mathbb{Q} \cap (0, T)$, $\Sigma^q_\#(\boldsymbol{\eta}_x \llcorner \Gamma_q)$ is either a Dirac mass or it is null. So, for $q \in \mathbb{Q} \cap (0, T)$ and $\delta \in (0, 1)$ fixed, it suffices to show that

$$\boldsymbol{\lambda}_x := \frac{1}{\boldsymbol{\eta}_x(\Gamma_q)} \Sigma^q_\#(\boldsymbol{\eta}_x \llcorner \Gamma_q) \in \mathscr{P}\big(C([0, q]; A)\big)$$

is a Dirac mass for μ_0-a.e. x satisfying $\boldsymbol{\eta}_x(\Gamma_q) \geq \delta$.

By construction the measures $\boldsymbol{\lambda}_x$ satisfy $\boldsymbol{\lambda}_x \leq \Sigma^q_\#(\boldsymbol{\eta}_y \llcorner \Gamma_q)/\delta$ and they are concentrated on curves $[0, q] \ni t \mapsto \eta(t)$ starting at x and solving the ODE $\dot\eta = \boldsymbol{b}_t(\eta)$ in $(0, q)$. Therefore

$$\boldsymbol{\lambda} := \int_{\{x \in A:\ \boldsymbol{\eta}_x(\Gamma_q) \geq \delta\}} \boldsymbol{\lambda}_x \, d\mu_0(x) \in \mathscr{P}\big(C([0, q]; A)\big)$$

satisfies all the assumptions of Theorem 1.12 with $T = q$ and $\Omega = A$, provided we check (1.12). To check this property with $C_0 = C/\delta$, for $t \in [0, q]$ and $\varphi \in C_c(A)$ nonnegative we use the fact that $\boldsymbol{\lambda}_y \leq \Sigma^q_\#(\boldsymbol{\eta}_y \llcorner \Gamma_q)/\delta$ and the fact that C is a compressibility constant of $\boldsymbol{\eta}$ to estimate

$$\int_{\mathbb{R}^d} \varphi \, d(e_t)_\# \boldsymbol{\lambda} \leq \frac{1}{\delta} \int_{\mathbb{R}^d} \varphi \, d(e_t)_\#(\boldsymbol{\eta} \llcorner \Gamma_q) \leq \frac{1}{\delta} \int_{\mathbb{R}^d} \varphi \, d(e_t)_\# \boldsymbol{\eta} \leq \frac{C}{\delta} \int_A \varphi \, dx.$$

Therefore Theorem 1.12 can be invoked: $\boldsymbol{\lambda}_x$ is a Dirac mass for μ_0-a.e. x and this gives that $\boldsymbol{\lambda}_x$ is a Dirac mass μ_0-a.e. in $\{\boldsymbol{\eta}_x(\Gamma_q) \geq \delta\}$. This concludes the proof of (i).

For (ii), we begin by defining $\boldsymbol{\eta}$ with the standard Cauchy-Lipschitz theory. More precisely, for $x \in A$ we let $\boldsymbol{X}(t, x)$ be the unique solution to the ODE $\dot\eta = \boldsymbol{b}_t(\eta)$ with $\eta(0) = x$ until the first time $T_A(x)$ that $\boldsymbol{X}(t, x)$ hits ∂A, and then we define $\boldsymbol{X}(t, x) = \boldsymbol{X}(t, T_A(x))$ for all $t \in [T_A(x), T]$. Finally, denoting by $\mathscr{L}^d_A$ the normalized Lebesgue measure in A, we define $\boldsymbol{\eta}$ as the law under $\mathscr{L}^d_A$ of the map $x \mapsto \boldsymbol{X}(\cdot, x)$. With this construction it is clear that condition (i) in Definition 2.10 holds.

Let us check condition (ii) as well, in the stronger form (2.19). Recall that $\boldsymbol{X}$ is smooth before the hitting time and that the map $t \mapsto J(t) := \det \nabla_x \boldsymbol{X}(t, x)$ is nonnegative and solves the ODE

$$\begin{cases} \dot J(t) = J(t) \operatorname{div} \boldsymbol{b}_t(\boldsymbol{X}(t, x)), \\ J(0) = 1. \end{cases} \tag{2.20}$$

Now, fix an open set $A' \Subset A$, and observe that (2.19) is equivalent to prove that for every $t \in [0, T]$

$$\int_{A' \cap \{x : \mathsf{h}_{A'}(X(\cdot,x)) > t\}} \varphi(X(t,x))dx \le e^{L(A',\boldsymbol{b})} \int_{A'} \varphi(x)dx \quad \text{for every } \varphi \in C_c(A').$$

Fix $\varphi \in C_c(A')$ nonnegative and notice that $\varphi(X(t,x)) = 0$ if $t \ge \mathsf{h}_{A'}(X(\cdot,x))$, hence $\operatorname{supp} \varphi \circ X(t,\cdot)$ is a compact subset of the open set $G_t := \{x : \mathsf{h}_{A'}(X(\cdot,x)) > t\}$. By the change of variables formula

$$\int_{\mathbb{R}^d} \varphi(X(t,x)) \det \nabla_x X(t,x)\, dx = \int_{\mathbb{R}^d} \varphi(x)\, dx,$$

in order to estimate from below the left-hand side it suffices to estimate from below $\det \nabla_x X(t,x)$ in G_t; using (2.20) and Gronwall's lemma, this estimate is provided by $e^{-L(A',\boldsymbol{b})}$. □

Remark 2.13. For the proof of Theorem 3.2 we record the following facts, proved but not stated in Proposition 2.12: if $\boldsymbol{\eta}$ is as in the statement of the proposition, then for $(e_0)_\# \boldsymbol{\eta}$-a.e. x the hitting time $\mathsf{h}_A(\eta)$ is equal to a positive constant $T_A(x)$ for $\boldsymbol{\eta}_x$-a.e. η; furthermore, $(e_t)_\# \boldsymbol{\eta}_x$ is a Dirac mass for all $t \in [0, T_A(x)]$.

Proof of Theorem 2.9. By the first part of Proposition 2.12, it suffices to build a regular generalized flow $\boldsymbol{\eta}$ in $\overline{A}$ relative to $\boldsymbol{c} = \chi_A \boldsymbol{b}$ with compressibility constant $e^{L(A,\boldsymbol{b})}/\mathscr{L}^d(A)$ such that $(e_0)_\# \boldsymbol{\eta} = \rho_0 \mathscr{L}^d$ with $\rho_0 > 0$ $\mathscr{L}^d$-a.e. in A. By the second part of the proposition, we have existence of $\boldsymbol{\eta}$ with $(e_0)_\# \boldsymbol{\eta}$ equal to the normalized Lebesgue measure $\mathscr{L}^d_A$ and satisfying (2.19) whenever $\boldsymbol{b} \in C^\infty([0,T] \times \overline{A}; \mathbb{R}^d)$.

Hence, to use this fact, extend $\boldsymbol{b}$ with the 0 value to $\mathbb{R} \times \mathbb{R}^d$ and let $\boldsymbol{b}_\varepsilon$ be mollified vector fields. We have that $L(A, \boldsymbol{b}_\varepsilon)$ are uniformly bounded (because $A \Subset \Omega$) and, in addition, the properties of convolution immediately yield

$$\limsup_{\varepsilon \downarrow 0} L(A', \boldsymbol{b}_\varepsilon) \le L(A, \boldsymbol{b}) \qquad \text{for any } A' \Subset A \text{ open.} \tag{2.21}$$

If $\boldsymbol{\eta}_\varepsilon$ are regular generalized flows associated to $\boldsymbol{c}_\varepsilon = \chi_A \boldsymbol{b}_\varepsilon$, we can apply Theorem 2.11 to get that any limit point $\boldsymbol{\eta}$ is a regular generalized flow associated to $\boldsymbol{c}$ and it satisfies $(e_0)_\# \boldsymbol{\eta} = \mathscr{L}^d_A$. In addition, given $A' \Subset A$ open we have

$$((e_t)_\# \boldsymbol{\eta}_\varepsilon \llcorner \{\mathsf{h}_{A'}(\cdot) > t\}) \llcorner A' \le \frac{e^{L(A',\boldsymbol{b}_\varepsilon)}}{\mathscr{L}^d(A)} \mathscr{L}^d \qquad \forall t \in [0,T],$$

thus (2.14) and (2.21) yield

$$((e_t)_\#\boldsymbol{\eta} \llcorner \{\mathsf{h}_{A'}(\cdot) > t\}) \llcorner A' \le \frac{e^{L(A,\boldsymbol{b})}}{\mathscr{L}^d(A)} \mathscr{L}^d \qquad \forall t \in [0, T].$$

Letting $A' \uparrow A$ gives that $e^{L(A,\boldsymbol{b})}/\mathscr{L}^d(A)$ is a compressibility constant for $\boldsymbol{\eta}$. □

Using a gluing procedure in space, we can now build the maximal regular flow in Ω using the flows provided by Theorem 2.9 in domains $\Omega_n \Subset \Omega_{n+1}$ with $\Omega_n \uparrow \Omega$.

Theorem 2.14. *Let $\boldsymbol{b} : (0, T) \times \Omega \to \mathbb{R}^d$ be a Borel vector field which satisfies (a-Ω) and (b-Ω). Then the maximal regular flow is unique, and existence is ensured under the additional assumption* (2.10). *In addition,*

(a) *for any $\Omega' \Subset \Omega$ the compressibility constant $C(\Omega', \boldsymbol{X})$ in Definition 2.4 can be taken to be $e^{L(\Omega',\boldsymbol{b})}$, where $L(\Omega', \boldsymbol{b})$ is the constant in* (2.10)*;*
(b) *if $\boldsymbol{Y}$ is a regular flow in B up to τ with values in Ω, then $T_{\Omega,\boldsymbol{X}} > \tau$ $\mathscr{L}^d$-a.e. in B and*

$$\boldsymbol{X}(\cdot, x) = \boldsymbol{Y}(\cdot, x) \quad \text{in } [0, \tau], \text{for } \mathscr{L}^d\text{-a.e. } x \in B. \tag{2.22}$$

Proof. Let us prove first the uniqueness of the maximal regular flow in Ω. Given regular maximal flows $\boldsymbol{X}^i$ in Ω, $i = 1, 2$, by Lemma 2.2 and Remark 2.5 we easily obtain

$$\boldsymbol{X}^1(\cdot, x) = \boldsymbol{X}^2(\cdot, x) \quad \text{in } [0, T_{\Omega,\boldsymbol{X}^1}(x) \wedge T_{\Omega,\boldsymbol{X}^2}(x)), \text{for } \mathscr{L}^d\text{-a.e. } x \in \Omega.$$

On the other hand, for $\mathscr{L}^d$-a.e. $x \in \{T_{\Omega,\boldsymbol{X}^1} > T_{\Omega,\boldsymbol{X}^2}\}$, the image of $[0, T_{\Omega,\boldsymbol{X}^2}(x)]$ through $V_\Omega(\boldsymbol{X}^1(\cdot, x))$ is bounded in $\mathbb{R}$, whereas the image of $[0, T_{\Omega,\boldsymbol{X}^2}(x))$ through $V_\Omega(\boldsymbol{X}^2(\cdot, x))$ is not. It follows that the set $\{T_{\Omega,\boldsymbol{X}^1} > T_{\Omega,\boldsymbol{X}^2}\}$ is $\mathscr{L}^d$-negligible. Reversing the roles of $\boldsymbol{X}^1$ and $\boldsymbol{X}^2$ we obtain that $T_{\Omega,\boldsymbol{X}^1} = T_{\Omega,\boldsymbol{X}^2}$ $\mathscr{L}^d$-a.e. in Ω.

In order to show existence we are going to use auxiliary flows $\boldsymbol{X}_n$ in Ω_n with hitting times $T_n : \Omega_n \to (0, T]$, *i.e.*,

(1) for $\mathscr{L}^d$-a.e. $x \in \Omega_n$, $\boldsymbol{X}_n(\cdot, x) \in AC([0, T_n(x)]; \mathbb{R}^d)$, $\boldsymbol{X}_n(0, x) = x$, $\boldsymbol{X}_n(t, x) \in \Omega_n$ for all $t \in [0, T_n(x))$, and $\boldsymbol{X}_n(T_n(x), x) \in \partial\Omega_n$ when $T_n(x) < T$, so that $\mathsf{h}_{\Omega_n}(\boldsymbol{X}_n(\cdot, x)) = T_n(x)$;
(2) for $\mathscr{L}^d$-a.e. $x \in \Omega_n$, $\boldsymbol{X}_n(\cdot, x)$ solves the ODE $\dot\gamma = \boldsymbol{b}_t(\gamma)$ in $(0, T_n(x))$;
(3) $\boldsymbol{X}_n(t, \cdot)_\#(\mathscr{L}^d \llcorner \{T_n > t\}) \le e^{L(\Omega_n,\boldsymbol{b})} \mathscr{L}^d \llcorner \Omega_n$ for all $t \in [0, T]$, where $L(\Omega_n, \boldsymbol{b})$ is given as in (2.10).

The existence of X_n, T_n as in (1), (2), (3) has been achieved in Theorem 2.9.

If $n \le m$, the uniqueness argument outlined at the beginning of this proof gives immediately that $T_n(x) \le T_m(x)$, and that $X_n(\cdot, x) \equiv X_m(\cdot, x)$ in $[0, T_n(x)]$ for $\mathscr{L}^d$-a.e. $x \in \Omega_n$. Hence the limits

$$T_{\Omega,X}(x) := \lim_{n\to\infty} T_n(x), \qquad X(t,x) = \lim_{n\to\infty} X_n(t,x) \quad t \in [0, T_{\Omega,X}(x)) \tag{2.23}$$

are well defined for $\mathscr{L}^d$-a.e. $x \in \Omega$. By construction

$$X(\cdot, x) = X_n(\cdot, x) \qquad \text{in } [0, T_n(x)), \text{for } \mathscr{L}^d\text{-a.e. } x \in \Omega_n. \tag{2.24}$$

We now check that X and $T_{\Omega,X}$ satisfy the conditions (i), (ii), (iii) of Definition 2.4. Property (i) is a direct consequence of property (2) of X_n, (2.23), and (2.24).

In connection with property (ii) of Definition 2.4, in the more specific form stated in (a) for any open set $\Omega' \Subset \Omega$, it suffices to check it for all open sets Ω_n: indeed, it is clear that in the uniqueness proof we need it only for a family of sets that invade Ω and, as soon as uniqueness is established, we can always assume in our construction that Ω' is one of the sets Ω_n. Now, given n, we first remark that property (1) of X_n yields $T_n(x) = \mathsf{h}_{\Omega_n}(X(\cdot, x))$ for $\mathscr{L}^d$-a.e. $x \in \Omega_n$; moreover (2.24) gives

$$X(t,\cdot)_\#(\mathscr{L}^d \llcorner \{T_n > t\}) = X_n(t,\cdot)_\#(\mathscr{L}^d \llcorner \{T_n > t\})$$

for all $t \in [0, T]$. Hence, we can now use property (3) of X_n to get

$$X(t,\cdot)_\#(\mathscr{L}^d \llcorner \{T_n > t\}) \le e^{L(\Omega_n, \boldsymbol{b})} \mathscr{L}^d \llcorner \Omega_n \quad \text{for every } t \in [0, T], \tag{2.25}$$

which together with the identity $T_n(x) = \mathsf{h}_{\Omega_n}(X(\cdot, x))$ for $\mathscr{L}^d$-a.e. $x \in \Omega_n$ concludes the verification of Definition 2.4(ii).

Now we check Definition 2.4(iii): we obtain that $\limsup V_\Omega(X(t,x)) = \infty$ as $t \uparrow T_{\Omega,X}(x)$ for $\mathscr{L}^d$-a.e. $x \in \Omega$ such that $T_{\Omega,X}(x) < T$ from the fact that $X(t, T_n(x)) \in \partial\Omega_n$, and the sets Ω_n contain eventually any set $K \Subset \Omega$. This completes the existence proof and the verification of the more specific property (a).

The proof of property (b) in the statement of the theorem follows at once from Lemma 2.2 and Remark 2.5. □

2.3. On the local character of the assumption (b-Ω)

Here we prove that the property (b-Ω) is local, in analogy with the other assumptions ((a-Ω) and the local bounds on distributional divergence) made throughout this Chapter. More precisely, the following assumption is equivalent to (b-Ω):

(b'-Ω) for any $t_0 \geq 0, x_0 \in \Omega$ there exists $\varepsilon := \varepsilon(t_0, x_0) > 0$ such that for any nonnegative $\bar\rho \in L^\infty(\mathbb{R}^d)$ with compact support contained in $B_\varepsilon(x_0) \subset \Omega$ and any closed interval $I = [a, b] \subset [t_0 - \varepsilon, t_0 + \varepsilon] \cap [0, T]$, the continuity equation

$$\frac{d}{dt}\rho_t + \operatorname{div}(\boldsymbol{b}\rho_t) = 0 \qquad \text{in } (a, b) \times \mathbb{R}^d$$

has at most one weakly* continuous solution $I \ni t \mapsto \rho_t \in \mathcal{L}_{I,\Omega}$ with $\rho_a = \bar\rho$ and ρ_t compactly supported in $B_\varepsilon(x_0)$ for every $t \in [a, b]$.

Lemma 2.15. *If the assumptions (a-Ω) and (b'-Ω) on the vector field* $\boldsymbol{b}$ *are satisfied, then (b-Ω) is satisfied.*

Proof. **Step 1.** Let $\boldsymbol{\eta} \in \mathscr{P}\big(C([a, b]; \mathbb{R}^d)\big)$, $0 \leq a < b \leq T$, be concentrated on absolutely continuous curves $\eta \in AC([a, b]; K)$ for some $K \subset \Omega$ compact, solving the ODE $\dot\eta = \boldsymbol{b}_t(\eta)$ $\mathscr{L}^1$-a.e. in (a, b), and such that $(e_t)_\# \boldsymbol{\eta} \leq C\mathscr{L}^d$ for any $t \in [0, T]$. We claim that the conditional probability measures $\boldsymbol{\eta}_x$ induced by the map e_a are Dirac masses for $(e_a)_\#\boldsymbol{\eta}$-a.e. x.

To this end, for $s, t \in [a, b], s < t$, we denote by $\Sigma^{s,t} : C([a, b]; \mathbb{R}^d) \to C([s, t]; \mathbb{R}^d)$ the map induced by restriction to $[s, t]$, namely $\Sigma^{s,t}(\eta) = \eta|_{[s,t]}$. For $(e_a)_\#\boldsymbol{\eta}$-a.e. $x \in \mathbb{R}^d$ we define $\tau(x)$ the first splitting time of $\boldsymbol{\eta}_x$, namely the infimum of all $t > a$ such that $(\Sigma^{a,t})_\#\boldsymbol{\eta}_x$ is not a Dirac mass. We agree that $\tau(x) = T$ if $\boldsymbol{\eta}_x$ is a Dirac mass. We also define the splitting point $B(x)$ as $\eta(\tau(x))$ for any $\eta \in \operatorname{supp}\boldsymbol{\eta}_x$. By contradiction, we assume that the set $\{x \in \mathbb{R}^d : \tau(x) < T\}$ has positive $(e_a)_\#\boldsymbol{\eta}$ measure.

For every $t_0 > 0$ and $x_0 \in \mathbb{R}^d$ let $\varepsilon(t_0, x_0) > 0$ be as in (b'-Ω). By a covering argument, we can take a finite cover of $[a, b] \times K$ with sets of the form

$$I_{t_0,x_0,\varepsilon(t_0,x_0)} = (t_0 - \varepsilon(t_0, x_0), t_0 + \varepsilon(t_0, x_0)) \times B_{\varepsilon(t_0,x_0)/2}(x_0).$$

We deduce that there exist $t_0 > 0$ and $x_0 \in \mathbb{R}^d$ such that the set

$$E_0 := \{x \in \mathbb{R}^d : \tau(x) < T,\ (\tau(x), B(x)) \in I_{t_0,x_0,\varepsilon(t_0,x_0)}\} \tag{2.26}$$

has positive $(e_a)_\#\boldsymbol{\eta}$ measure.

For every $p, q \in \mathbb{Q}$ with $a \leq p < q \leq b$ we define the open set

$$E_{p,q} := \{\eta \in C([a, b]; \mathbb{R}^d) : \eta([p, q]) \subset B_{\varepsilon(t_0,x_0)/2}(x_0)\}.$$

We claim that there exist a set $E_1 \subset E_0$ and $p, q \in \mathbb{Q} \cap [a, b]$, $p < q$ such that $(e_a)_\#\boldsymbol{\eta}(E_1) > 0$ and for every $x \in E_1$ the measure $\Sigma^{p,q}_\#(1_{E_{p,q}}\boldsymbol{\eta}_x)$ is not a Dirac delta.

To this end, it is enough to show that for $\mathscr{L}^d$-a.e. $x \in E_0$ there exist $p_x, q_x \in \mathbb{Q} \cap [a,b]$, $p_x < q_x$ such that $\Sigma_{\#}^{p_x,q_x}(1_{E_{p_x,q_x}} \boldsymbol{\eta}_x)$ is not a Dirac delta.

Let us consider $\eta_1 \in \operatorname{supp} \boldsymbol{\eta}_x$; it satisfies $\eta_1(\tau(x)) = B(x) \in B_{\varepsilon(t_0,x_0)/2}(x_0)$. Let p_x, q_x be chosen such that $\eta_1([p_x, q_x]) \subseteq B_{\varepsilon(t_0,x_0)/2}(x_0)$. By definition of $\tau(x)$ we know that $\Sigma_{\#}^{p_x,q_x} \boldsymbol{\eta}_x$ is not a Dirac delta. Hence there exists $\eta_2 \in C([a,b]; \mathbb{R}^d)$ such that $\eta_2 \in \operatorname{supp}(\boldsymbol{\eta}_x)$, $\eta_2(\tau(x)) = B(x)$, $\eta_1(t) \neq \eta_2(t)$ for every $t \in [a, \tau(x)]$, $\eta_1(t) \neq \eta_2(t)$ for some $t \in [\tau(x), q_x]$. Up to reducing q_x, we can assume that $\Sigma^{p_x,q_x}(\eta_1)$, $\Sigma^{p_x,q_x}(\eta_2)$ are curves whose image is contained in $B_{\varepsilon(t_0,x_0)/2}(x_0)$, so that $\eta_1, \eta_2 \in E_{p_x,q_x}$, and which do not coincide. Moreover, since $\operatorname{supp}(\Sigma_{\#}^{p_x,q_x} \boldsymbol{\eta}_x) = \Sigma^{p_x,q_x}(\operatorname{supp} \boldsymbol{\eta}_x)$, we deduce that both $\Sigma^{p_x,q_x}(\eta_1)$ and $\Sigma^{p_x,q_x}(\eta_2)$ belong to the support of $\Sigma_{\#}^{p_x,q_x}(\boldsymbol{\eta}_x)$ and hence $\Sigma_{\#}^{p_x,q_x}(1_{E_{p_x,q_x}} \boldsymbol{\eta}_x) = 1_{\Sigma^{p_x,q_x}(E_{p_x,q_x})} \Sigma_{\#}^{p_x,q_x} \boldsymbol{\eta}_x$ is not a Dirac delta.

Let $\delta > 0$ be small enough so that $E_\delta = E_1 \cap \{x : \boldsymbol{\eta}_x(E_{p,q}) \geq \delta\}$ has positive $(e_a)_\# \boldsymbol{\eta}$-measure. We introduce the probability measure $\tilde{\boldsymbol{\eta}} \in \mathscr{P}\big(C([a,b]; \mathbb{R}^d)\big)$

$$\tilde{\boldsymbol{\eta}} := ((e_a)_\# \boldsymbol{\eta} \llcorner E_\delta) \otimes \left(\frac{1_{E_{p,q}}}{\boldsymbol{\eta}_x(E_{p,q})} \boldsymbol{\eta}_x \right) = ((e_a)_\# \boldsymbol{\eta} \llcorner E_\delta) \otimes \tilde{\boldsymbol{\eta}}_x,$$

which is nonnegative, and less than or equal to $\boldsymbol{\eta}/\delta$. Moreover $\Sigma_{\#}^{p,q} \tilde{\boldsymbol{\eta}} \in \mathscr{P}\big(C([p,q]; \mathbb{R}^d)\big)$ is concentrated on curves in $B_{\varepsilon(t_0,x_0)/2}(x_0)$, and

$$\Sigma_{\#}^{p,q} \tilde{\boldsymbol{\eta}}_x = \frac{\Sigma_{\#}^{p,q}(1_{E_{p,q}} \boldsymbol{\eta}_x)}{\boldsymbol{\eta}_x(E_{p,q})} \text{ is not a Dirac mass for } (e_a)_\# \boldsymbol{\eta}\text{-a.e. } x \in E_\delta.$$

Applying Theorem 1.12 with $\boldsymbol{\lambda} = \Sigma_{\#}^{p,q} \tilde{\boldsymbol{\eta}}$, $\Omega = B_{\varepsilon(t_0,x_0)}(x_0)$, in the time interval $[p,q]$, and thanks to the local uniqueness of bounded, nonnegative solutions of the continuity equation in $I_{t_0,x_0,\varepsilon(t_0,x_0)}$, which in turn follows from (b'-Ω), we deduce that the disintegration $\Sigma_{\#}^{p,q} \tilde{\boldsymbol{\eta}}_x$ of $\Sigma_{\#}^{p,q} \tilde{\boldsymbol{\eta}}$ induced by e_a is a Dirac mass for $(e_a)_\# \boldsymbol{\eta}$-a.e. $x \in E_\delta$. By the uniqueness of the disintegration, we obtain a contradiction.

Step 2. Let μ^1 and μ^2 be two solutions of the continuity equation as in (b) with the same initial datum. Let $\boldsymbol{\eta}^1, \boldsymbol{\eta}^2 \in \mathscr{P}\big(C([a,b]; \mathbb{R}^d)\big)$ be the representation of μ^1 and μ^2 obtained through the superposition principle; they are concentrated on absolutely continuous integral curves of $\boldsymbol{b}$ and they satisfy $\mu_t^i = (e_t)_\# \boldsymbol{\eta}^i$ for any $t \in [0,T]$, $i = 1,2$. Since there exists a compact set $K \subset \Omega$ such that μ_t^i is concentrated on K for every $t \in [0,T]$, $\boldsymbol{\eta}^i$ is concentrated on absolutely continuous curves contained in K for $i = 1,2$. Then by the linearity of the continuity equation $(e_t)_\#[(\boldsymbol{\eta}_1 + \boldsymbol{\eta}_2)/2] = (\mu_t^1 + \mu_t^2)/2$ is still a solution to the continuity equation; by

Step 1 we obtain that $(\boldsymbol{\eta}_x^1 + \boldsymbol{\eta}_x^2)/2$ are Dirac masses for μ_0-a.e. x. This shows that $\boldsymbol{\eta}_x^1 = \boldsymbol{\eta}_x^2$ for μ_0-a.e. x and therefore that $\mu_t^1 = \mu_t^2$ for every $t \in [0, T]$. □

Chapter 3
Main properties of maximal regular flows and analysis of blow-up

The chapter is devoted to the properties of the maximal regular flow built in Chapter 2 under suitable assumptions on the vector field. Since these hypotheses are the natural setting to study the semigroup and stability properties, as well as the proper blow-up of the trajectories, we recall them here. For $T \in (0,\infty)$ we consider a Borel vector field $\boldsymbol{b} : (0,T) \times \Omega \to \mathbb{R}^d$ satisfying:

(a-Ω) $\int_0^T \int_{\Omega'} |\boldsymbol{b}_t(x)|\,dxdt < \infty$ for any $\Omega' \Subset \Omega$;
(b-Ω) for any nonnegative $\bar\rho \in L^\infty_+(\Omega)$ with compact support in Ω and any closed interval $I = [a,b] \subset [0,T]$, the continuity equation

$$\frac{d}{dt}\rho_t + \operatorname{div}(\boldsymbol{b}\rho_t) = 0 \qquad \text{in } (a,b)\times\Omega$$

has at most one weakly* continuous solution $I \ni t \mapsto \rho_t \in \mathcal{L}_{I,\Omega}$ (defined in (1.9)) with $\rho_a = \bar\rho$.

We further assume that the spatial divergence $\operatorname{div} \boldsymbol{b}_t(\cdot)$ in the sense of distributions satisfies

$$\forall\, \Omega' \Subset \Omega,\ \operatorname{div} \boldsymbol{b}_t(\cdot) \geq m(t) \ \text{ in } \Omega', \text{with } L(\Omega', \boldsymbol{b}) := \int_0^T |m(t)|\,dt < \infty. \tag{3.1}$$

In Section 3.1 and 3.2, we prove a natural semigroup property for $\boldsymbol{X}$ and for $T_{\Omega,\boldsymbol{X}}$ and the stability properties of $\boldsymbol{X}$ before the blow-up time $T_{\boldsymbol{X}}$ with respect to perturbations of $\boldsymbol{b}$. Finally, we discuss some additional properties which depend on *global* bounds on the divergence, more precisely on (2.4). The first property, presented in Section 3.3 and well known in the classical setting, is properness of the blow-up, namely this enforcement of (2.2):

$$\lim_{t \uparrow T_{\Omega,\boldsymbol{X}}(x)} V_\Omega(\boldsymbol{X}(t,x)) = \infty \qquad \text{for } \mathscr{L}^d\text{-a.e. } x \in \{T_{\Omega,\boldsymbol{X}} < T\}. \tag{3.2}$$

In other terms, for any $\Omega' \Subset \Omega$ we have that $\boldsymbol{X}(t,x) \notin \Omega'$ for t sufficiently close to $T_{\Omega,\boldsymbol{X}}(x)$.

In $\Omega = \mathbb{R}^d$, $d \geq 2$, we also provide an example of an autonomous Sobolev vector field showing that (2.2) cannot be improved to (3.2) when only local bounds on divergence are present. We also discuss the 2-dimensional case for BV_{loc} vector fields. The second property is the continuity of $\boldsymbol{X}(\cdot,x)$ up to $T_{\Omega,\boldsymbol{X}}(x)$, discussed in Section 3.4, and sufficient conditions for $T_{\Omega,\boldsymbol{X}}(x) = T$.

3.1. Semigroup property

In order to discuss the semigroup property, we double the time variable and denote by

$$\boldsymbol{X}(t,s,x), \qquad t \geq s,$$

the maximal flow with s as initial time, so that $\boldsymbol{X}(t,0,x) = \boldsymbol{X}(t,x)$ and $\boldsymbol{X}(s,s,x) = x$. The maximal time of $\boldsymbol{X}(\cdot,s,x)$ will be denoted by $T_{\Omega,\boldsymbol{X},s}(x)$.

In the smooth setting, it is easily seen that $\boldsymbol{X}(\cdot,x)$ solves the ODE $\dot{x}(t) = \boldsymbol{b}_t(x(t))$ and its value at time s is $\boldsymbol{X}(s,x)$, hence it coincides with the unique trajectory $\boldsymbol{X}(\cdot,s,\boldsymbol{X}(s,x))$. If $\boldsymbol{b}$ and $\boldsymbol{X}$ are as in Theorem 1.5, and a two-sided bound on the divergence of $\boldsymbol{b}$ is assumed, an analogous argument (based also, this time, on the compressibility condition) allows to show that for every $s \in [0,T]$, for $\mathscr{L}^d$-a.e. $x \in \mathbb{R}^d$

$$\boldsymbol{X}\big(\cdot,s,\boldsymbol{X}(s,x)\big) = \boldsymbol{X}(\cdot,x) \text{ in } [s,T]. \tag{3.3}$$

In the context of maximal regular flows, namely under the assumptions of Theorem 2.14, the semigroup property is a natural extension of (3.3) involving also the existence times.

The proof of the semigroup property and of the identity $T_{\Omega,\boldsymbol{X},s}(\boldsymbol{X}(s,x)) = T_{\Omega,\boldsymbol{X}}(x)$ satisfied by the maximal existence time follows the classical scheme. It is however a bit more involved than usual because we are assuming only one-sided bounds on the divergence of $\boldsymbol{b}$, therefore the inverse of the map $\boldsymbol{X}(s,\cdot)$ (which corresponds to a flow with reversed time) is a priori not defined. For this reason, using disintegrations, we define in the proof a kind of multi-valued inverse of $\boldsymbol{X}(s,\cdot)$.

Theorem 3.1 (Semigroup property). *Under assumptions (a-Ω), (b-Ω), and* (3.1) *on $\boldsymbol{b}$, for all $s \in [0,T]$ the maximal regular flow $\boldsymbol{X}$ satisfies*

$$T_{s,\Omega,\boldsymbol{X}}\big(\boldsymbol{X}(s,x)\big) = T_{\Omega,\boldsymbol{X}}(x) \qquad \textit{for } \mathscr{L}^d\textit{-a.e. } x \in \{T_{\Omega,\boldsymbol{X}} > s\}, \tag{3.4}$$

$$\boldsymbol{X}\big(\cdot,s,\boldsymbol{X}(s,x)\big) = \boldsymbol{X}(\cdot,x) \textit{ in } [s,T_{\Omega,\boldsymbol{X}}(x)), \textit{for } \mathscr{L}^d\textit{-a.e. } x \in \{T_{\Omega,\boldsymbol{X}} > s\}. \tag{3.5}$$

Proof. Let us fix $s \geq 0$ and assume without loss of generality that $\mathscr{L}^d(\{T_{\Omega,X} > s\}) > 0$. Let us fix a Borel $B_s \subset \{T_{\Omega,X} > s\}$ with positive and finite measure, and let $\mathscr{L}^d_s$ denote the renormalized Lebesgue measure on B_s, namely $\mathscr{L}^d_s := \mathscr{L}^d \llcorner B_s / \mathscr{L}^d(B_s)$. We denote by ρ_s the bounded density of the probability measure $X(s,\cdot)_\# \mathscr{L}^d_s$ with respect to $\mathscr{L}^d$. We can disintegrate the probability measure $\pi := (Id \times X(s,\cdot))_\# \mathscr{L}^d_s$ with respect to ρ_s, getting a family $\{\pi_y\}$ of probability measures in $\mathbb{R}^d$ such that $\pi = \int \pi_y \otimes \delta_y \, \rho_s(y)\, dy$. Notice that in the case when $X(s,\cdot)$ is (essentially) injective, π_y is the Dirac mass at $(X(s,\cdot))^{-1}(y)$ for $X(s,\cdot)_\# \mathscr{L}^d_s$-a.e. y.

For $\varepsilon > 0$, let us set

$$\pi_\varepsilon := \int_{\{\rho_s \geq \varepsilon\}} \pi_y \otimes \delta_y \, dy \in \mathscr{P}\left(\mathbb{R}^{2d}\right).$$

Since $\varepsilon \pi_\varepsilon \leq \pi$, the first marginal $\tilde{\rho}_\varepsilon$ of π_ε is bounded from above by $\mathscr{L}^d_s / \varepsilon$, therefore it has a bounded density $\tilde{\rho}_\varepsilon$ with respect to $\mathscr{L}^d$. Moreover, since $\pi \leq \|\rho_s\|_{L^\infty(\mathbb{R}^d)} \sup_{\varepsilon > 0} \pi_\varepsilon$ and the first marginal of π is $\mathscr{L}^d_s$, we obtain

$$\sup_{\varepsilon > 0} \tilde{\rho}_\varepsilon(x) > 0 \qquad \text{for } \mathscr{L}^d\text{-a.e. } x \in B_s. \tag{3.6}$$

Now, for $\tau > s$ and $\varepsilon > 0$ fixed, let $B^\tau_s := \{T_{\Omega,X} > \tau\}$ and define a generalized flow $\boldsymbol{\eta}_{\tau,\varepsilon} \in \mathscr{P}\left(C([s,\tau];\mathbb{R}^d)\right)$ by

$$\boldsymbol{\eta}_{\tau,\varepsilon} := \int_{(x,y) \in B^\tau_s \times \{\rho_s \geq \varepsilon\}} \delta_{X(\cdot,x)} \, d\pi_y(x)\, dy = \int_{B^\tau_s} \delta_{X(\cdot,x)} \, \tilde{\rho}_\varepsilon(x)\, dx. \tag{3.7}$$

For any $r \in [s,\tau]$ and any $\phi \in C_b(\mathbb{R}^d)$ nonnegative there holds

$$\int_{\mathbb{R}^d} \phi \, d[(e_r)_\# \boldsymbol{\eta}_{\tau,\varepsilon}] = \int_{B^\tau_s} \phi(X(r,x)) \tilde{\rho}_\varepsilon(x)\, dx \leq L \|\tilde{\rho}_\varepsilon\|_\infty \int_{\mathbb{R}^d} \phi(z)\, dz.$$

Evaluating at $r = s$, a similar computation gives

$$(e_s)_\# \boldsymbol{\eta}_{\tau,\varepsilon} = X(s,\cdot)_\# (\chi_{B^\tau_s} \tilde{\rho}_\varepsilon).$$

By Theorem 1.12 (applied in the time interval $[s,\tau]$ instead of $[0,T]$) it follows that

$$\boldsymbol{\eta}_{\tau,\varepsilon} = \int \delta_{\eta_z} \, d[(e_s)_\# \boldsymbol{\eta}_{\tau,\varepsilon}](z). \tag{3.8}$$

Now, it is clear that $W(\cdot,z) := \eta_z(\cdot)$ is a regular flow in $[s,\tau]$, hence (by uniqueness) $\eta_z = X(\cdot,s,z)$ for $(e_s)_\# \boldsymbol{\eta}_{\tau,\varepsilon}$-a.e. z. Returning to (3.8) we get

$$\boldsymbol{\eta}_{\tau,\varepsilon} = \int \delta_{X(\cdot,s,z)} \, d[(e_s)_\# \boldsymbol{\eta}_{\tau,\varepsilon}](z) = \int_{B^\tau_s} \delta_{X(\cdot,s,X(s,x))} \tilde{\rho}_\varepsilon(x)\, dx, \tag{3.9}$$

where in the second equality we used the formula for $(e_s)_\#\boldsymbol{\eta}_{\tau,\varepsilon}$. Comparing formulas (3.7) and (3.9), and taking (3.6) into account, we find that $T_{s,\Omega,\boldsymbol{X}}(\boldsymbol{X}(s,x)) \geq \tau$ and that $\boldsymbol{X}\big(\cdot, s, \boldsymbol{X}(s,x)\big) \equiv \boldsymbol{X}(\cdot, x)$ in $[s,\tau]$, for $\mathscr{L}^d$-a.e. $x \in B_s^\tau$. Since $\tau > s$ is arbitrary, it follows that $T_{s,\Omega,\boldsymbol{X}}(\boldsymbol{X}(s,x)) \geq T_{\Omega,\boldsymbol{X}}(x)$ and that $\boldsymbol{X}\big(t,s,\boldsymbol{X}(s,x)\big) = \boldsymbol{X}(t,x)$ $\mathscr{L}^d$-a.e. in B_s.

If $T_{\Omega,\boldsymbol{X}}(x) < T$, by the semigroup identity it follows that

$$\limsup_{t \uparrow T_{\Omega,\boldsymbol{X}}(x)} V_\Omega\left(\boldsymbol{X}\big(t,s,\boldsymbol{X}(s,x)\big)\right) = \limsup_{t \uparrow T_{\Omega,\boldsymbol{X}}(x)} V_\Omega(\boldsymbol{X}(t,x)) = \infty,$$

and hence

$$T_{s,\Omega,\boldsymbol{X}}(\boldsymbol{X}(s,x)) = T_{\Omega,\boldsymbol{X}}(x) \qquad \text{for } \mathscr{L}^d\text{-a.e. } x \in B_s. \tag{3.10}$$

Eventually we use the arbitrariness of B_s to conclude (3.4) and (3.5). □

3.2. Stability

This Section provides the stability of maximal regular flows in Ω. This result, which is usually related to an analogous stability property for solutions of the continuity equation, plays an important role in applications, since it allows for instance to build weak solutions of nonlinear systems of PDEs by approximation (see for instance Theorem 8.8 and Theorem 9.4).

A classical stability result for regular Lagrangian flows in $\mathbb{R}^d$ is the following: if $\{\boldsymbol{b}^n\}_{n\in\mathbb{N}}$, $\boldsymbol{b}$ satisfy the assumptions of Theorem 1.5 and

$$\begin{aligned} &\boldsymbol{b}^n \to \boldsymbol{b} \quad \text{in } L^1((0,T)\times\mathbb{R}^d),\\ &\sup_{n\in\mathbb{N}} \|\boldsymbol{b}^n\|_{L^\infty((0,T)\times\mathbb{R}^d)} + \|(\operatorname{div}\boldsymbol{b}^n)_-\|_{L^\infty((0,T)\times\mathbb{R}^d)} < \infty, \end{aligned}$$

then the regular Lagrangian flows $\boldsymbol{X}^n$ of $\boldsymbol{b}^n$ converge to the regular Lagrangian flow $\boldsymbol{X}$ of $\boldsymbol{b}$ in the sense

$$\lim_{n\to\infty} \int_{\mathbb{R}^d} \min\Big\{ \max_{t\in[0,T]} |\boldsymbol{X}^n(t,x) - \boldsymbol{X}(t,x)|, e^{-|x|^2}\Big\}\,dx = 0.$$

The proof of this statement can be found in [13, Theorem 33]. A similar result under purely local assumptions on the vector field requires to localize also the thesis of the stability property, since no control can be expected at the blow-up time (see Remark 3.4 below). We state the result when the vector fields converge strongly in space and weakly in time, in analogy with the classical theory (see also Remark 3.3 below).

Theorem 3.2 (Stability of maximal regular flows in Ω). *Let $\Omega \subset \mathbb{R}^d$ be an open set. Let $\boldsymbol{X}^n$ be maximal regular flows in Ω relative to locally integrable Borel vector fields $\boldsymbol{b}^n : (0,T) \times \Omega \to \mathbb{R}^d$. Assume that:*

(a) *for any $A \Subset \Omega$ open the compressibility constants $C(A, \boldsymbol{X}^n)$ in Definition 2.4 are uniformly bounded;*
(b) *for any $A \Subset \Omega$ open, setting $A^\varepsilon := \{x \in A : \mathrm{dist}(x, \mathbb{R}^d \setminus A) \geq \varepsilon\}$ for $\varepsilon > 0$, there holds, uniformly with respect to n,*

$$\lim_{h\to 0} \left|\chi_{A^{|h|}}(x+h)\boldsymbol{b}_t^n(x+h) - \chi_A(x)\boldsymbol{b}_t^n(x)\right| = 0 \quad \text{in } L^1((0,T)\times A); \tag{3.11}$$

(c) *there exists a Borel vector field $\boldsymbol{b} : (0,T) \times \Omega \to \mathbb{R}^d$ satisfying (a-Ω) and (b-Ω) such that*

$$\boldsymbol{b}^n \rightharpoonup \boldsymbol{b} \ \text{ weakly in } L^1((0,T)\times A; \mathbb{R}^d) \ \text{ for all } A \Subset \Omega \text{ open}. \tag{3.12}$$

Then there exists a unique maximal regular flow $\boldsymbol{X}$ for $\boldsymbol{b}$ and, for every $t \in [0,T]$ and any open set $A \Subset \Omega$, we have

$$\lim_{n\to\infty} \left\| \max_{s\in[0,t]} |\boldsymbol{X}_A^n(s,\cdot) - \boldsymbol{X}(s,\cdot)| \wedge 1 \right\|_{L^1(\{x:\, \mathsf{h}_A(\boldsymbol{X}(\cdot,x))>t\})} = 0, \tag{3.13}$$

where

$$\boldsymbol{X}_A^n(t,x) := \begin{cases} \boldsymbol{X}^n(t,x) & \text{for } t \in [0, \mathsf{h}_A(\boldsymbol{X}^n(\cdot,x))], \\ \boldsymbol{X}^n(\mathsf{h}_A(\boldsymbol{X}^n(\cdot,x)),x) & \text{for } t \in [\mathsf{h}_A(\boldsymbol{X}^n(\cdot,x)), T]. \end{cases}$$

Remark 3.3. The convergence (3.12) and (3.11) of $\boldsymbol{b}^n$ to $\boldsymbol{b}$ is implied by the strong convergence of $\boldsymbol{b}^n$ to $\boldsymbol{b}$ in space-time. It is however quite natural to state the convergence in these terms in view of some applications. For example, the weak convergence of (3.12) and the boundedness in a fractional Sobolev space $\boldsymbol{b}^n \in L^1((0,T); W^{m,p}(\mathbb{R}^d))$, $p > 1, m > 0$, is enough to guarantee that (3.11) holds. The same kind of convergence appears in [81, Theorem II.7] to prove convergence of distributional solutions of the continuity equation, and in [67, Remark 2.11] in the context of quantitative estimates on the flows of Sobolev vector fields.

Remark 3.4. The convergence of the flows in (3.13) is localized to the trajectories of $\boldsymbol{b}$ which are inside A in $[0,t]$. This is indeed natural: even with smooth vector fields one can construct examples where the existence time of $\boldsymbol{X}(\cdot,x)$ is strictly smaller than the existence time of $\boldsymbol{X}^n(\cdot,x)$ and the convergence of $\boldsymbol{X}^n(\cdot,x)$ to $\boldsymbol{X}(\cdot,x)$, or to its constant extension beyond the existence time $T_{\Omega,\boldsymbol{X}}(x)$, fails after $T_{\Omega,\boldsymbol{X}}(x)$ (see Figure 3.1).

The stability of maximal flows in Theorem 3.2 implies a lower semicontinuity property of hitting times.

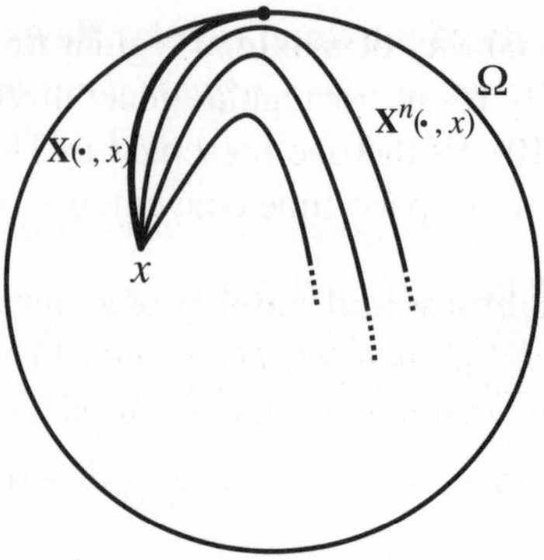

Figure 3.1. One can build a sequence of smooth vector fields $\boldsymbol{b}^n$ whose trajectories $\boldsymbol{X}^n(\cdot, x)$ starting from a point x is drawn in the figure. These trajectories fail to converge to the constant extension of $\boldsymbol{X}(\cdot, x)$ after $T_{\Omega,\boldsymbol{X}}(x)$.

Corollary 3.5 (Semicontinuity of hitting times). *With the same notation and assumptions of Theorem 3.2, for every $t \in [0, T]$ we have that*

$$\lim_{n\to\infty} \mathscr{L}^d\big(\{x : \mathsf{h}_A(\boldsymbol{X}^n(\cdot, x)) \le t < \mathsf{h}_A(\boldsymbol{X}(\cdot, x))\}\big) = 0. \tag{3.14}$$

In particular, there exists a subsequence $n(k) \to \infty$ (which depends, in particular, on A) such that

$$\mathsf{h}_A(\boldsymbol{X}(\cdot, x)) \le \liminf_{k\to\infty} \mathsf{h}_A(\boldsymbol{X}^{n(k)}(\cdot, x)) \quad \mathscr{L}^d\text{-a.e. in } A. \tag{3.15}$$

Proof. For every x such that $\mathsf{h}_A(\boldsymbol{X}^n(\cdot, x)) \le t < \mathsf{h}_A(\boldsymbol{X}(\cdot, x))$ we have that

$$\max_{s\in[0,t]} |\boldsymbol{X}^n_A(s, x) - \boldsymbol{X}(s, x)| \ge \operatorname{dist}(\partial A, \boldsymbol{X}([0, t], x)) > 0.$$

It implies, together with (3.13), that (3.14) holds.

Up to a subsequence and with a diagonal argument, by (3.14) we deduce that for every $t \in \mathbb{Q} \cap [0, T]$ the functions $1_{\{\mathsf{h}_A(\boldsymbol{X}^{n(k)}(\cdot,x))\le t\}}$ converge pointwise a.e. to 0 in $\{\mathsf{h}_A(\boldsymbol{X}(\cdot, x)) > t\}$ and therefore for $\mathscr{L}^d$-a.e. x such that $t < \mathsf{h}_A(\boldsymbol{X}(\cdot, x))$ we have $\mathsf{h}_A(\boldsymbol{X}^n(\cdot, x)) > t$ for n large enough. This implies that for every $t \in \mathbb{Q} \cap [0, T]$, for $\mathscr{L}^d$-a.e. x such that $t < \mathsf{h}_A(\boldsymbol{X}(\cdot, x))$ we have

$$t \le \liminf_{k\to\infty} \mathsf{h}_A(\boldsymbol{X}^{n(k)}(\cdot, x)) \quad \mathscr{L}^d\text{-a.e. in } A,$$

which implies (3.15). ☐

The proof of the stability of maximal regular flows in Ω is based on a tightness and stability result for regular generalized flows in $\overline{A}$ (according to Definition 2.10), as the one presented in Theorem 2.11 under the assumption of the strong space-time convergence of the vector fields.

Proposition 3.6 (Tightness and stability of generalized regular flows). *Let $A \subset \mathbb{R}^d$ be a bounded open set. The result of Theorem* 2.11 *holds true also if we replace the strong convergence of the vector fields* (2.13) *with the assumptions*

$$\lim_{h\to 0} \chi_{A^{|h|}}(x+h)\boldsymbol{c}^n_t(x+h) \\ = \chi_A(x)\boldsymbol{c}^n_t(x) \text{ in } L^1((0,T)\times A), \text{ uniformly with respect to } n, \tag{3.16}$$

$$\boldsymbol{c}^n \rightharpoonup \boldsymbol{c} \qquad \text{weakly in } L^1((0,T)\times A), \tag{3.17}$$

where $A^\varepsilon := \{x \in A : \operatorname{dist}(x, \mathbb{R}^d \setminus A) \geq \varepsilon\}$ *for* $\varepsilon > 0$ *(compare with* (3.11) *and* (3.12)*).*

Proof. The tightness was based on Dunford-Pettis' theorem and it can be repeated in this context thanks to (3.17): in particular, there exists a modulus of integrability F such that

$$\sup_{n\in\mathbb{N}} \int \int_0^T F(|\dot\eta(t)|)\,dt\,d\boldsymbol{\eta}^n < \infty. \tag{3.18}$$

We show that $\boldsymbol{\eta}$ is concentrated on integral curves of $\boldsymbol{c}$, namely

$$\int \left| \eta(t) - \eta(0) - \int_0^t \boldsymbol{c}_s(\eta(s))\,ds \right| d\boldsymbol{\eta}(\eta) = 0 \tag{3.19}$$

for any $t \in [0,T]$. To this end we consider $\boldsymbol{c}^\varepsilon := (\boldsymbol{c}\chi_{A^\varepsilon}) * \rho_\varepsilon$, where $\rho_\varepsilon(x) := \varepsilon^{-d}\rho(x/\varepsilon)$, $\rho \in C^\infty_c(\mathbb{R}^d)$ nonnegative, is a standard convolution kernel in the space variable with compact support in the unit ball. Notice that $\boldsymbol{c}^\varepsilon \in L^1((0,T); C^\infty_c(A;\mathbb{R}^d))$ and that $|\boldsymbol{c}^\varepsilon - \boldsymbol{c}| \to 0$ in $L^1((0,T)\times A)$ as $\varepsilon \to 0$. Similarly, for every $n \in \mathbb{N}$ we set $\boldsymbol{c}^{n,\varepsilon} := (\boldsymbol{c}^n\chi_{A^\varepsilon}) * \rho_\varepsilon$. We first prove that, for every $\varepsilon > 0$,

$$\int \left| \eta(t) - \eta(0) - \int_0^t \boldsymbol{c}^\varepsilon_s(\eta(s))\,ds \right| d\boldsymbol{\eta}(\eta) \leq \omega(\varepsilon), \tag{3.20}$$

where $\omega : (0,\infty) \to (0,\infty)$ is a nondecreasing function which goes to 0 as $\varepsilon \to 0$ to be chosen later.

Since the integrand is a continuous (possibly unbounded) function of $\eta \in C([0,T];\mathbb{R}^d)$ and $\boldsymbol{\eta}^n$ is concentrated on integral curves of $\boldsymbol{c}^n$, by the triangular inequality we have the estimate

$$\begin{aligned}
&\int \left| \eta(t) - \eta(0) - \int_0^t \boldsymbol{c}_s^\varepsilon(\eta(s))\,ds \right| d\boldsymbol{\eta}(\eta) \\
&\le \liminf_{n\to\infty} \int \left| \eta(t) - \eta(0) - \int_0^t \boldsymbol{c}_s^\varepsilon(\eta(s))\,ds \right| d\boldsymbol{\eta}^n(\eta) \\
&\le \liminf_{n\to\infty} \left[\int \left| \int_0^t [\boldsymbol{c}_s^n - \boldsymbol{c}_s^{n,\varepsilon}](\eta(s))\,ds \right| d\boldsymbol{\eta}^n(\eta) \right. \\
&\qquad\qquad \left. + \int \left| \int_0^t [\boldsymbol{c}_s^{n,\varepsilon} - \boldsymbol{c}_s^\varepsilon](\eta(s))\,ds \right| d\boldsymbol{\eta}^n(\eta). \right]
\end{aligned}$$

To estimate the first term in the right-hand side of (3.21), we notice that

$$\sup_{n\in\mathbb{N}} \|\boldsymbol{c}^{n,\varepsilon} - \boldsymbol{c}^n\|_{L^1((0,T)\times A)} \le \omega(\varepsilon)$$

and $\omega(\varepsilon) \to 0$ as $\varepsilon \to 0$. Indeed, consider a nondecreasing function $\omega_0 : (0,\infty) \to (0,\infty)$ which goes to 0 as $\varepsilon \to 0$ and such that

$$\|\chi_{A^{|h|}}(x-h)\boldsymbol{c}_t^n(x-h) - \chi_A(x)\boldsymbol{c}_t^n(x)\|_{L^1((0,T)\times A)} \le \omega_0(|h|) \tag{3.21}$$

for every $n \in \mathbb{N}$, which exists thanks to (3.16). We notice that

$$\begin{aligned}
&\int_0^T \int_A |\boldsymbol{c}^{n,\varepsilon} - \boldsymbol{c}^n|\,dx\,dt \\
&\le \int_{\mathbb{R}^d} \rho_\varepsilon(z) \int_0^T \int_A |\chi_{A^\varepsilon}(x-z)\boldsymbol{c}_t^n(x-z) - \boldsymbol{c}_t^n(x)|\,dx\,dt\,dz \\
&\le \int_{\mathbb{R}^d} \rho_\varepsilon(z) \int_0^T \int_A [\chi_{A^{|z|}}(x-z) - \chi_{A^\varepsilon}(x-z)]|\boldsymbol{c}_t^n(x-z)|\,dx\,dt\,dz \\
&\quad + \int_{\mathbb{R}^d} \rho_\varepsilon(z) \int_0^T \int_A |\chi_{A^{|z|}}(x-z)\boldsymbol{c}_t^n(x-z) - \boldsymbol{c}_t^n(x)|\,dx\,dt\,dz \\
&\le \int_{\mathbb{R}^d} \rho_\varepsilon(z) \int_0^T \int_{\mathbb{R}^d} [\chi_A(x) - \chi_{A^\varepsilon}(x)]|\boldsymbol{c}_t^n(x)|\,dx\,dt\,dz + \omega_0(\varepsilon)
\end{aligned}$$

and the first term converges to 0 uniformly in n thanks to (3.17), Dunford-Pettis' theorem and since $A^\varepsilon \uparrow A$ as $\varepsilon \to 0$.

Hence, using the fact that $\boldsymbol{c}^n = 0$ on ∂A and the definition (2.12) of compressibility constant C_n for $\boldsymbol{\eta}^n$ we get

$$\begin{aligned}
\int \left| \int_0^t [\boldsymbol{c}_s^n - \boldsymbol{c}_s^{n,\varepsilon}](\eta(s))\,ds \right| d\boldsymbol{\eta}^n(\eta) &\le C_n \int_\Omega \int_0^t |\boldsymbol{c}^n - \boldsymbol{c}^{n,\varepsilon}|\,ds\,dx \\
&\le \sup_n C_n\,\omega(\varepsilon).
\end{aligned} \tag{3.22}$$

We now estimate the second term in the right-hand side of (3.21). To this end, for every $k > 0$ we consider the set of curves

$$\Gamma_k := \left\{ \eta \in AC([0,T];\overline{A}) : \int_0^T F(|\dot\eta(t)|)\,dt \le k \right\}.$$

We notice that all curves in Γ_k have a uniform modulus of continuity that we denote by $\tilde\omega_k$. By Chebyshev's inequality and (3.18) we deduce that

$$\boldsymbol{\eta}^n(C([0,T];\overline{A}) \setminus \Gamma_k) \le \frac{C}{k}$$

for some constant $C > 0$, hence in the complement of Γ_k we estimate the integrand with its L^∞ norm:

$$\begin{aligned}\int_{\Gamma_k^c} \left| \int_0^t [\boldsymbol{c}_s^{n,\varepsilon} - \boldsymbol{c}_s^{\varepsilon}](\eta(s))\,ds \right| d\boldsymbol{\eta}^n(\eta) &\le \boldsymbol{\eta}^n(\Gamma_k^c) \int_0^T \|\boldsymbol{c}_s^{n,\varepsilon} - \boldsymbol{c}_s^{\varepsilon}\|_{L^\infty(A)}\,ds \\ &\le \frac{C}{k} \|\boldsymbol{c}^n - \boldsymbol{c}\|_{L^1((0,T)\times A)} \|\rho_\varepsilon\|_{L^\infty(A)}.\end{aligned} \tag{3.23}$$

Hence, choosing k large enough we can make this term as small as we wish uniformly with respect to n, since $\|\boldsymbol{c}^n - \boldsymbol{c}\|_{L^1((0,T)\times A)} \le \|\boldsymbol{c}^n\|_{L^1((0,T)\times A)} + \|\boldsymbol{c}\|_{L^1((0,T)\times A)}$ is bounded.

In Γ_k, for any $N \in \mathbb{N}$ we can use the triangular inequality, the fact that $\boldsymbol{c}^{n,\varepsilon}$ and $\boldsymbol{c}^{\varepsilon}$ are null on $(0,T) \times \partial A$, and the bounded compression condition $(e_{i/N})_\# \boldsymbol{\eta}^n \llcorner A \le C_n \mathscr{L}^d$ for every $i = 1, \dots, N$, to get

$$\begin{aligned}&\int_{\Gamma_k} \left| \int_0^t [\boldsymbol{c}_s^{n,\varepsilon} - \boldsymbol{c}_s^{\varepsilon}](\eta(s))\,ds \right| d\boldsymbol{\eta}^n(\eta) \\ &\le \sum_{i=1}^N \int_{\Gamma_k} \left| \int_{t_{i-1}^N}^{t_i^N} [\boldsymbol{c}_s^{n,\varepsilon} - \boldsymbol{c}_s^{\varepsilon}](\eta(s))\,ds \right| d\boldsymbol{\eta}^n(\eta) \\ &\le \sum_{i=1}^N \int_{\Gamma_k} \left| \int_{t_{i-1}^N}^{t_i^N} [\boldsymbol{c}_s^{n,\varepsilon} - \boldsymbol{c}_s^{\varepsilon}]\big(\eta(t_i^N)\big)\,ds \right| d\boldsymbol{\eta}^n(\eta) \\ &\quad + \tilde\omega_k\left(\frac{t}{N}\right) \sum_{i=1}^N \int_{t_{i-1}^N}^{t_i^N} \|\nabla[\boldsymbol{c}_s^{n,\varepsilon} - \boldsymbol{c}_s^{\varepsilon}](\cdot)\|_{L^\infty(A)}\,ds \\ &\le C_n \sum_{i=1}^N \int_A \left| \int_{t_{i-1}^N}^{t_i^N} [\boldsymbol{c}^{n,\varepsilon} - \boldsymbol{c}^{\varepsilon}]\,ds \right| dx \\ &\quad + \tilde\omega_k\left(\frac{t}{N}\right) \|\boldsymbol{c}^n - \boldsymbol{c}\|_{L^1((0,T)\times A)} \|\nabla\rho_\varepsilon\|_{L^\infty(\mathbb{R}^d)},\end{aligned}$$

where $t_i^N = it/N$. Choosing N large enough we can make the second term in the right-hand side as small as we want, uniformly in n. Letting $n \to \infty$ in (3.24), each term in the first sum in the right-hand side converges to 0 pointwise in x by the weak convergence (3.12) tested with the function $\varphi_s^x(y) = 1_{[t_{i-1}^N, t_i^N]}(s)\rho_\varepsilon(x-y)$, namely, for every $x \in A$,

$$\lim_{n\to\infty} \int_{t_{i-1}^N}^{t_i^N} [\boldsymbol{c}_s^{n,\varepsilon}(x) - \boldsymbol{c}_s^\varepsilon(x)]\, ds = \lim_{n\to\infty} \int_{t_{i-1}^N}^{t_i^N} [\boldsymbol{c}_s^n(y) - \boldsymbol{c}_s(y)]\rho_\varepsilon(x-y)\, ds = 0.$$

These functions are bounded by $\|\boldsymbol{c}^n - \boldsymbol{c}\|_{L^1((0,T)\times A)}\|\rho_\varepsilon\|_{L^\infty(\mathbb{R}^d)}$, thus by dominated convergence the first sum in the right-hand side of (3.24) converges to 0. It follows that, given ε and k, by choosing N sufficiently large we can make also this term as small as we wish, hence (3.20) follows from (3.21). We now let $\varepsilon \to 0$ in (3.20) and notice that, since $\boldsymbol{\eta}$ satisfies (2.12) with $C = \liminf_n C_n$ and $\boldsymbol{c}^\varepsilon \to \boldsymbol{c}$ in $L^1((0,T)\times A)$,

$$\lim_{\varepsilon\to 0} \int \left| \int_0^t [\boldsymbol{c}_s - \boldsymbol{c}_s^\varepsilon](\eta(s))\, ds \right| d\boldsymbol{\eta}(\eta) \le C \lim_{\varepsilon\to 0} \int_A \int_0^t |\boldsymbol{c} - \boldsymbol{c}^\varepsilon|\, ds\, dx = 0,$$

proving the validity of (3.19). □

The following lemma is standard in optimal transport theory (see [6, Lemma 22] or [124, Corollary 5.23]), but we prove it for completeness.

Lemma 3.7. *Let X_1, X_2 be Polish metric spaces, let $\mu \in \mathscr{P}(X_1)$, and let $F_n : X_1 \to X_2$ be a sequence of Borel functions. If*

$$(\mathrm{Id}, F_n)_\# \mu \rightharpoonup (\mathrm{Id}, F)_\# \mu \qquad \textit{narrowly in } \mathscr{P}(X_1 \times X_2), \tag{3.24}$$

then F_n converge to F in μ-measure, namely

$$\lim_{n\to\infty} \mu(\{d_{X_2}(F_n, F) > \varepsilon\}) = 0 \qquad \forall \varepsilon > 0.$$

Proof. Let us fix $\varepsilon \in (0,1)$. For every $\delta > 0$ we consider a continuous map $\tilde F$ which coincides with F up to a set of μ-measure δ. Taking the bounded continuous function $\phi(x,y) = \min\{d_{X_2}(y, \tilde F(x)), 1\}$ for $(x,y) \in X_1 \times X_2$ as a test function in (3.24) we deduce that

$$\begin{aligned}\varepsilon \limsup_{n\to\infty} \mu(\{d_{X_2}(F_n, \tilde F) > \varepsilon\}) &\le \lim_{n\to\infty} \int_{X_1} \min\{d_{X_2}(F_n(x), \tilde F(x)), 1\}\, d\mu(x) \\ &= \int_{X_1} \min\{d_{X_2}(F(x), \tilde F(x)), 1\}\, d\mu(x) \le \delta.\end{aligned}$$

Therefore

$$\limsup_{n\to\infty}\mu\left(\left\{d_{X_2}(F_n,F)>\varepsilon\right\}\right)\le\mu\left(\left\{F\neq\tilde F\right\}\right) + \limsup_{n\to\infty}\mu\left(\left\{d_{X_2}(F_n,\tilde F)>\varepsilon\right\}\right) \le \delta+\frac{\delta}{\varepsilon}$$

which can be made arbitrarily small by taking δ small. □

Proof of Theorem 3.2. Fix $A \Subset \Omega$ open, denote by $\mathscr{L}^d_A$ the normalized Lebesgue measure on A, and define $\boldsymbol{X}^n_A$ as in the statement of the theorem. Then the laws $\boldsymbol{\eta}^n$ of $x \mapsto \boldsymbol{X}^n_A(\cdot, x)$ under $\mathscr{L}^d_A$ define regular generalized flows in $\overline{A}$ relative to $\boldsymbol{c}^n = \chi_A \boldsymbol{b}^n$, according to Definition 2.10, with compressibility constants $C_n = C(A, \boldsymbol{X}^n)$.

Hence we can apply Proposition 3.6 to obtain that, up to a subsequence, $\boldsymbol{\eta}^n$ weakly converge to a generalized flow $\boldsymbol{\eta}$ in $\overline{A}$ relative to the vector field $\boldsymbol{c} = \chi_A \boldsymbol{b}$, with compressibility constant $C = \liminf_n C_n$. Let $\boldsymbol{\eta}_x$ be the conditional probability measures induced by the map e_0, and let $\boldsymbol{X}_A$ and T_A be given by Proposition 2.12; recall that $\boldsymbol{X}_A(\cdot, x)$ is an integral curve of $\boldsymbol{b}$ in $[0, T_A(x)]$, that $\boldsymbol{X}_A([0, T_A(x)), x) \subset A$, and that $\boldsymbol{X}_A(T_A(x), x) \in \partial A$ if $T_A(x) < T$; as explained in Remark 2.13, for $\mathscr{L}^d_A$-almost every x the hitting time $\mathsf{h}_A(\eta)$ is equal to $T_A(x)$ for $\boldsymbol{\eta}_x$-a.e. η, and $(e_t)_\#\boldsymbol{\eta}_x = \delta_{\boldsymbol{X}_A(t,x)}$ for all $t \in [0, T_A(x)]$. For every $t \in [0, T]$ we set $E_{t,A} := \{T_A(x) > t\}$; since

$$\boldsymbol{X}_A(s,\cdot)_\#(\mathscr{L}^d \llcorner E_{t,A}) = (e_s)_\# \int_{E_{t,A}} \delta_{\boldsymbol{X}_A(\cdot,x)}\, d\mathscr{L}^d \le (e_s)_\#\boldsymbol{\eta} \le C\mathscr{L}^d \quad \forall s\in[0,t],$$

we obtain that $\boldsymbol{X}_A$ is a regular flow for $\boldsymbol{b}$ on $[0, t] \times E_t$. Applying Theorem 2.14(b) to $\boldsymbol{X}_{A_1}$ and $\boldsymbol{X}_{A_2}$ with $A_1 \subset A_2$ we deduce that $\boldsymbol{X}_{A_1} = \boldsymbol{X}_{A_2}$ on E_{t,A_1}, and this allows us (by a gluing procedure) to obtain a maximal regular flow for $\boldsymbol{b}$.

To prove the last statement, we apply Lemma 3.7 with $X_1 = \mathbb{R}^d$, $\mu = (\mathscr{L}^d \llcorner \{T_A > t\})/\mathscr{L}^d(\{T_A > t\})$, $X_2 = C([0, t]; \overline{A})$, $F_n(x) = \boldsymbol{X}^n_A(\cdot, x)$, $F(x) = \boldsymbol{X}_A(\cdot, x)$. More precisely, we consider the laws $\tilde{\boldsymbol{\eta}}^n \in \mathscr{P}\big(C([0, t]; \mathbb{R}^d)\big)$ of $x \mapsto \boldsymbol{X}^n_A(\cdot, x)$ under μ; with the same argument as above, we know that $\tilde{\boldsymbol{\eta}}^n$ weakly converge to $\tilde{\boldsymbol{\eta}}$ and that the disintegration $\tilde{\boldsymbol{\eta}}_x$ coincides with $\delta_{\boldsymbol{X}_A(\cdot,x)}$ for μ-a.e. $x \in \mathbb{R}^d$ (notice that $\boldsymbol{X}_A(\cdot, x)$ is defined in $[0, t]$ for μ-a.e. x). The assumption (3.24) is satisfied, since for every bounded continuous function $\varphi : \mathbb{R}^d \times C([0, T]; \overline{A}) \to \mathbb{R}$ we have

$$\int \varphi(x, \gamma)\, d(\mathrm{Id}, \boldsymbol{X}^n_A(\cdot, x))_\#\mu(x, \gamma) = \int \varphi(\gamma(0), \gamma)\, d\tilde{\boldsymbol{\eta}}^n(\gamma)$$

(and similarly with $\tilde{\eta}$) and the weak convergence of $\tilde{\eta}^n$ to $\tilde{\eta}$ shows that

$$\lim_{n\to\infty}\int \varphi(x,\gamma)d(\mathrm{Id},\boldsymbol{X}_A^n(\cdot,x))_{\#}\mu(x,\gamma)=\int \varphi(x,\gamma)d(\mathrm{Id},\boldsymbol{X}_A(\cdot,x))_{\#}\mu(x,\gamma).$$

We deduce the convergence in μ-measure of $\boldsymbol{X}_A^n$ to $\boldsymbol{X}_A$ in $C([0,t];\overline{A})$, *i.e.*,

$$\lim_{n\to\infty}\mathscr{L}^d\left(\left\{x\in\{T_A>t\}:\ \sup_{s\in[0,t]}|\boldsymbol{X}_A^n(s,x)-\boldsymbol{X}_A(s,x)|>\varepsilon\right\}\right)=0\quad\forall\varepsilon>0,$$

from which (3.13) follows easily. □

3.3. Proper blow-up of trajectories under global bounds on divergence

Recall that the blow-up time $T_{\Omega,\boldsymbol{X}}(x)$ for maximal regular flows is characterized by the property $\limsup_{t\uparrow T_{\Omega,\boldsymbol{X}}(x)} V_\Omega(\boldsymbol{X}(t,x))=\infty$ when $T_{\Omega,\boldsymbol{X}}(x)<T$. We say that $\boldsymbol{X}(\cdot,x)$ blows up *properly* (*i.e.* with no oscillations) if the stronger condition

$$\lim_{t\uparrow T_{\Omega,\boldsymbol{X}}(x)} V_\Omega(\boldsymbol{X}(t,x))=\infty \tag{3.25}$$

holds. This property says in particular that the modulus of every unbounded trajectory must converge to infinity. On the other hand, (3.25) does not guarantee that, even in a bounded domain Ω, bounded trajectories have a limit as t approaches the blow-up time (the limit belongs to $\partial\Omega$ if it exists). This fact may happen even with smooth vector fields (see Figure 3.2); we show in Theorem 3.12 that this cannot happen if we assume global integrability of $\boldsymbol{b}$.

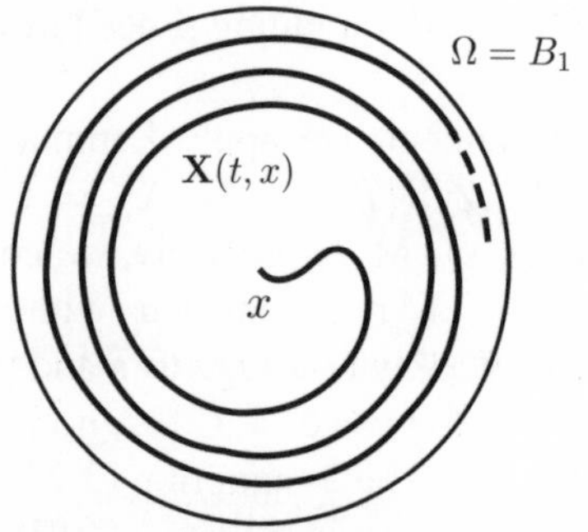

Figure 3.2. The picture shows a bounded trajectory of a smooth vector field in $\Omega=B_1$ such that (3.25) holds but the trajectory blows up in finite time without having a limit.

In the following theorem we prove the proper blow-up of trajectories when a global bounded compression condition on $\boldsymbol{X}$ is available, see (3.27) below. Thanks to the properties of the maximal regular flow the global bounded compression condition is fulfilled, for instance, in all cases when the divergence bounds $L(\Omega')$ in (3.1) are uniformly bounded. More precisely

$$\operatorname{div} \boldsymbol{b}_t(\cdot) \geq m(t) \quad \text{in } \Omega, \text{with } L(\Omega) := \int_0^T |m(t)|\, dt < \infty \tag{3.26}$$

implies (3.27) with $C_* \leq e^{L(\Omega)}$.

Theorem 3.8. *Let $\boldsymbol{X}$ be a maximal regular flow relative to a Borel vector field $\boldsymbol{b}$ satisfying (a-Ω) and (b-Ω), and assume that the bounded compression condition is global, namely there exists a constant $C_* \geq 0$ satisfying*

$$\boldsymbol{X}(t,\cdot)_\#(\mathscr{L}^d \llcorner \{T_{\Omega,\boldsymbol{X}} > t\}) \leq C_* \mathscr{L}^d \qquad \forall t \in [0,T). \tag{3.27}$$

Then

$$\liminf_{t \uparrow T_{\Omega,\boldsymbol{X}}(x)} |\boldsymbol{X}(t,x)| = \infty$$

$$\text{for } \mathscr{L}^d\text{-a.e. } x \in \mathbb{R}^d \text{ such that } \limsup_{t \uparrow T_{\Omega,\boldsymbol{X}}(x)} |\boldsymbol{X}(t,x)| = \infty,$$

and in particular $\lim_{t\uparrow T_{\Omega,\boldsymbol{X}}(x)} V_\Omega(\boldsymbol{X}(t,x)) = \infty$ for $\mathscr{L}^d$-a.e. x with $T_{\Omega,\boldsymbol{X}}(x) < T$.

Proof. Let Ω_n be open sets with $\Omega_n \Subset \Omega_{n+1} \Subset \Omega$, with $\cup_n \Omega_n = \Omega$. We consider cut-off functions $\psi_n \in C_c^\infty(\Omega_{n+1})$ with $0 \leq \psi_n \leq 1$ and $\psi_n \equiv 1$ on a neighborhood of $\overline{\Omega}_n$.

Since $\boldsymbol{X}(\cdot, x)$ is an integral curve of $\boldsymbol{b}$ for $\mathscr{L}^d$-a.e $x \in \Omega$ we can use (3.27) to estimate

$$\begin{aligned}
&\int_\Omega \int_0^{T_{\Omega,\boldsymbol{X}}(x)} \left| \frac{d}{dt} \psi_n(\boldsymbol{X}(t,x)) \right| dt\, dx \\
&\leq \int_\Omega \int_0^{T_{\Omega,\boldsymbol{X}}(x)} |\nabla \psi_n(\boldsymbol{X}(t,x))|\, |\boldsymbol{b}_t(\boldsymbol{X}(t,x))|\, dt\, dx \\
&= \int_0^T \int_{\{T_{\Omega,\boldsymbol{X}} > t\}} |\nabla \psi_n(\boldsymbol{X}(t,x))|\, |\boldsymbol{b}_t(\boldsymbol{X}(t,x))|\, dx\, dt \qquad (3.28) \\
&\leq C_* \int_0^T \int_{\mathbb{R}^d} |\nabla \psi_n(y)| |\boldsymbol{b}_t(y)|\, dy\, dt \\
&\leq C \|\nabla \psi_n\|_{L^\infty(\Omega)} \int_0^T \int_{\Omega_{n+1}} |\boldsymbol{b}_t(x)|\, dx\, dt.
\end{aligned}$$

Hence $\psi_n(\boldsymbol{X}(\cdot, x))$ is the restriction of an absolutely continuous map in $[0, T_{\Omega,\boldsymbol{X}}(x)]$ (and therefore uniformly continuous in $[0, T_{\Omega,\boldsymbol{X}}(x))$) for $\mathscr{L}^d$-a.e. $x \in \Omega$.

Let us fix $x \in \Omega$ such that $\limsup_{t\uparrow T_{\Omega,\boldsymbol{X}}(x)} V_\Omega(\boldsymbol{X}(t,x)) = \infty$ and $\psi_n(\boldsymbol{X}(\cdot, x))$ is uniformly continuous in $[0, T_{\Omega,\boldsymbol{X}}(x))$ for every $n \in \mathbb{N}$. The lim sup condition yields that the limit of all $\psi_n(\boldsymbol{X}(t,x))$ as $t \uparrow T_{\Omega,\boldsymbol{X}}(x)$ must be 0. On the other hand, if the lim inf of $V_\Omega(\boldsymbol{X}(t,x))$ as $t \uparrow T_{\Omega,\boldsymbol{X}}(x)$ were finite, we could find an integer n and $t_k \uparrow T_{\Omega,\boldsymbol{X}}(x)$ with $\boldsymbol{X}(t_k, x) \in \Omega_n$ for all k. Since $\psi_{n+1}(\boldsymbol{X}(t_k, x)) = 1$ we obtain a contradiction. □

Remark 3.9. Under the assumptions of the previous theorem applied with $\Omega = \mathbb{R}^d$, given any probability measure $\mu_0 \leq C\mathscr{L}^d$ for some $C > 0$, it can be easily shown that the measure

$$\mu_t := \boldsymbol{X}(t, \cdot)_\#(\mu_0 \llcorner \{T_{\boldsymbol{X}} > t\}), \qquad t \in [0, T] \tag{3.29}$$

is a bounded (by (3.27)), weakly* continuous, distributional solution to the continuity equation. We notice that the same statement is not true if we assume only a local bound on div $\boldsymbol{b}$, since the measure (3.29) can be locally unbounded, as in the example of Proposition 3.10, and therefore we cannot write the distributional formulation of the continuity equation.

To see that (3.29) is a distributional solution of the continuity equation, we consider $\varphi \in C^\infty_c(\mathbb{R}^d)$ and we define the function $g_t(x)$ as $\varphi(\boldsymbol{X}(t,x))$ if $t < T_{\boldsymbol{X}}(x)$ or $t = T_{\boldsymbol{X}}(x) = T$, and $g_t(x) = 0$ otherwise. By Theorem 3.8 we notice that $g_t(x)$ is absolutely continuous with respect to t for $\mathscr{L}^d$-a.e. $x \in \mathbb{R}^d$ and that $\frac{d}{dt} g_t(x) = 1_{\{T_{\boldsymbol{X}}(x)>t\}} \nabla\varphi(\boldsymbol{X}(t,x))\boldsymbol{b}_t(\boldsymbol{X}(t,x))$ for $\mathscr{L}^1$-a.e. $t \in (0, T)$, for $\mathscr{L}^d$-a.e. $x \in \mathbb{R}^d$. We deduce that the function $t \to \int_{\{T_{\boldsymbol{X}}>t\}} \varphi(\boldsymbol{X}(t,x))\, d\mu_0(x)$ is absolutely continuous and its derivative is given by

$$\begin{aligned}\frac{d}{dt}\int_{\{T_{\boldsymbol{X}}>t\}} \varphi(\boldsymbol{X}(t,x))\, d\mu_0(x) &= \frac{d}{dt}\int_{\mathbb{R}^d} g_t(x)\, d\mu_0(x)\\ &= \int_{\{T_{\boldsymbol{X}}>t\}} \nabla\varphi(\boldsymbol{X}(t,x))\boldsymbol{b}_t(\boldsymbol{X}(t,x))\, d\mu_0(x).\end{aligned}$$

The proper blow-up may fail for the maximal regular flow due only to the lack of a global bound on the divergence of $\boldsymbol{b}$, as shown in the next example.

In the following we denote by $\mathbf{v}_1, \ldots, \mathbf{v}_d$ the canonical basis of $\mathbb{R}^d$ and $B_r^{(d-1)}(x') \subset \mathbb{R}^{d-1}$ the ball of center $x' \in \mathbb{R}^{d-1}$ and radius r. We denote each point $x \in \mathbb{R}^d$ as $x = (x', x_d)$, where x' are the first $d-1$ coordinates of x. For simplicity we write $T_{\boldsymbol{X}}$ for $T_{\mathbb{R}^d,\boldsymbol{X}}$.

Proposition 3.10. *Let $d \geq 3$. There exist an autonomous vector field $\boldsymbol{b} : \mathbb{R}^d \to \mathbb{R}^d$ and a Borel set of positive measure $\Sigma \subset \mathbb{R}^d$ such that $\boldsymbol{b} \in W^{1,p}_{\mathrm{loc}}(\mathbb{R}^d; \mathbb{R}^d)$ for some $p > 1$, $\operatorname{div} \boldsymbol{b} \in L^\infty_{\mathrm{loc}}(\mathbb{R}^d)$, and*

$$T_X(x) \leq 2, \qquad \liminf_{t \uparrow T_X(x)} |X(t,x)| = 0, \qquad \limsup_{t \uparrow T_X(x)} |X(t,x)| = \infty \tag{3.30}$$

for every $x \in \Sigma$.

Proof. We build a vector field whose trajectories are represented in Figure 3.3. Let $\{a_k\}_{k \in \mathbb{N}}$ be a fastly decaying sequence to be chosen later. For every $k = 1, 2, \ldots$ we define the cylinders

$$E_k = \begin{cases} B^{(d-1)}_{a_k}(2^{-k}\mathbf{v}_1) \times [-2^{k-1}, 2^k] & \text{if } k \text{ is odd} \\ B^{(d-1)}_{a_k}(2^{-k}\mathbf{v}_1) \times [-2^k, 2^{k-1}] & \text{if } k \text{ is even.} \end{cases}$$

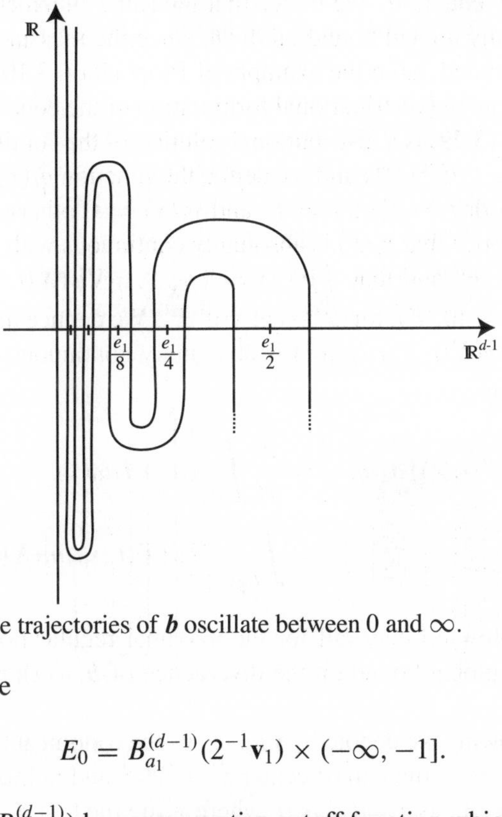

Figure 3.3. The trajectories of $\boldsymbol{b}$ oscillate between 0 and ∞.

We also define

$$E_0 = B^{(d-1)}_{a_1}(2^{-1}\mathbf{v}_1) \times (-\infty, -1].$$

Let $\varphi \in C^\infty_0(B^{(d-1)}_1)$ be a nonnegative cutoff function which is equal to 1 in $B_{1/2}$. In every E_k the vector field $\boldsymbol{b}$ points in the d-th direction and it depends only on the first $d-1$ variables

$$\boldsymbol{b}(x) := \begin{cases} (-1)^{k+1}4^k\varphi\Big(\dfrac{x'-2^{-k}\mathbf{v}_1}{a_k}\Big)\mathbf{v}_d & \forall x \in E_k,\ k \geq 1 \\ 4\varphi\Big(\dfrac{x'-2^{-1}\mathbf{v}_1}{a_1}\Big)\mathbf{v}_d & \forall x \in E_0. \end{cases} \tag{3.31}$$

Notice that div $\boldsymbol{b} = 0$ in every E_k and that $\boldsymbol{b}$ is 0 on the lateral boundary of every cylinder E_k since φ is compactly supported.

For every $k \geq 1$ we define the cylinders $E'_k \subset \mathbb{R}^d$ as

$$E'_k = \begin{cases} B^{(d-1)}_{a_k/2}(2^{-k}\mathbf{v}_1) \times [-2^{k-1}, 2^k] & \text{if } k \text{ is odd} \\ B^{(d-1)}_{a_k/2}(2^{-k}\mathbf{v}_1) \times [-2^k, 2^{k-1}] & \text{if } k \text{ is even.} \end{cases}$$

For every $k \in \mathbb{N}$ we define a handle F_k which connects E_k with E_{k+1} as in Figure 3.4. It is made of a family of smooth, nonintersecting curves of length less than 1 which connect the top of E_k to the top of E_{k+1} and E'_k with E'_{k+1}. We denote by F'_k the handle between E'_k and E'_{k+1}, as in Figure 3.4.

The vector field $\boldsymbol{b}$ is extended to be 0 outside $\cup_{k=0}^{\infty}(E_k \cup F_k)$. It is extended inside every F_k by choosing a smooth extension in a neighborhood of each handle, whose trajectories are the ones described by the handle. The modulus of $\boldsymbol{b}$ is chosen to be between 4^k and 4^{k+1} in F'_k (notice that $|\boldsymbol{b}(x)| = 4^k$ on the top of E'_k thanks to (3.31)).

With this choice, every trajectory in F'_k is not longer than 1 and the vector field $\boldsymbol{b}$ is of size 4^k. We deduce that the handle is covered in time less than 4^{-k}.

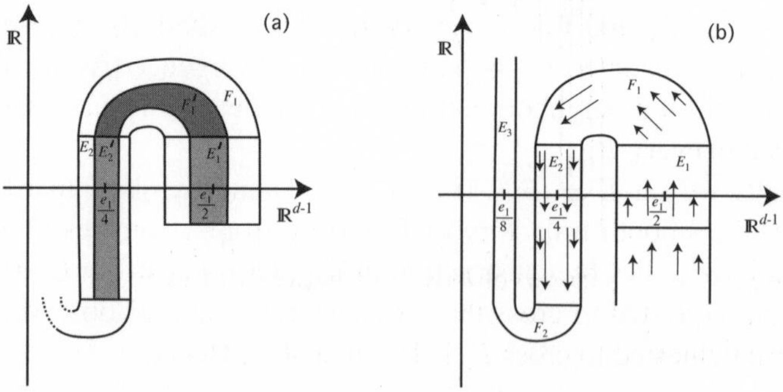

Figure 3.4. The sets E_k, F_k, E'_k, and F'_k and the vector field $\boldsymbol{b}$.

By the construction it is clear that $\boldsymbol{b}$ is smooth in $\mathbb{R}^d \setminus \mathbb{R}\mathbf{v}_d$. We show that $\boldsymbol{b} \in W^{1,p}_{\text{loc}}(\mathbb{R}^d; \mathbb{R}^d)$ for some $p > 1$ by estimating the $W^{1,p}$ norm of $\boldsymbol{b}$ in every ball B_R. With this estimate, one can easily see that $\boldsymbol{b}$ is the limit of smooth vector fields with bounded $W^{1,p}$ norms on B_R; it is enough to consider the restriction of $\boldsymbol{b}$ to the first n sets $E_k \cup F_k$.

Fix $R > 0$. The $W^{1,p}$ norm of $\boldsymbol{b}$ in B_R is estimated by

$$\|\boldsymbol{b}\|_{W^{1,p}(B_R)} \leq \|\boldsymbol{b}\|_{W^{1,p}(E_0 \cap B_R)} + \sum_{k=1}^{\infty} \|\boldsymbol{b}\|_{W^{1,p}(F_k \cap B_R)} + \sum_{k=1}^{\infty} \|\boldsymbol{b}\|_{W^{1,p}(E_k)}. \tag{3.32}$$

The first term is obviously finite (depending on R); since B_R intersects at most finitely many F_k, the second sum in the right-hand side of (3.32) has only finitely many nonzero terms. As regards the third sum, we compute the $W^{1,p}$ norm of $\boldsymbol{b}$ in each set E_k. For every $k \in \mathbb{N}$

$$\begin{aligned} \|\boldsymbol{b}\|_{L^p(E_k)} &\leq 4^k (2R)^{1/p} \left\| \varphi\left(\frac{x' - 2^{-k}\mathbf{v}_1}{a_k}\right) \right\|_{L^p\left(B^{(d-1)}_{a_k}(2^{-k}\mathbf{v}_1)\right)} \\ &= 4^k (2R a_k^{d-1})^{1/p} \|\varphi\|_{L^p(B_1^{(d-1)})} \end{aligned}$$

and similarly

$$\begin{aligned} \|\nabla \boldsymbol{b}\|_{L^p(E_k)} &\leq \frac{4^k (2R)^{1/p}}{a_k} \left\| \nabla\varphi\left(\frac{x' - 2^{-k}\mathbf{v}_1}{a_k}\right) \right\|_{L^p\left(B^{(d-1)}_{a_k}(2^{-k}\mathbf{v}_1)\right)} \\ &= \frac{4^k (2R a_k^{d-1})^{1/p}}{a_k} \|\nabla\varphi\|_{L^p(B_1^{(d-1)})}. \end{aligned} \tag{3.33}$$

Since $a_k \leq 1$, the series in the right-hand side of (3.32) is estimated by

$$\sum_{k=1}^{\infty} \|\boldsymbol{b}\|_{W^{1,p}(E_k)} \leq C(R, \varphi) \sum_{k=1}^{\infty} 4^k a_k^{(d-1)/p-1}$$

and it is convergent for every $p < d - 1$ provided that we take $a_k \leq 8^{-pk/(d-1-p)}$. Hence $\boldsymbol{b} \in W^{1,p}(B_R; \mathbb{R}^d)$ for every $R > 0$.

To check that $\operatorname{div} \boldsymbol{b} \in L^\infty_{\text{loc}}(\mathbb{R}^d)$, we notice that $\boldsymbol{b}$ is divergence free in $\mathbb{R}^d \setminus \cup_{k=0}^{\infty} F_k$ and that for every $R > 0$ the ball B_R intersects only finitely many handles F_k; in particular $\boldsymbol{b}$ is divergence free in B_1. Since $\boldsymbol{b}$ is smooth in a neighborhood of each handle, we deduce that $\operatorname{div} \boldsymbol{b}$ is bounded in every B_R.

Finally we set $\Sigma = B_{a_1/2}(\mathbf{v}_1/2) \times [0, 1]$ and we show that for every $x \in \Sigma$ the smooth trajectory of $\boldsymbol{b}$ starting from x satisfies (3.30). The trajectory of x lies by construction in $\cup_{k=0}^{\infty}(E'_k \cup F'_k)$. For every $k \in \mathbb{N}$, the time requested to cross the set E'_k is $2^k/4^k$ and, as observed before, the time requested to cross F'_k is less than 4^{-k}. Hence

$$T_X(x) \leq \sum_{k=1}^{\infty} \frac{2^k + 1}{4^k} \leq 2 \qquad \forall x \in \Sigma.$$

The other properties in (3.30) are satisfied by construction. □

In dimension $d = 2$, thanks to the smoothness of the vector field built in the previous example outside the x_2-axis, there exists only an integral curve of $\boldsymbol{b}$ for every $x \in \mathbb{R}^2 \setminus \{x_1 = 0\}$. Hence, thanks to the superposition principle the previous example satisfies the assumption (b-Ω) on $\boldsymbol{b}$ and therefore provides a two-dimensional counterexample to the proper blow-up of trajectories. On the other hand, the vector field built in the previous example is not in $BV_{\mathrm{loc}}(\mathbb{R}^2; \mathbb{R}^2)$. We show indeed in the next proposition that for any autonomous BV_{loc} vector field in dimension $d = 2$ the behavior of the previous example (see Figure 3.3) cannot happen and the trajectories must blow up properly. It looks likely that, with $d = 2$ and a non-autonomous vector field, one can build an example following the lines of the example in Proposition 3.10.

Proposition 3.11. *Let* $\boldsymbol{b} \in BV_{\mathrm{loc}}(\mathbb{R}^2; \mathbb{R}^2)$, $\operatorname{div} \boldsymbol{b} \in L^\infty_{\mathrm{loc}}(\mathbb{R}^2)$. *Then*

$$\lim_{t \uparrow T_{\boldsymbol{X}}(x)} |\boldsymbol{X}(t,x)| = \infty \text{ for } \mathscr{L}^2\text{-a.e. } x \in \mathbb{R}^2 \text{ such that } \limsup_{t \uparrow T_{\boldsymbol{X}}(x)} |\boldsymbol{X}(t,x)| = \infty. \tag{3.34}$$

Proof. **Step 1.** Let $R > 0$. We prove that for every vector field $\boldsymbol{b} \in BV_{\mathrm{loc}}(\mathbb{R}^2; \mathbb{R}^2)$

$$\int_R^{R+1} \operatorname*{ess\,sup}_{x \in \partial B_r} |\boldsymbol{b}(x)| \, dr \le \frac{1}{2\pi R} \int_{B_{R+1} \setminus B_R} |\boldsymbol{b}(x)| \, dx + |D\boldsymbol{b}|(B_{R+1} \setminus B_R). \tag{3.35}$$

For this, let $\boldsymbol{b}_\varepsilon$ be a sequence of smooth vector fields which approximate $\boldsymbol{b}$ in $BV(B_{R+1} \setminus B_R)$, namely

$$\lim_{\varepsilon \to 0} |\boldsymbol{b}_\varepsilon - \boldsymbol{b}| = 0 \qquad \text{in } L^1(B_{R+1} \setminus B_R),$$

$$\lim_{\varepsilon \to 0} \int_{B_{R+1} \setminus B_R} |\nabla \boldsymbol{b}_\varepsilon(x)| \, dx = |D\boldsymbol{b}|(B_{R+1} \setminus B_R).$$

Up to a subsequence (not relabeled) we deduce that for $\mathscr{L}^1$-a.e. $r \in (R, R+1)$

$$\lim_{\varepsilon \to 0} \boldsymbol{b}_\varepsilon = \boldsymbol{b} \qquad \text{in } L^1(\partial B_r; \mathbb{R}^2).$$

Since we can control the supremum of the one dimensional restriction of $\boldsymbol{b}_\varepsilon$ to ∂B_r through the L^1 norm of $\boldsymbol{b}_\varepsilon$ and the total variation we have that

$$\sup_{x \in \partial B_r} |\boldsymbol{b}_\varepsilon(x)| \le \frac{1}{2\pi r} \int_{\partial B_r} |\boldsymbol{b}_\varepsilon(x)| \, dx + \int_{\partial B_r} |\nabla \boldsymbol{b}_\varepsilon(x)| \, dx.$$

Hence, integrating with respect to r in $(R, R+1)$, (3.35) holds for $\boldsymbol{b}_\varepsilon$:

$$\int_R^{R+1} \sup_{x \in \partial B_r} |\boldsymbol{b}_\varepsilon(x)| \, dr \le \frac{1}{2\pi R} \int_{B_{R+1} \setminus B_R} |\boldsymbol{b}_\varepsilon(x)| \, dx + \int_{B_{R+1} \setminus B_R} |\nabla \boldsymbol{b}_\varepsilon(x)| \, dx.$$

Taking the lim inf in both sides as ε goes to 0, by Fatou lemma we deduce that

$$\begin{aligned}\int_R^{R+1} \operatorname*{ess\,sup}_{x\in\partial B_r} |\boldsymbol{b}(x)|\,dr &\le \int_{B_{R+1}\setminus B_R} \liminf_{\varepsilon\to 0} \sup_{x\in\partial B_r} |\boldsymbol{b}_\varepsilon(x)|\,dr \\ &\le \liminf_{\varepsilon\to 0} \int_R^{R+1} \sup_{x\in\partial B_r} |\boldsymbol{b}_\varepsilon(x)|\,dr \\ &\le \lim_{\varepsilon\to 0}\left(\frac{1}{2\pi R}\int_{B_{R+1}\setminus B_R} |\boldsymbol{b}_\varepsilon(x)|\,dx \right.\\ &\qquad \left. + \int_{B_{R+1}\setminus B_R} |D\boldsymbol{b}_\varepsilon(x)|\,dx\right) \\ &= \frac{1}{2\pi R}\int_{B_{R+1}\setminus B_R} |\boldsymbol{b}(x)|\,dx + |D\boldsymbol{b}|(B_{R+1}\setminus B_R).\end{aligned}$$

Step 2. Let $R > 0$ and let $\boldsymbol{c} : \mathbb{R}^2 \to \mathbb{R}^2$ be a Borel vector field such that

$$f(r) := \sup_{x\in\partial B_r} |\boldsymbol{c}(x)| \in L^1(R, R+1).$$

Let $\gamma : [0,\tau] \to \overline{B}_{R+1} \setminus B_R$ be an absolutely continuous integral curve of $\boldsymbol{c}$ (namely $\dot\gamma = \boldsymbol{c}(\gamma)$ $\mathscr{L}^1$-a.e. in $(0,\tau)$) such that $\gamma(0) \in \partial B_R$ and $\gamma(\tau) \in \partial B_{R+1}$. We claim that

$$\tau \ge \left(\int_R^{R+1} f(r)\,dr\right)^{-1}. \tag{3.36}$$

To prove this, we define the nondecreasing function $\sigma : [0,\tau] \to \mathbb{R}$

$$\sigma(t) = \max_{s\in[0,t]} |\gamma(s)| \qquad \forall t \in [0,\tau]; \tag{3.37}$$

we have that $\sigma(0) = R$ and $\sigma(\tau) = R+1$. For every $s, t \in [0,\tau]$ with $s < t$ there holds

$$0 \le \sigma(t) - \sigma(s) \le \sup_{r\in(s,t]} (|\gamma(r)| - |\gamma(s)|)^+ \le \int_s^t \left|\frac{d}{dr}|\gamma(r)|\right| dr$$
$$\le \int_s^t |\dot\gamma(r)|\,dr.$$

Thus σ is absolutely continuous and $\dot\sigma \le |\dot\gamma|$ $\mathscr{L}^1$-a.e in $(0,\tau)$. In addition, for every $t \in (0,\tau)$ such that $\sigma(t) \ne |\gamma(t)|$ the function σ is constant in a neighborhood of t, hence $\dot\sigma \le \chi_{\{\sigma=|\gamma|\}}|\dot\gamma|$ $\mathscr{L}^1$-a.e. in $(0,\tau)$. Therefore

$$\dot\sigma(t) \le 1_{\{\sigma=|\gamma|\}}(t)|\dot\gamma(t)| = 1_{\{\sigma=|\gamma|\}}(t)|\boldsymbol{c}(\gamma(t))| \le f(\sigma(t)) \quad \text{for } \mathscr{L}^1\text{-a.e. } t \in (0,\tau).$$

By Hölder inequality and the change of variable formula we deduce that

$$\begin{aligned} 1 \le [\sigma(\tau)-\sigma(0)]^2 &\le \left(\int_0^\tau \dot\sigma(t)\,dt\right)^2 \le \tau\int_0^\tau [\dot\sigma(t)]^2\,dt \\ &\le \tau\int_0^\tau \dot\sigma(t) f(\sigma(t))\,dt = \tau\int_R^{R+1} f(\sigma)\,d\sigma, \end{aligned}$$

which proves (3.36).

Step 3. We conclude the proof. Using the invariance of the concept of maximal regular flow (see Remark 2.6) we can work with a well-chosen representative which allows us to apply the estimate in Step 2. For this specific representation of $\boldsymbol{b}$, we show that *every* integral unbounded trajectory blows up properly.

For $\mathscr{L}^d$-a.e. $r > 0$ the restriction $\boldsymbol{b}_r(x) = \boldsymbol{b}(rx)$, $x \in \mathbb{S}^1$, of the vector field $\boldsymbol{b}$ to ∂B_r is BV. We remind that every 1-dimensional BV function has a precise representative given at every point by the average of the right approximate limit and of the left approximate limit, which exist everywhere. We define the Borel vector field $\boldsymbol{c} : \mathbb{R}^2 \to \mathbb{R}$ as

$$\boldsymbol{c}(rx) = \text{the precise representative of } \boldsymbol{b}_r \text{ at } x \qquad \forall x \in \mathbb{S}^1$$

for all r such that $\boldsymbol{b}_r \in BV(\mathbb{S}^1)$, and 0 otherwise. Notice that, by Fubini theorem, $\boldsymbol{c}$ coincides $\mathscr{L}^2$-a.e. with $\boldsymbol{b}$, and that $\sup|\boldsymbol{c}(r\cdot)| \le \operatorname{ess\,sup}|\boldsymbol{b}(r\cdot)|$ for all $r > 0$.

Let us assume by contradiction the existence of $\bar x \in \mathbb{R}^d$ such that $\boldsymbol{X}(\cdot,\bar x)$ is an integral curve of the precise representative $\boldsymbol{c}$ and

$$\liminf_{t\uparrow T_{\boldsymbol{X}}(\bar x)} |\boldsymbol{X}(t,\bar x)| < \infty, \qquad \limsup_{t\uparrow T_{\boldsymbol{X}}(\bar x)} |\boldsymbol{X}(t,\bar x)| = \infty. \tag{3.38}$$

We fix $R > 0$ greater than the lim inf in (3.38), as in Figure 3.5 and we define $f(r) := \sup_{x\in\partial B_r}|\boldsymbol{c}(x)|, r \in [R, R+1]$. Thanks to (3.35) applied to $\boldsymbol{c}$, we deduce that $f \in L^1(R, R+1)$. Therefore we can apply Step 2 to deduce that every transition from inside B_R to outside B_{R+1} requires at least time $1/\|f\|_{L^1(R,R+1)} > 0$. Hence the trajectory $\boldsymbol{X}(\cdot,\bar x)$ can cross the set $B_{R+1}\setminus B_R$ only finitely many times in finite time, a contradiction. □

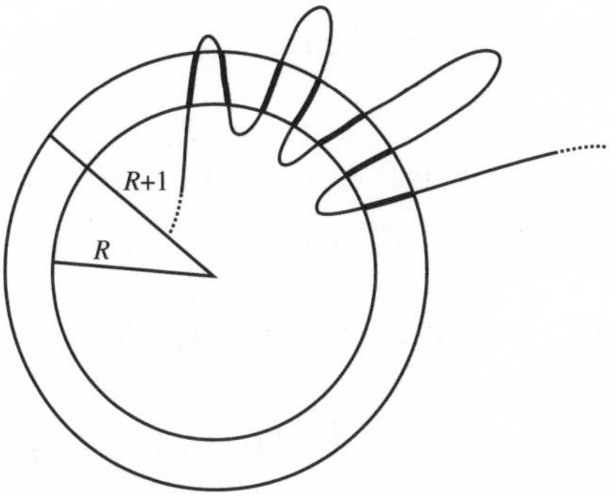

Figure 3.5. For an autonomous vector field $\boldsymbol{b}$ in the plane, we consider an integral curve of a suitable representative of $\boldsymbol{b}$, namely a vector field which coincides $\mathscr{L}^2$-a.e. with $\boldsymbol{b}$. Given $R > 0$, the time needed for the integral curve to cross the annulus $B_{R+1} \setminus B_R$ is greater or equal than the constant $\| \operatorname{ess\,sup}_{\partial B_r} |\boldsymbol{b}| \|^{-1}_{L^1(R,R+1)}$ (see (3.36) below). For this reason, every trajectory can cross only finitely many times the annulus in finite time and therefore every unbounded trajectory must blow up properly, as in (3.30).

3.4. No blow-up criteria under global bounds on divergence

If one is interested in estimating the blow-up time $T_{\Omega,X}$ of the maximal regular flow, or even if one wants to rule out the blow-up, one may easily adapt to this framework the classical criterion based on the existence of a Lyapunov function $\Psi : \mathbb{R}^d \to [0, \infty]$ satisfying $\Psi(z) \to \infty$ as $|z| \to \infty$ and

$$\frac{d}{dt}\Psi(x(t)) \leq C_\Psi\big(1 + \Psi(x(t))\big)$$

along absolutely continuous solutions to $\dot{x} = \boldsymbol{b}_t(x)$. On the other hand, in some cases, by a suitable approximation argument one can exhibit a solution $\mu_t = \rho_t \mathscr{L}^d$ to the continuity equation with velocity field $\boldsymbol{b}$ with $|\boldsymbol{b}_t|\rho_t$ integrable. As in [19, Proposition 8.1.8] (where locally Lipschitz vector fields were considered) we can use the existence of this solution to rule out the blow-up.

In the next theorem we provide a sufficient condition for the continuity of $\boldsymbol{X}$ at the blow-up time, using a global version of (a-Ω) and the global bounded compression condition (3.27), implied by the global bound on divergence (3.26).

Theorem 3.12. *Let $\boldsymbol{b} \in L^1((0,T) \times \Omega; \mathbb{R}^d)$ satisfy (b-Ω) and assume that the maximal regular flow $\boldsymbol{X}$ satisfies* (3.27). *Then $\boldsymbol{X}(\cdot, x)$ is absolutely continuous in $[0, T_{\Omega,X}(x)]$ for $\mathscr{L}^d$-a.e. $x \in \Omega$, and the limit of $\boldsymbol{X}(t,x)$ as $t \uparrow T_{\Omega,X}(x)$ belongs to $\partial\Omega$ whenever $T_{\Omega,X}(x) < T$.*

Proof. By (3.27) we have that

$$\begin{aligned}\int_\Omega \int_0^{T_{\Omega,\boldsymbol{X}}(x)} |\dot{\boldsymbol{X}}(t,x)|\,dt\,dx &= \int_\Omega \int_0^{T_{\Omega,\boldsymbol{X}}(x)} |\boldsymbol{b}_t(\boldsymbol{X}(t,x))|\,dt\,dx \\ &= \int_0^T \int_{\{T_{\Omega,\boldsymbol{X}}>t\}} |\boldsymbol{b}_t(\boldsymbol{X}(t,x))|\,dx\,dt \\ &\le C_* \int_0^T \int_\Omega |\boldsymbol{b}_t(z)|\,dz\,dt.\end{aligned}$$

Hence $\boldsymbol{X}$ satisfies (3.27). Then $\boldsymbol{X}(\cdot,x)$ is absolutely continuous in $[0, T_{\Omega,\boldsymbol{X}}(x)]$ for $\mathscr{L}^d$-a.e. $x \in \Omega$. Since the $\limsup V_\Omega(\boldsymbol{X}(t,x))$ as $t \uparrow T_{\Omega,\boldsymbol{X}}$ is ∞ whenever $T_{\Omega,\boldsymbol{X}}(x) < T$, we obtain that in this case the limit of $\boldsymbol{X}(t,x)$ as $t \to T_{\Omega,\boldsymbol{X}}(x)$ belongs to $\partial\Omega$. □

In the case $\Omega = \mathbb{R}^d$ we now prove a simple criterion for global existence, which allows us to recover the classical result in the DiPerna-Lions theory on the existence of a global flow under the growth condition

$$\frac{|\boldsymbol{b}_t(x)|}{1+|x|} \in L^1((0,T);L^1(\mathbb{R}^d)) + L^1((0,T);L^\infty(\mathbb{R}^d)). \tag{3.39}$$

As in the previous section, we will use in the next theorem the simplified notation $T_{\boldsymbol{X}}$ for $T_{\mathbb{R}^d,\boldsymbol{X}}$.

Theorem 3.13 (No blow-up criterion). *Let $\boldsymbol{b} : (0,T) \times \mathbb{R}^d \to \mathbb{R}^d$ be a Borel vector field which satisfies (a-$\mathbb{R}^d$) and (b-$\mathbb{R}^d$), and assume that the maximal regular flow $\boldsymbol{X}$ satisfies* (3.27)*. Assume that $\rho_t \in L^\infty\big((0,T); L^\infty_+(\mathbb{R}^d)\big)$ is a weakly* continuous solution of the continuity equation satisfying the integrability condition*

$$\int_0^T \int_{\mathbb{R}^d} \frac{|\boldsymbol{b}_t(x)|}{1+|x|} \rho_t(x)\,dx\,dt < \infty. \tag{3.40}$$

Then $T_{\boldsymbol{X}}(x) = T$ and $\boldsymbol{X}(\cdot,x) \in AC([0,T];\mathbb{R}^d)$ for $\rho_0\mathscr{L}^d$-a.e. $x \in \mathbb{R}^d$. In addition, if the growth condition (3.39) *holds, then ρ_t satisfying* (3.40) *exist for any $\rho_0 \in L^1 \cap L^\infty(\mathbb{R}^d)$ nonnegative, so that $\boldsymbol{X}$ is defined in the whole* $[0,T] \times \mathbb{R}^d$.

Proof. For the first part of the statement we apply Theorem 1.6 to deduce that ρ_t is the marginal at time t of a measure $\boldsymbol{\eta} \in \mathscr{M}_+\big(C([0,T];\mathbb{R}^d)\big)$ concentrated on absolutely continuous curves η in $[0,T]$ solving the ODE $\dot{\eta} = \boldsymbol{b}_t(\eta)$. We then apply Theorem 1.12 to obtain that the conditional probability measures $\boldsymbol{\eta}_x$ induced by the map e_0 are Dirac masses for

$(e_0)_\#\eta$-a.e. x, hence (by uniqueness of the maximal regular flow) ρ_t is transported by $\boldsymbol{X}$. Notice that, as a consequence of the fact that η is concentrated on absolutely continuous curves in $[0, T]$, the flow is globally defined on $[0, T]$, thus $T_{\boldsymbol{X}}(x) = T$.

For the second part, under assumption (3.39) the existence of a nonnegative and weakly* continuous solution of the continuity equation ρ_t in $L^\infty\big((0, T); L^1 \cap L^\infty(\mathbb{R}^d)\big)$ can be achieved by a simple smoothing argument. So, the bound in $L^1 \cap L^\infty$ on ρ_t can be combined with (3.39) to obtain (3.40). □

Remark 3.14. We remark that if only a local bound on the divergence is assumed as in Section 2.2, the growth assumption (3.39) is not enough to guarantee that the trajectories of the regular flow do not blow up. On the other hand, it can be easily seen that if we assume that $\boldsymbol{b}$ satisfies (a-$\mathbb{R}^d$), (b-$\mathbb{R}^d$), (3.1) and $|\boldsymbol{b}_t(x)|/(1 + |x|) \in L^1((0, T); L^\infty(\mathbb{R}^d))$, every integral curve of $\boldsymbol{b}$ cannot blow up in finite time and therefore the maximal regular flow satisfies $T_{\boldsymbol{X}}(x) = T$ and $\boldsymbol{X}(\cdot, x) \in AC([0, T]; \mathbb{R}^d)$ for $\mathscr{L}^d$-a.e. $x \in \mathbb{R}^d$.

Theorem 3.13 is useful in applications when one constructs solutions by approximation. For instance, for the Vlasov-Poisson system in dimension $d = 2$ and 3, this result can be used to show that trajectories which transport a bounded solution with finite energy do not explode in the phase space (see Theorem 8.2).

3.5. Forward and backward Maximal Regular Flows with divergence free vector fields

The theory of maximal regular flows developed in the previous and in this Chapter applies to study the Lagrangian structure of transport equations, which in turn give information on solutions of nonlinear PDEs such as the Vlasov-Poisson system. In order to avoid unnecessary complications, we develop this theory under the assumption that the vector field is divergence-free, which is satisfied in the application. In the following, we give a notion of maximal regular flow and we state an existence and uniqueness result that fully suits the application to the Vlasov-Poisson system, since it deals with forward and backward flows starting at any time.

Let $T \in (0, \infty)$ and let $\boldsymbol{b} : (0, T) \times \mathbb{R}^d \to \mathbb{R}^d$ be a Borel vector field. The following definition of maximal regular flow has an initial condition at time $s \in (0, T)$. Since this definition works only with a global bound on the divergence of $\boldsymbol{b}$, it appears simplified with respect to Definition 2.4 (specifically, compare (ii) and (iii)). A posteriori, however, in the cases of interest (namely, when assumptions (a), (b), and 3.43 below are satisfied), the two definitions are fully equivalent.

Definition 3.15 (Maximal Regular Flow). For every $s \in (0, T)$ we say that a Borel map $\boldsymbol{X}(\cdot, s, \cdot)$ is a *Maximal Regular Flow* starting at time s if there exist two Borel maps $T^+_{s,\boldsymbol{X}} : \mathbb{R}^d \to (s, T]$, $T^-_{s,\boldsymbol{X}} : \mathbb{R}^d \to [0, s)$ such that $\boldsymbol{X}(\cdot, x)$ is defined in $(T^-_{s,\boldsymbol{X}}(x), T^+_{s,\boldsymbol{X}}(x))$ and the following two properties hold:

(i) for $\mathscr{L}^d$-a.e. $x \in \mathbb{R}^d$, $\boldsymbol{X}(\cdot, x) \in AC_{\mathrm{loc}}((T^-_{s,\boldsymbol{X}}(x), T^+_{s,\boldsymbol{X}}(x)); \mathbb{R}^d)$ and solves the ODE $\dot{x}(t) = \boldsymbol{b}_t(x(t))$ $\mathscr{L}^1$-a.e. in $(T^-_{s,\boldsymbol{X}}(x), T^+_{s,\boldsymbol{X}}(x))$, with the initial condition $\boldsymbol{X}(s, s, x) = x$;
(ii) there exists a constant $C = C(s, \boldsymbol{X})$ such that

$$\boldsymbol{X}(t, s, \cdot)_\#\big(\mathscr{L}^d \llcorner \{T^-_{s,\boldsymbol{X}} < t < T^+_{s,\boldsymbol{X}}\}\big) \le C \mathscr{L}^d \quad \forall t \in [0, T]. \tag{3.41}$$

(iii) for $\mathscr{L}^d$-a.e. $x \in \mathbb{R}^d$, either $T^+_{s,\boldsymbol{X}}(x) = T$ (respectively $T^-_{s,\boldsymbol{X}}(x) = 0$) and $\boldsymbol{X}(\cdot, s, x)$ can be continuously extended up to $t = T$ (respectively $t = 0$) so that $\boldsymbol{X}(\cdot, s, x) \in C([s, T]; \mathbb{R}^d)$ (respectively $\boldsymbol{X}(\cdot, s, x) \in C([0, s]; \mathbb{R}^d)$), or

$$\lim_{t \uparrow T^+_{s,\boldsymbol{X}}(x)} |\boldsymbol{X}(t, s, x)| = \infty \ \text{(respectively} \lim_{t \downarrow T^-_{s,\boldsymbol{X}}(x)} |\boldsymbol{X}(t, s, x)| = \infty). \tag{3.42}$$

In particular, $T^+_{s,\boldsymbol{X}}(x) < T$ (respectively $T^-_{s,\boldsymbol{X}}(x) > 0$) implies (3.42).

The definition of Maximal Regular Flow can be extended up to the extreme times $s = 0, s = T$, setting $T^-_{0,\boldsymbol{X}} \equiv 0$ and $T^+_{T,\boldsymbol{X}} \equiv T$.

A Maximal Regular Flow has been built in Theorem 2.14 under general local assumptions on $\boldsymbol{b}$. Before stating the result, we recall the assumptions of this Section, which are a particular case of the assumptions of this Chapter. For $T \in (0, \infty)$ we are given a Borel vector field $\boldsymbol{b} : (0, T) \times \mathbb{R}^d \to \mathbb{R}^d$ satisfying:

(a) $\int_0^T \int_\Omega |\boldsymbol{b}_t(x)|\, dxdt < \infty$ for any $\Omega \Subset \mathbb{R}^d$;
(b) for any nonnegative $\bar{\rho} \in L^\infty_+(\mathbb{R}^d)$ with compact support and any closed interval $I = [a, b] \subset [0, T]$, the continuity equation

$$\frac{d}{dt}\rho_t + \mathrm{div}\,(\boldsymbol{b}\rho_t) = 0 \qquad \text{in } (a, b) \times \mathbb{R}^d$$

has at most one weakly* continuous solution $I \ni t \mapsto \rho_t \in \mathcal{L}_{I,\mathbb{R}^d}$ (defined in (1.9)) with $\rho_a = \bar{\rho}$.

Since the vector fields that arise in the applications we have in mind are divergence-free, we assume throughout the Section that our velocity field $\boldsymbol{b}$ satisfies

$$\operatorname{div} \boldsymbol{b} = 0 \qquad \text{in } (0, T) \times \mathbb{R}^d \qquad \text{in the sense of distributions.} \tag{3.43}$$

Equivalently, $\operatorname{div} \boldsymbol{b}_t = 0$ in the sense of distributions for $\mathscr{L}^1$-a.e. $t \in (0, T)$.

The existence and uniqueness of the Maximal Regular Flow after time s, as well as the semigroup property, were proved in Theorems 2.14 and 3.1 assuming a one sided bound (specifically a lower bound) on the divergence. We recall that, in this context, uniqueness should be understood as follows: if $\boldsymbol{X}$ and $\boldsymbol{Y}$ are Maximal Regular Flows, for all $s \in [0, T]$ one has

$$\begin{cases} T^{\pm}_{s,\boldsymbol{X}}(x) = T^{\pm}_{s,\boldsymbol{Y}}(x) & \text{for } \mathscr{L}^d\text{-a.e. } x \in \mathbb{R}^d \\ \boldsymbol{X}(\cdot, s, x) = \boldsymbol{Y}(\cdot, s, x) \text{ in } (T^{-}_{s,\boldsymbol{X}}(x), T^{+}_{s,\boldsymbol{X}}(x)) & \text{for } \mathscr{L}^d\text{-a.e. } x \in \mathbb{R}^d. \end{cases} \tag{3.44}$$

Under our assumptions on the divergence, by simply reversing the time variable the Maximal Regular Flow can be built both forward and backward in time, so we state the result in the time-reversible case.

Theorem 3.16 (Existence, uniqueness, and semigroup property). *Let us consider a Borel vector field* $\boldsymbol{b} : (0, T) \times \mathbb{R}^d \to \mathbb{R}^d$ *which satisfies* (a) *and* (b). *Then the Maximal Regular Flow starting from any* $s \in [0, T]$ *is unique according to* (3.44)*, and existence is ensured under the additional assumption* (3.43). *In addition, still assuming* (3.43)*, for all* $s \in [0, T]$ *the following properties hold:*

(i) *the compressibility constant* $C(s, \boldsymbol{X})$ *in Definition 3.15 equals* 1 *and for every* $t \in [0, T]$

$$\begin{aligned} &\boldsymbol{X}(t, s, \cdot)_{\#}\big(\mathscr{L}^d \llcorner \{T^{-}_{s,\boldsymbol{X}} < t < T^{+}_{s,\boldsymbol{X}}\}\big) \\ &\quad = \mathscr{L}^d \llcorner \big(\boldsymbol{X}(t, s, \cdot)(\{T^{-}_{s,\boldsymbol{X}} < t < T^{+}_{s,\boldsymbol{X}}\})\big); \end{aligned} \tag{3.45}$$

(ii) *if* $\tau_1 \in [0, s]$, $\tau_2 \in [s, T]$, *and* $\boldsymbol{Y}$ *is a Regular Flow in* $[\tau_1, \tau_2] \times B$, *then* $T^{+}_{s,\boldsymbol{X}} > \tau_2$, $T^{-}_{s,\boldsymbol{X}} < \tau_1$ $\mathscr{L}^d$*-a.e. in* B*; moreover*

$$\boldsymbol{X}(\cdot, s, x) = \boldsymbol{Y}(\cdot, \boldsymbol{X}(\tau_1, s, x)) \quad \text{in } [\tau_1, \tau_2], \text{for } \mathscr{L}^d\text{-a.e. } x \in B;$$

(iii) *the Maximal Regular Flow satisfies the semigroup property, namely for all* $s,\ s' \in [0, T]$

$$T^{\pm}_{s',\boldsymbol{X}}(\boldsymbol{X}(s', s, x)) = T^{\pm}_{s,\boldsymbol{X}}(x), \ \text{for } \mathscr{L}^d\text{-a.e. } x \in \{T^{+}_{s,\boldsymbol{X}} > s' > T^{-}_{s,\boldsymbol{X}}\}, \tag{3.46}$$

and, for $\mathscr{L}^d$*-a.e.* $x \in \{T^+_{s,X} > s' > T^-_{s,X}\}$,

$$X\big(t, s', X(s', s, x)\big) = X(t, s, x) \;\; \forall t \in (T^-_{s,X}(x), T^+_{s,X}(x)). \quad (3.47)$$

We finally mention that Theorem 3.13 provides, also in the context of the previous theorem, a simple condition for global existence of the maximal flow.

Chapter 4
Lagrangian structure of transport equations

When considering a fast growing, smooth vector field $\boldsymbol{b} : [0, T] \times \mathbb{R}^d \to \mathbb{R}^d$, we know from the Cauchy-Lipschitz theorem (see Theorem 1.1) that, starting at every time $s \in [0, T]$, we can build, forward and backward in time, the unique maximal regular flow $\boldsymbol{X}(\cdot, s, x)$ starting at time s from position x. This construction provides a set of curves that "foliate" the space-time (see Figure 4.1). Correspondingly, every smooth solution $u_t : [0, T] \times \mathbb{R}^d \to \mathbb{R}$ of the transport equation is transported by this set of curves, meaning that $u_t(\boldsymbol{X}(t, s, x))$ is constant with respect to t in the existence interval of the curve $\boldsymbol{X}(\cdot, s, x)$ for every $s \in [0, T]$ and $x \in \mathbb{R}^d$.

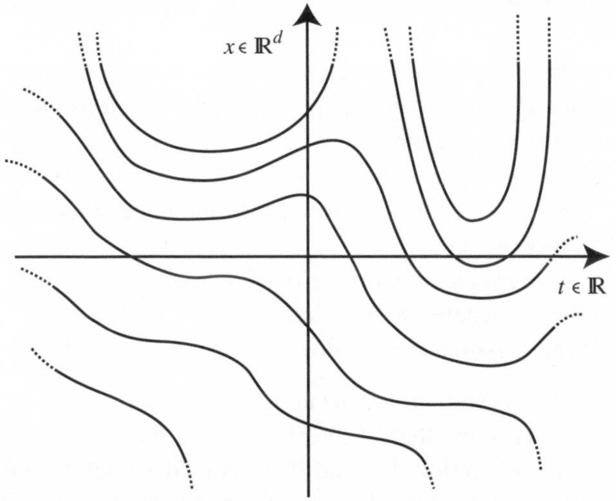

Figure 4.1. In the case of a smooth vector field $\boldsymbol{b} : \mathbb{R} \times \mathbb{R}^d \to \mathbb{R}^d$, the maximal flows give a set of smooth curves that "foliate" the space-time. In the non-smooth setting a similar phenomenon occurs with the maximal regular flows introduced in Chapters 2 and 3, but measurability issues arise in considering them as a set of curves in space-time.

A similar description is not available in the literature in the context of non-smooth vector fields, since up to now global assumptions were always made on $\boldsymbol{b}$ to prevent the blow-up of the flow. The aim of this Chapter is to develop the abstract theory that connects the notion of Maximal Regular Flow and Lagrangian/renormalized solutions for the continuity/transport equation under purely local assumptions on the vector field and on the solution.

The whole content of this Chapter will be applied in Chapter 8 to show that the Eulerian description of weak solutions of the Vlasov-Poisson system corresponds to a Lagrangian evolution of particles. In view of the applications, we present the theory in this Chapter only for divergence-free vector fields, remarking that some statements would require more technical tools to be extended to the case of vector fields with bounded divergence. We warn the reader that, since the theory is completely general, we shall consider flows of vector fields in $\mathbb{R}^d$ and denote by x a point in $\mathbb{R}^d$. Then, for the applications to kinetic equations in the phase-space $\mathbb{R}^{2d}$, one should apply these results replacing d with $2d$ and x with (x, v).

In the following, we consider four different notions of solutions of the continuity/transport equation. The first two are of Eulerian nature, whereas the remaining two are Lagrangian.

- *Distributional solutions* have been introduced in Definition 1.2 and regard the PDE point of view. This is the weakest possible notion of solution.
- *Renormalized solutions* (see Definition 4.4 below) are more rigid and encode in a PDE language the property that solutions are transported along curves in space-time.
- *Generalized flows* are weighted collections of integral curves of the vector field $\boldsymbol{b}$ (which may go to infinity and come back in finite time, see Definition 4.1 below).
- *Solutions transported by the maximal regular flow* are particular generalized flows, where the only integral curves allowed are the trajectories of the maximal regular flow (see Definition 4.2 below).

If the vector field is smooth, all the notions of solutions are equivalent. In the following, we show that, if the vector field $\boldsymbol{b}$ is divergence-free and satisfies (a)-(b) of Section 3.5, and if we consider bounded nonnegative solutions, the equivalence still holds. Although the concepts of distributional and renormalized solutions are completely local, in the literature they have been related in [81] and [5] only by means of global assumptions on the vector field, that we avoid in the following. For instance, the superposition principle, presented in Theorem 1.6, relates distributional solutions and generalized flows under the global assumption (1.5); Theorem 1.12 relates generalized flows to solutions transported by the regular lagrangian flow assuming that no blow-up is allowed in the curves on which the generalized flow is concentrated. In this Chapter, we present the connection between the previous definitions under local assumptions on the vector field and on the solution.

- Distributional solutions vs generalized flows: in Section 4.3 we prove that, under general assumptions both on the vector field and on the solution (in particular, no regularity or boundedness is assumed) every distributional solution can be represented as a generalized flow. This is the local counterpart of the superposition principle, Theorem 1.6.
- Generalized flows vs solutions transported by the maximal regular flow: in Section 4.1 we show that the well posedness of the continuity equation with vector field $\boldsymbol{b}$, which in turn follows usually from the regularity of $\boldsymbol{b}$ (see Remark 1.9), implies that generalized regular flows (here, "regular" avoids concentration, see Definition 4.1 below) are transported by the maximal regular flow.
- Solutions transported by the maximal regular flow vs renormalized solutions: finally, this connection is exploited in Section 4.2.

4.1. Generalized flows and Maximal Regular Flows

We denote by $\mathring{\mathbb{R}}^d = \mathbb{R}^d \cup \{\infty\}$ the one-point compactification of $\mathbb{R}^d$ and we recall the definition of generalized flow and of regular generalized flow in our context. This is a generalization of Definition 2.10, which was used in open bounded sets, and it allows the integral curves of $\boldsymbol{b}$, on which the generalized flow is concentrated, to go to infinity and come back.

Definition 4.1 (Generalized flow). Let $\boldsymbol{b} : (0,T) \times \mathbb{R}^d \to \mathbb{R}^d$ be a Borel vector field. The measure $\boldsymbol{\eta} \in \mathscr{M}_+\big(C([0,T]; \mathring{\mathbb{R}}^d)\big)$ is said to be a *generalized flow* of $\boldsymbol{b}$ if $\boldsymbol{\eta}$ is concentrated on the set

$$\Gamma := \Big\{\eta \in C([0,T]; \mathring{\mathbb{R}}^d) : \ \eta \in AC_{\mathrm{loc}}(\{\eta \neq \infty\}; \mathbb{R}^d) \text{ and } \dot\eta(t) = \boldsymbol{b}_t(\eta(t)) \text{ for } \mathscr{L}^1\text{-a.e. } t \in \{\eta \neq \infty\}\Big\}. \quad (4.1)$$

We say that a generalized flow $\boldsymbol{\eta}$ is *regular* if there exists $L_0 \geq 0$ satisfying

$$(e_t)_\# \boldsymbol{\eta} \llcorner \mathbb{R}^d \leq L_0 \mathscr{L}^d \qquad \forall t \in [0,T]. \quad (4.2)$$

In connection with this definition, let us provide a sketch of proof of the fact that the set Γ in (4.1) is Borel in $C([0,T]; \mathring{\mathbb{R}}^d)$.

First of all one notices that for all intervals $[a,b] \subset [0,T]$ the set $\{\eta : \eta([a,b]) \subset \mathbb{R}^d\}$ is Borel. Then, considering the absolute continuity of a curve η in the integral form

$$|\eta(t) - \eta(s)| \leq \int_s^t |\boldsymbol{b}_r(\eta(r))|\, dr \qquad \forall\, s, t \in [a, b],\ s \leq t$$

it is sufficient to verify (arguing componentwise and splitting in positive and negative part) that for any nonnegative Borel function $\boldsymbol{c}$ and for any $s, t \in [0, T]$ with $s \leq t$ fixed, the function

$$\eta \mapsto \int_s^t \boldsymbol{c}_r(\eta(r))\, dr$$

is Borel in $\{\eta : \ \eta([a, b]) \subset \mathbb{R}^d\}$. This follows by a monotone class argument, since the property is obviously true for continuous functions and it is stable under equibounded and monotone convergence. Finally, as soon as the absolute continuity property is secured, also the verification of the Borel regularity of the class

$$\begin{aligned} &\Gamma \cap \{\eta : \eta([a, b]) \subset \mathbb{R}^d\} \\ &= \Big\{\eta \in C([0, T]); \mathring{\mathbb{R}}^d) : \eta \in AC([a, b]; \mathbb{R}^d), \\ &\qquad \dot{\eta}(t) = \boldsymbol{b}_t(\eta(t))\ \mathscr{L}^1\text{-a.e. in } (a, b)\Big\} \end{aligned}$$

can be achieved following similar lines. Finally, by letting the endpoints a, b vary in a countable dense set we obtain that Γ is Borel.

In the case of a smooth, bounded vector field, a particular class of generalized flows is the one generated by transporting a given measure $\mu_0 \in \mathscr{P}\left(\mathbb{R}^d\right)$ along the integral lines of the flow:

$$\boldsymbol{\eta} = \int_{\mathbb{R}^d} \delta_{\boldsymbol{X}(\cdot, x)}\, d\mu_0(x),$$

where $\delta_{\boldsymbol{X}(\cdot,x)} \in \mathscr{P}\big(AC([0, T]; \mathbb{R}^d)\big)$ denotes the Dirac delta on the path $\boldsymbol{X}(\cdot, x)$.

In the next definition we propose a generalization of this construction involving Maximal Regular Flows.

Definition 4.2 (Measures transported by the Maximal Regular Flow). Let $\boldsymbol{b} : (0, T) \times \mathbb{R}^d \to \mathbb{R}^d$ be a Borel vector field having a Maximal Regular Flow $\boldsymbol{X}$ (according to Definition 3.15) and let $\boldsymbol{\eta} \in \mathscr{M}_+\big(C([0, T]; \mathring{\mathbb{R}}^d)\big)$ with $(e_t)_\# \boldsymbol{\eta} \ll \mathscr{L}^d$ for all $t \in [0, T]$. We say that $\boldsymbol{\eta}$ is transported by $\boldsymbol{X}$ if, for all $s \in [0, T]$, $\boldsymbol{\eta}$ is concentrated on

$$\begin{aligned} \big\{\eta \in C([0, T]; \mathring{\mathbb{R}}^d) : \eta(s) = \infty \text{ or } \eta(\cdot) \\ = \boldsymbol{X}(\cdot, s, \eta(s)) \text{ in } (T^-_{s,\boldsymbol{X}}(\eta(s)), T^+_{s,\boldsymbol{X}}(\eta(s)))\big\}. \end{aligned} \tag{4.3}$$

Correspondingly, let $\rho_t \in L^\infty((0,T);L^1_+(\mathbb{R}^d))$ be a distributional solution of the continuity equation, weakly continuous on $[0,T]$ in duality with $C_c(\mathbb{R}^d)$. We say that ρ_t is a Lagrangian solution if there exists $\boldsymbol{\eta} \in \mathscr{M}_+\big(C([0,T];\mathring{\mathbb{R}}^d)\big)$ transported by $\boldsymbol{X}$ with $(e_t)_\#\boldsymbol{\eta} = \rho_t\,\mathscr{L}^d$ for every $t \in [0,T]$.

The absolute continuity assumption $(e_t)_\#\boldsymbol{\eta} \ll \mathscr{L}^d$ on the marginals of $\boldsymbol{\eta}$ is needed to ensure that this notion is invariant with respect to the uniqueness property in (3.44). In other words, if $\boldsymbol{X}$ and $\boldsymbol{Y}$ are related as in (3.44), then $\boldsymbol{\eta}$ is transported by $\boldsymbol{X}$ if and only if $\boldsymbol{\eta}$ is transported by $\boldsymbol{Y}$. Indeed, given $s \in [0,T]$ the symmetric difference between the set in (4.3) and the corresponding set with $\boldsymbol{Y}$ in place of $\boldsymbol{X}$ is contained in

$$\Gamma_{s,\boldsymbol{X},\boldsymbol{Y}} = \{\eta \in C([0,T];\mathring{\mathbb{R}}^d) : \eta(s) \in E_s\},$$

where

$$E_s = \big\{x \in \mathbb{R}^d : T^-_{s,\boldsymbol{X}}(x) \neq T^-_{s,\boldsymbol{Y}}(x) \text{ or } T^+_{s,\boldsymbol{X}}(x) \neq T^+_{s,\boldsymbol{Y}}(x) \text{ or } \boldsymbol{X}(\cdot,s,x) \neq \boldsymbol{Y}(\cdot,s,x) \text{ in } (T^-_{s,\boldsymbol{X}}(x), T^+_{s,\boldsymbol{X}}(x))\big\}.$$

Our goal is to show that the set $\Gamma_{s,\boldsymbol{X},\boldsymbol{Y}}$ is $\boldsymbol{\eta}$-negligible. This follows by the uniqueness property (3.44), which says that $\mathscr{L}^d(E_s) = 0$, and by $(e_t)_\#\boldsymbol{\eta} \ll \mathscr{L}^d$, which implies that $\boldsymbol{\eta}(\Gamma_{s,\boldsymbol{X},\boldsymbol{Y}}) = (e_t)_\#\boldsymbol{\eta}(E_s) = 0$.

It is easily seen that if $\boldsymbol{\eta}$ is transported by a Maximal Regular Flow, then $\boldsymbol{\eta}$ is a generalized flow according to Definition 4.1, but in connection with the proof of the renormalization property we are more interested to the converse statement. As shown in the next theorem, this holds for regular generalized flows and for divergence-free vector fields satisfying (a)-(b) of Section 3.5.

Theorem 4.3 (Regular generalized flows are transported by $\boldsymbol{X}$). *Let $\boldsymbol{b} : (0,T)\times\mathbb{R}^d \to \mathbb{R}^d$ be a divergence-free vector field which satisfies* (a)-(b) *of Section* 3.5 *and let $\boldsymbol{X}$ be its Maximal Regular Flow (according to Definition* 3.15*). Let $\boldsymbol{\eta} \in \mathscr{M}_+\big(C([0,T];\mathring{\mathbb{R}}^d)\big)$ be a regular generalized flow according to Definition* 4.1.

Consider $s \in [0,T]$ and a Borel family $\{\boldsymbol{\eta}^s_x\} \subset \mathscr{P}\big(C([0,T];\mathring{\mathbb{R}}^d)\big)$, $x \in \mathring{\mathbb{R}}^d$, of conditional probability measures representing $\boldsymbol{\eta}$ with respect to the marginal $(e_s)_\#\boldsymbol{\eta}$, i.e., *$\int \boldsymbol{\eta}^s_x\, d(e_s)_\#\boldsymbol{\eta}(x) = \boldsymbol{\eta}$. Then for $(e_s)_\#\boldsymbol{\eta}$-almost every $x \in \mathbb{R}^d$ we have that $\boldsymbol{\eta}^s_x$ is concentrated on the set*

$$\Gamma_s := \big\{\eta \in C([0,T];\mathring{\mathbb{R}}^d) \,:\, \eta(s) = x,\ \eta(\cdot) = \boldsymbol{X}(\cdot,s,\eta(s)) \text{ in } (T^-_{s,\boldsymbol{X}}(\eta(s)), T^+_{s,\boldsymbol{X}}(\eta(s)))\big\}. \tag{4.4}$$

In particular $\boldsymbol{\eta}$ is transported by $\boldsymbol{X}$.

Proof. First of all we notice that the set Γ_s in (4.4) is Borel. Indeed, the maps $\eta \mapsto T^{\pm}_{s,\boldsymbol{X}}(\eta(s))$ are Borel because $T^{\pm}_{\boldsymbol{X}}$ are Borel in $\mathbb{R}^d$, and the map $\eta \mapsto \boldsymbol{X}(t, s, \eta(s))$ is Borel as well for any $t \in [0, T]$. Therefore, choosing a countable dense set of times $t \in [0, T]$ the Borel regularity of Γ_s is achieved.

The fact that $\boldsymbol{\eta}^s_x$ is concentrated on the set $\{\eta : \eta(s) = x\}$ is immediate from the definition of $\boldsymbol{\eta}^s_x$. We now show that for $(e_s)_\#\boldsymbol{\eta}$-almost every $x \in \mathbb{R}^d$ the measure $\boldsymbol{\eta}^s_x$ is concentrated on the set

$$\left\{\eta \in C([0, T]; \mathring{\mathbb{R}}^d) : \eta(\cdot) = \boldsymbol{X}(\cdot, s, x) \text{ in } [s, T^+_{s,\boldsymbol{X}}(x))\right\}. \tag{4.5}$$

Notice that applying the same result after reversing the time variable, this proves the concentration on the set Γ_s in (4.4).

For $r \in (s, T]$ we denote by $\Sigma^{s,r} : C([0, T]; \mathring{\mathbb{R}}^d) \to C([s, r]; \mathring{\mathbb{R}}^d)$ the map induced by restriction to $[s, r]$, namely $\Sigma^{s,r}(\eta) := \eta|_{[s,r]}$.

For every $R > 0, r \in (s, T]$, let us consider

$$\boldsymbol{\eta}^{R,r} := \Sigma^{s,r}_\# \Big(\boldsymbol{\eta} \llcorner \big\{\eta : \eta(t) \in B_R \text{ for every } t \in [s, r]\big\}\Big).$$

By construction $\boldsymbol{\eta}^{R,r}$ is a regular generalized flow relative to $\boldsymbol{b}$ with compact support, hence our regularity assumption on $\boldsymbol{b}$ allows us to apply Theorem 1.12 to deduce that

$$\boldsymbol{\eta}^{R,r} = \int \delta_{\boldsymbol{Y}(\cdot,x)}\, d[(e_s)_\#\boldsymbol{\eta}^{R,r}](x), \tag{4.6}$$

where $\boldsymbol{Y}(\cdot, x)$ is an integral curve of $\boldsymbol{b}$ in $[s, r]$ for $(e_s)_\#\boldsymbol{\eta}$-a.e. $x \in \mathbb{R}^d$. Let us denote by $\rho_{R,r}$ the density of $(e_s)_\#\boldsymbol{\eta}^{R,r}$ with respect to $\mathscr{L}^d$, which is bounded by L_0 thanks to (4.2). For every $\delta > 0$ we have that

$$\begin{aligned}
\boldsymbol{Y}(t,\cdot)_\#\big(\mathscr{L}^d \llcorner \{\rho_{R,r} > \delta\}\big) &= (e_t)_\# \int_{\{\rho_{R,r}>\delta\}} \delta_{\boldsymbol{Y}(\cdot,x)}\, d\mathscr{L}^d(x) \\
&\le \frac{1}{\delta}(e_t)_\# \int_{\{\rho_{R,r}>\delta\}} \delta_{\boldsymbol{Y}(\cdot,x)}\, d[(e_s)_\#\boldsymbol{\eta}^{R,r}](x) \qquad (4.7)\\
&\le \frac{1}{\delta}(e_t)_\#\boldsymbol{\eta}^{R,r} \le \frac{1}{\delta}(e_t)_\#\boldsymbol{\eta} \llcorner \mathbb{R}^d \le \frac{L_0}{\delta}\mathscr{L}^d,
\end{aligned}$$

hence $\boldsymbol{Y}(\cdot, x)$ is a Regular Flow of $\boldsymbol{b}$ in $[s, r] \times \{\rho_{R,r} > \delta\}$ according to Definition 2.1. By Theorem 3.16(ii) we deduce that $\boldsymbol{Y}(\cdot, x) = \boldsymbol{X}(\cdot, s, x)$ for $\mathscr{L}^d$-a.e. $x \in \{\rho_{R,s} > \delta\}$ and therefore, letting $\delta \to 0$,

$$\boldsymbol{Y}(\cdot, x) = \boldsymbol{X}(\cdot, s, x) \qquad \text{in } [s, r] \text{ for } (e_s)_\#\boldsymbol{\eta}^{R,s}\text{-a.e. } x \in \mathbb{R}^d. \tag{4.8}$$

Letting $R \to \infty$ we have that $\boldsymbol{\eta}^{R,r} \to \boldsymbol{\sigma}^r$ increasingly, where

$$\boldsymbol{\sigma}^r := \Sigma_\#^{s,r}\Big(\boldsymbol{\eta} \llcorner \{\eta : \eta(t) \neq \infty \text{ for every } t \in [s,r]\}\Big).$$

By (4.6) and (4.8) we deduce that for every $r \in (s,T]$

$$\boldsymbol{\sigma}^r = \int \delta_{\boldsymbol{X}(\cdot,s,x)}\, d[(e_s)_\# \boldsymbol{\sigma}^r](x). \tag{4.9}$$

Arguing by contradiction, let us assume that there exists a Borel set $E \subset \mathbb{R}^d$ such that $(e_s)_\# \boldsymbol{\eta}(E) > 0$ and $\boldsymbol{\eta}_x^s$ is not concentrated on the set (4.5) for every $x \in E$, namely

$$\boldsymbol{\eta}_x^s\Big(\Big\{\eta \in C([0,T]; \mathring{\mathbb{R}}^d) : \eta \neq \boldsymbol{X}(\cdot,s,x) \text{ as a curve in } [s, T^+_{s,\boldsymbol{X}}(x))\Big\}\Big) > 0.$$

Since this can be rewritten as

$$\boldsymbol{\eta}_x^s\left(\bigcup_{r \in \mathbb{Q} \cap (s, T^+_{s,\boldsymbol{X}}(x))} \left\{\eta \in C([0,T];\mathring{\mathbb{R}}^d) : \eta \neq \boldsymbol{X}(\cdot,s,x) \text{ in } [s,r], \eta([s,r]) \subset \mathbb{R}^d\right\}\right) > 0,$$

for every $x \in E$ there exists $r_x \in \mathbb{Q} \cap (s, T^+_{s,\boldsymbol{X}}(x))$ such that

$$\boldsymbol{\eta}_x^s\Big(\Big\{\eta \in C([0,T]; \mathring{\mathbb{R}}^d) : \eta \neq \boldsymbol{X}(\cdot,s,x) \\ \text{as a curve in } [s,r_x],\ \eta([s,r_x]) \subset \mathbb{R}^d\Big\}\Big) > 0.$$

In other words, for every $x \in E$ there exists a rational number r_x such that

$$\Sigma_\#^{s,r_x}\Big(\boldsymbol{\eta}_x^s \llcorner \{\eta : \eta(t) \neq \infty \text{ for every } t \in [s,r_x]\}\Big)$$

is nonzero and not multiple of $\delta_{\boldsymbol{X}(\cdot,s,x)}$.

Therefore, there exist a Borel set $E' \subset E$ of positive $(e_s)_\# \boldsymbol{\eta}$-measure and $r \in (s,T] \cap \mathbb{Q}$ such that for every $x \in E'$

$$\Sigma_\#^{s,r}\Big(\boldsymbol{\eta}_x^s \llcorner \{\eta : \eta(t) \neq \infty \text{ for every } t \in [s,r]\}\Big)$$

is nonzero and not multiple of $\delta_{\boldsymbol{X}(\cdot,s,x)}$.

By (4.9) and $(e_s)_\# \boldsymbol{\sigma}^r \leq (e_s)_\# \boldsymbol{\eta}$ we have that

$$\int \delta_{\boldsymbol{X}(\cdot,s,x)}\, d(e_s)_\# \boldsymbol{\eta}(x) \geq \boldsymbol{\sigma}^r$$
$$= \int \Sigma_\#^{s,r}\Big(\boldsymbol{\eta}_x^s \llcorner \{\eta : \eta(t) \neq \infty \text{ for every } t \in [s,r]\}\Big)\, d(e_s)_\# \boldsymbol{\eta}(x).$$

This yields $\delta_{X(\cdot,s,x)} \geq \Sigma_{\#}^{s,r}\Big(\boldsymbol{\eta}_x^s \llcorner \{\eta : \eta(t) \neq \infty \text{ for every } t \in [s,r]\}\Big)$ for $(e_s)_\# \boldsymbol{\eta}$-a.e. x, and therefore a contradiction with the existence of E'. This proves that $\boldsymbol{\eta}_x^s$ is concentrated on the set defined in (4.5), as desired.

Finally, in order to prove that $\boldsymbol{\eta}$ is transported by $\boldsymbol{X}$ we apply the definition of disintegration and the fact that for $(e_s)_\# \boldsymbol{\eta}$-a.e. $x \in \mathbb{R}^d$ the measure $\boldsymbol{\eta}_x^s$ is concentrated on the set Γ_s in (4.4) to obtain that $\boldsymbol{\eta}(\Gamma) = \int \boldsymbol{\eta}_x^s(\Gamma)\, d(e_s)_\# \boldsymbol{\eta}(x) = 1$, where Γ is the set in (4.3). □

4.2. Generalized flows transported by the maximal regular flow and renormalized solutions

We now recall the concept of renormalized solution to a continuity equation. This was already introduced in Section 1.1, but we prefer to reintroduce it here in its formulation adapted to the particular situation of a divergence-free vector field, for the convenience of the reader. To fix the ideas we consider the interval $(0, T)$ and 0 as initial time, but the definition can be immediately adapted to general intervals, forward and backward in time.

Definition 4.4 (Renormalized solutions). Let $\boldsymbol{b} \in L^1_{\mathrm{loc}}((0,T)\times\mathbb{R}^d;\mathbb{R}^d)$ be a Borel and divergence-free vector field. A Borel function $\rho : (0,T)\times \mathbb{R}^d \to \mathbb{R}$ is a renormalized solution of the continuity equation relative to $\boldsymbol{b}$ if

$$\partial_t \beta(\rho) + \nabla\cdot(\boldsymbol{b}\beta(\rho)) = 0 \text{ in } (0,T)\times\mathbb{R}^d \quad \forall \beta \in C^1 \cap L^\infty(\mathbb{R}) \tag{4.10}$$

in the sense of distributions. Analogously, we say that ρ is a renormalized solutions starting from a Borel function $\rho_0 : \mathbb{R}^d \to \mathbb{R}$ if

$$\begin{aligned}\int_{\mathbb{R}^d} \phi_0(x)\beta(\rho_0(x))\,dx + \int_0^T \int_{\mathbb{R}^d} [\partial_t \phi_t(x) \\ + \nabla\phi_t(x)\cdot \boldsymbol{b}_t(x)]\beta(\rho_t(x))\,dx\,dt = 0\end{aligned} \tag{4.11}$$

for all $\phi \in C_c^\infty([0,T)\times\mathbb{R}^d)$ and all $\beta \in C^1\cap L^\infty(\mathbb{R})$.

Remark 4.5 (Equivalent formulations). The definition is equivalent to test (4.10) with compactly supported functions in the space variable (see for instance [19, Section 8.1]); in other words, (4.11) holds if and only if for every $\varphi \in C_c^\infty(\mathbb{R}^d)$ the function $\int_{\mathbb{R}^d} \varphi(x)\beta(\rho_t(x))\,dx$ coincides $\mathscr{L}^1$-a.e. with an absolutely continuous function $t \mapsto A(t)$ such that $A(0) = \int_{\mathbb{R}^d} \varphi(x)\beta(\rho_0(x))\,dx$ and

$$\frac{d}{dt}A(t) = \int_{\mathbb{R}^d} \nabla\varphi(x)\cdot \boldsymbol{b}_t(x)\beta(\rho_t(x))\,dx \quad \text{for } \mathscr{L}^1\text{-a.e. } t\in(0,T). \tag{4.12}$$

Moreover, by an easy approximation argument, the same holds for every Lipschitz, compactly supported $\varphi : \mathbb{R}^d \to \mathbb{R}$. This way, possibly splitting φ in positive and negative parts, only nonnegative test functions need to be considered. Analogously, by writing every $\beta \in C^1(\mathbb{R}^d)$ as the sum of a C^1 nondecreasing function and of a C^1 nonincreasing function, we can use the linearity of the equation with respect to $\beta(\rho_t)$ to reduce to the case of $\beta \in C^1 \cap L^\infty(\mathbb{R})$ nondecreasing.

In the next theorem we show first that, flowing an initial datum $\rho_0 \in L^1(\mathbb{R}^d)$ through the maximal flow, we obtain a renormalized solution of the continuity equation. This is, in turn, a key tool to prove the second part of the lemma, namely that any solution transported by the maximal regular flow flow induces, with its marginals, renormalized solutions. The proof of these facts heavily relies on the incompressibility of the flow and therefore on the assumption that the vector field is divergence-free. A generalization of this lemma to the case of vector fields with bounded divergence is possible, but rather technical and long. We notice that the assumptions (a) and (b), as well as the one on the divergence of the vector field $\boldsymbol{b}$, are used only for the existence and uniqueness of a maximal regular flow which preserves the Lebesgue measure on its domain of definition, through Theorem 3.16.

To fix the ideas, in part (i) of the theorem below we consider only 0 as initial time. An analogous statement can be given for any other initial time $s \in [0, T]$, considering intervals $[0, s]$ or $[s, T]$, with no additional assumption on $\boldsymbol{b}$.

Theorem 4.6. *Let* $\boldsymbol{b} : (0, T) \times \mathbb{R}^d \to \mathbb{R}^d$ *be a divergence-free vector field which satisfies* (a)-(b) *of Section* 3.5. *Let* $\boldsymbol{X}(t, s, x)$ *be the maximal regular flow of* $\boldsymbol{b}$ *according to Definition* 3.15.

(i) *If* $\rho_0 \in L^1(\mathbb{R}^d)$, *we define* $\rho_t \in L^1(\mathbb{R}^d)$ *by*

$$\rho_t := \boldsymbol{X}(t, 0, \cdot)_\#(\rho_0 \llcorner \{T^+_{0,\boldsymbol{X}} > t\}) \qquad t \in [0, T).$$

Then ρ_t *is a renormalized solution of the continuity equation starting from* ρ_0. *In addition the map* $t \mapsto \rho_t$ *is strongly continuous on* $[0, T)$ *with respect to the* L^1_{loc} *convergence, and even strongly* L^1 *continuous on* $[0, T)$ *from the right.*

(ii) *If* $\boldsymbol{\eta} \in \mathscr{M}_+\big(C([0, T]; \mathring{\mathbb{R}}^d)\big)$ *is transported by* $\boldsymbol{X}$, *and* $(e_t)_\#\boldsymbol{\eta} \llcorner \mathbb{R}^d \ll \mathscr{L}^d$ *for every* $t \in [0, T]$, *then the density* ρ_t *of* $(e_t)_\#\boldsymbol{\eta} \llcorner \mathbb{R}^d$ *with respect to* $\mathscr{L}^d$ *is strongly continuous on* $[0, T)$ *with respect to the* L^1_{loc} *convergence and it is a renormalized solution of the continuity equation.*

Proof. We split the proof in four steps.

Step 1: proof of (i), renormalization property of ρ_t. In the proof of (i) we set for simplicity $X(t,x) = X(t,0,x)$ and $T^+_{0,X} = T_X$. We first notice that by the incompressibility of the flow (3.45) and by the definition of ρ_t, for every $t \in [0,T)$ and $\varphi \in C_c(\mathbb{R}^d)$ one has

$$\begin{aligned}\int_{\{T_X>t\}} \varphi(X(t,x))\rho_t(X(t,x))\,dx &= \int_{X(t,\cdot)(\{T_X>t\})} \varphi\rho_t\,dx\\ &= \int_{\{T_X>t\}} \varphi(X(t,x))\rho_0\,dx.\end{aligned}$$

Hence, for any $t \in [0,T)$ it holds

$$\rho_t(X(t,x)) = \rho_0(x) \qquad \text{for } \mathscr{L}^d\text{-a.e. } x \in \{T_X > t\}. \tag{4.13}$$

Let $\beta \in C^1 \cap L^\infty(\mathbb{R})$. By the incompressibility of the flow (3.45) and by (4.13) we have that

$$\int_{\mathbb{R}^d} \varphi\beta(\rho_t)\,dx = \int_{X(t,\cdot)(\{T_X>t\})} \varphi\beta(\rho_t)\,dx = \int_{\{T_X>t\}} \varphi(X(t,\cdot))\beta(\rho_0)\,dx \tag{4.14}$$

for any $\varphi \in C_c(\mathbb{R}^d)$. In addition, the blow-up property (3.42) ensures that the map $t \mapsto \varphi(X(t,x))$ can be continuously extended to be identically 0 on the time interval $[T_X(x), T)$ (in the case of blow-up before time T); in addition, for the same reason, if $\varphi \in C^1_c(\mathbb{R}^d)$ the extended map is absolutely continuous in $[0,T]$ and

$$\frac{d}{dt}\varphi(X(t,x)) = \chi_{[0,T_X(x))}(t)\nabla\varphi(X(t,x)) \cdot \boldsymbol{b}_t(X(t,x)) \tag{4.15}$$

for $\mathscr{L}^1$-a.e. $t \in (0,T)$.

Therefore, using (4.14) and integrating (4.15), for all $\varphi \in C^1_c(\mathbb{R}^d)$ we find that

$$\begin{aligned}\frac{d}{dt}\int_{\mathbb{R}^d} \varphi\beta(\rho_t)\,dx &= \int_{\{T_X>t\}} \nabla\varphi(X(t,\cdot)) \cdot \boldsymbol{b}_t(X(t,\cdot))\beta(\rho_0)\,dx\\ &= \int_{\mathbb{R}^d} \nabla\varphi \cdot \boldsymbol{b}_t\beta(\rho_t)\,dx,\end{aligned}$$

for $\mathscr{L}^1$-a.e. $t \in (0,T)$, which proves the renormalization property.

Step 2: proof of (i), strong continuity of ρ_t. We notice that, as a consequence of the possibility of continuously extending the map $t \mapsto$

$\varphi(\boldsymbol{X}(\cdot,x))$ after $T_{\boldsymbol{X}}(x)$ for $\varphi \in C_c(\mathbb{R}^d)$, the map $[0,T) \ni t \mapsto \rho_t$ is weakly continuous in duality with $C_c(\mathbb{R}^d)$. Let us prove now the strong continuity of $t \mapsto \rho_t$. We start with the proof for $t = 0$. Fix $\epsilon > 0$, let $\psi \in C_c(\mathbb{R}^d)$ with $\|\psi - \rho_0\|_1 < \epsilon$, and notice that the positivity $\mathscr{L}^d$-a.e. in $\mathbb{R}^d$ of $T_{\boldsymbol{X}}$ gives

$$\int_{\mathbb{R}^d} |\rho_t(x) - \psi(x)|\,dx \le \int_{\boldsymbol{X}(t,\cdot)(\{T_{\boldsymbol{X}}>t\})} |\rho_t(x) - \psi(x)|\,dx + \int_{\boldsymbol{X}(t,\cdot)(\{0<T_{\boldsymbol{X}}\le t\})} |\psi(x)|\,dx$$

and that the second summand in the right hand side is infinitesimal. Changing variables and using (4.13) together with the incompressibility of the flow, it follows that

$$\int_{\boldsymbol{X}(t,\cdot)(\{T_{\boldsymbol{X}}>t\})} |\rho_t(x) - \psi(x)|\,dx = \int_{\{T_{\boldsymbol{X}}>t\}} |\rho_0(x) - \psi(\boldsymbol{X}(t,x))|\,dx,$$

therefore

$$\limsup_{t\downarrow 0} \int_{\mathbb{R}^d} |\rho_t - \psi|\,dx \le \limsup_{t\downarrow 0} \int_{\{T_{\boldsymbol{X}}>t\}} |\rho_0(x) - \psi(\boldsymbol{X}(t,x))|\,dx \le \int_{\mathbb{R}^d} |\rho_0 - \psi|\,dx.$$

This proves that $\limsup_t \|\rho_t - \rho_0\|_1 \le 2\epsilon$ and, by the arbitrariness of ϵ, the desired strong continuity for $t = 0$.

We now notice that the same argument together with the semigroup property of Theorem 3.16(iii) shows that the map $t \mapsto \rho_t$ is strongly continuous from the right in L^1. In addition, reversing the time variable and using again the semigroup property, we deduce the identity $\rho_t(x) = \rho_s(\boldsymbol{X}(t,s,x))1_{\{T_{\boldsymbol{X}}>t\}}(\boldsymbol{X}(0,s,x))$, therefore

$$\lim_{s\uparrow t} \int_{\mathbb{R}^d} |\rho_t(x) - \rho_s(x)1_{\{T_{\boldsymbol{X}}>t\}}(\boldsymbol{X}(0,s,x))|\,dx = 0 \qquad \forall t \in (0,T).$$

Hence, in order to prove that the map $t \mapsto \rho_t$ is strongly continuous in L^1_{loc}, we are left to show that for every $R > 0$ and $t \in (0,T)$ one has

$$\lim_{s\uparrow t} \int_{B_R} |\rho_s(x) - \rho_s(x)1_{\{T_{\boldsymbol{X}}>t\}}(\boldsymbol{X}(0,s,x))|\,dx = 0. \tag{4.16}$$

For this, we observe that by (4.13) and the incompressibility of the flow, we have that

$$\begin{aligned}
&\int_{B_R} |\rho_s(x) - \rho_s(x) 1_{\{T_{\boldsymbol{X}} > t\}}(\boldsymbol{X}(0, s, x))|\, dx \\
&\quad = \int_{B_R} |\rho_s|(x) 1_{\{T_{\boldsymbol{X}} \le t\}}(\boldsymbol{X}(0, s, x))\, dx \qquad (4.17)\\
&\quad = \int_{\mathbb{R}^d} |\rho_0|(y) 1_{\{T_{\boldsymbol{X}} \le t\}}(y) 1_{B_R}(\boldsymbol{X}(s, 0, y))\, dy.
\end{aligned}$$

Since trajectories go to infinity when the time approaches $T_{\boldsymbol{X}}$ (see (3.42)), it follows that

$$1_{\{T_{\boldsymbol{X}} \le t\}}(y) 1_{B_R}(\boldsymbol{X}(s, 0, y)) \to 0 \quad \text{for } \mathscr{L}^d\text{-a.e. } y \text{ as } s \uparrow t,$$

so (4.16) follows by dominated convergence. This concludes the proof of (i).

Step 3: proof of (ii), renormalization property of ρ_t.

To prove (ii), we begin by showing that ρ_t is a renormalized solution of the continuity equation. By Remark 4.5 it is enough to prove that, given a bounded nondecreasing $\beta \in C^1(\mathbb{R})$ and a nonnegative $\varphi \in C_c^\infty(\mathbb{R}^d)$, the function $t \mapsto \int_{\mathbb{R}^d} \varphi\beta(\rho_t)\, dx$ is absolutely continuous in $[0, T]$ and

$$\frac{d}{dt}\int_{\mathbb{R}^d} \varphi\beta(\rho_t)\, dx = \int_{\mathbb{R}^d} \nabla\varphi \cdot \boldsymbol{b}_t \beta(\rho_t)\, dx \qquad \text{for } \mathscr{L}^1\text{-a.e. } t \in (0, T). \tag{4.18}$$

To show that the map is absolutely continuous, let us consider $s, t \in [0, T]$ and let $\tilde\rho_r^t$ be the evolution of ρ_t through the flow $\boldsymbol{X}(\cdot, t, x)$, namely

$$\tilde\rho_r^t := \boldsymbol{X}(r, t, \cdot)_\#(\rho_t \llcorner \{T^+_{t,\boldsymbol{X}} > r > T^-_{t,\boldsymbol{X}}\}) \qquad \text{for every } r \in [0, T]. \tag{4.19}$$

Since, by our assumption, $\boldsymbol{\eta}$ is transported by $\boldsymbol{X}$, we can prove that

$$\tilde\rho_r^t \le \rho_r \qquad \text{for every } r \in [0, T]. \tag{4.20}$$

Indeed, with the notation of the statement of Theorem 4.3, since $\delta_{\boldsymbol{X}(r,t,x)} = (e_r)_\# \boldsymbol{\eta}_x^t$ for ρ_t-a.e. $x \in \{T^+_{t,\boldsymbol{X}} > r > T^-_{t,\boldsymbol{X}}\}$, for every $r \in [0, T]$ one has

$$\begin{aligned}
\tilde\rho_r^t \mathscr{L}^d &= \int_{\{T^-_{t,\boldsymbol{X}} < s\}} \delta_{\boldsymbol{X}(s,t,x)} \rho_t(x)\, dx \le \int_{\mathbb{R}^d} (e_r)_\# \boldsymbol{\eta}_x^t\, \rho_t(x)\, dx \\
&= (e_r)_\# \int_{\mathbb{R}^d} \boldsymbol{\eta}_x^t\, \rho_t(x)\, dx = (e_r)_\# \boldsymbol{\eta} = \rho_r \mathscr{L}^d.
\end{aligned}$$

Combining (4.20), the equality $\tilde{\rho}^t_t = \rho_t$, the monotonicity of β, and statement (i), we deduce that

$$\begin{aligned}\int_{\mathbb{R}^d} [\beta(\rho_t) - \beta(\rho_s)]\varphi\, dx &\le \int_{\mathbb{R}^d} [\beta(\tilde{\rho}^t_t) - \beta(\tilde{\rho}^t_s)]\varphi\, dx \\ &= \int_s^t \int_{\mathbb{R}^d} \beta(\tilde{\rho}^t_r)\nabla\varphi \cdot \boldsymbol{b}_r\, dx\, dr\end{aligned} \tag{4.21}$$

and similarly

$$\begin{aligned}\int_{\mathbb{R}^d} [\beta(\rho_t) - \beta(\rho_s)]\varphi\, dx &\ge \int_{\mathbb{R}^d} [\beta(\tilde{\rho}^s_t) - \beta(\tilde{\rho}^s_s)]\varphi\, dx \\ &= \int_s^t \int_{\mathbb{R}^d} \beta(\tilde{\rho}^s_r)\nabla\varphi \cdot \boldsymbol{b}_r\, dx\, dr.\end{aligned} \tag{4.22}$$

We deduce that

$$\left|\int_{\mathbb{R}^d} [\beta(\rho_t) - \beta(\rho_s)]\varphi\, dx\right| \le \|\beta\|_\infty \int_{\mathbb{R}^d} \int_s^t |\nabla\varphi||\boldsymbol{b}_r|\, dr\, dx,$$

which shows that the function $t \mapsto \int_{\mathbb{R}^d} \varphi\beta(\rho_t)\, dx$ is absolutely continuous in $[0, T]$.

In order to prove (4.18) it is sufficient to notice that (4.21) and the strong continuity of $r \mapsto \tilde{\rho}^t_r$ at $r = t$ (ensured by statement (i)) give

$$\int_{\mathbb{R}^d} [\beta(\rho_t) - \beta(\rho_s)]\varphi\, dx \le (t-s)\int_{\mathbb{R}^d} \beta(\rho_t)\nabla\varphi \cdot \boldsymbol{b}_t\, dx + o(t-s),$$

hence (4.18) holds at any differentiability point of $t \mapsto \int_{\mathbb{R}^d} \varphi\beta(\rho_t)\, dx$.

Step 4: proof of (ii), strong continuity of ρ_t. We now show that ρ_t is strongly continuous on $[0, T)$ with respect to the L^1_{loc} convergence; more precisely we show that, for every $t \in [0, T)$ and for every $r > 0$,

$$\lim_{s \uparrow t} \int_{B_r} |\rho_s - \rho_t|\, dx = 0 \tag{4.23}$$

(reversing the time variable, the same argument gives the right-continuity). To this end, let us define $\tilde{\rho}^t$ as in (4.19) for every $t \in [0, T]$; we claim that

$$\tilde{\rho}^t_s = \rho_s \llcorner \{T^+_{s,X} > t\} \qquad \text{for every } s \in [0, t]\,. \tag{4.24}$$

Indeed, let us fix $s, t \in [0, T]$ and $s \le t$. Denoting with $\boldsymbol{\eta}^t_x$ the disintegration of $\boldsymbol{\eta}$ with respect to the map e_t, recalling that $\boldsymbol{\eta}^t_x$ is concentrated on curves $\eta \in C([0,T];\mathring{\mathbb{R}}^d)$ with $\eta(t) = x$, by Theorem 4.3, we have that for $\mathscr{L}^d$-a.e. $x \in \mathbb{R}^d$

$$
\begin{aligned}
&1_{\{T^-_{t,\boldsymbol{X}}<s\}}(x)\delta_{\boldsymbol{X}(s,t,x)} \\
&\quad = (e_s)_\#\Big(\boldsymbol{\eta}^t_x \llcorner \big\{\eta \in C([0,T];\mathring{\mathbb{R}}^d) : \eta(t) = x \text{ and } T^-_{t,\boldsymbol{X}}(x) < s\big\}\Big) \\
&\quad = (e_s)_\#\Big(\boldsymbol{\eta}^t_x \llcorner \big\{\eta \in C([0,T];\mathring{\mathbb{R}}^d) : \eta(t) \neq \infty \text{ and } T^-_{t,\boldsymbol{X}}(\eta(t)) < s\big\}\Big).
\end{aligned}
$$

Hence we can rewrite $\tilde{\rho}^t_s$ in terms of $\boldsymbol{\eta}$

$$
\begin{aligned}
\tilde{\rho}^t_s\mathscr{L}^d &= \int_{\{T^-_{t,\boldsymbol{X}}<s\}} \delta_{\boldsymbol{X}(s,t,x)}\rho_t(x)\,dx \\
&= \int_{\mathbb{R}^d} (e_s)_\#\Big(\boldsymbol{\eta}^t_x \llcorner \big\{\eta \in C([0,T];\mathring{\mathbb{R}}^d) : \eta(t) \neq \infty \\
&\qquad\qquad \text{and } T^-_{t,\boldsymbol{X}}(\eta(t)) < s\big\}\Big)\,\rho_t(x)\,dx \qquad (4.25)\\
&= (e_s)_\#\Big(\boldsymbol{\eta} \llcorner \big\{\eta \in C([0,T];\mathring{\mathbb{R}}^d) : \eta(t) \neq \infty \text{ and } T^-_{t,\boldsymbol{X}}(\eta(t)) < s\big\}\Big).
\end{aligned}
$$

By the semigroup property (Theorem 3.16(iii)) there exists a set $E_{s,t} \subseteq \mathbb{R}^d$ of $\mathscr{L}^d$-measure 0 such that

$$
T^\pm_{s,\boldsymbol{X}}(\boldsymbol{X}(s,t,x)) = T^\pm_{t,\boldsymbol{X}}(x) \qquad \text{for every } x \in \{T^+_{t,\boldsymbol{X}} > s > T^-_{t,\boldsymbol{X}}\} \setminus E_{s,t},
$$

$$
T^\pm_{t,\boldsymbol{X}}(\boldsymbol{X}(t,s,x)) = T^\pm_{s,\boldsymbol{X}}(x) \qquad \text{for every } x \in \{T^+_{s,\boldsymbol{X}} > t > T^-_{s,\boldsymbol{X}}\} \setminus E_{s,t},
$$

for every $x \in \{T^+_{t,\boldsymbol{X}} > s > T^-_{t,\boldsymbol{X}}\} \setminus E_{s,t}$

$$
\boldsymbol{X}\big(\cdot, s, \boldsymbol{X}(s,t,x)\big) = \boldsymbol{X}(\cdot,t,x) \qquad \text{in } (T^-_{t,\boldsymbol{X}}(x), T^+_{t,\boldsymbol{X}}(x)),
$$

and for every $x \in \{T^+_{s,\boldsymbol{X}} > t > T^-_{s,\boldsymbol{X}}\} \setminus E_{s,t}$

$$
\boldsymbol{X}\big(\cdot, t, \boldsymbol{X}(t,s,x)\big) = \boldsymbol{X}(\cdot,s,x) \qquad \text{in } (T^-_{s,\boldsymbol{X}}(x), T^+_{s,\boldsymbol{X}}(x)).
$$

Since $(e_s)_\#\boldsymbol{\eta}\llcorner\mathbb{R}^d$ is absolutely continuous with respect to $\mathscr{L}^d$ (so that the set of curves η such that $\eta(s) \in E_{s,t}$ is $\boldsymbol{\eta}$-negligible) and $\boldsymbol{\eta}$ is transported by the maximal regular flow, we have the following equalities, which hold up to a set of $\boldsymbol{\eta}$-measure 0:

$$
\begin{aligned}
&\big\{\eta \in C([0,T];\mathring{\mathbb{R}}^d) : \eta(s) \neq \infty \text{ and } T^+_{s,\boldsymbol{X}}(\eta(s)) > t\big\} \\
&= \big\{\eta \in C([0,T];\mathring{\mathbb{R}}^d) : \eta(s) \neq \infty,\ \eta(s) \notin E_{s,t},\ T^+_{s,\boldsymbol{X}}(\eta(s)) > t \\
&\qquad \text{and } \eta(\cdot) = \boldsymbol{X}(\cdot, s, \eta(s)) \text{ in } (T^-_{s,\boldsymbol{X}}(\eta(s)), T^+_{s,\boldsymbol{X}}(\eta(s))\big\} \\
&= \big\{\eta \in C([0,T];\mathring{\mathbb{R}}^d) : \eta(t) \neq \infty,\ \eta(t) \notin E_{s,t},\ T^-_{t,\boldsymbol{X}}(\eta(t)) < s \\
&\qquad \text{and } \eta(\cdot) = \boldsymbol{X}(\cdot, t, \eta(t)) \text{ in } (T^-_{t,\boldsymbol{X}}(\eta(t)), T^+_{t,\boldsymbol{X}}(\eta(t))\big\} \\
&= \big\{\eta \in C([0,T];\mathring{\mathbb{R}}^d) : \eta(t) \neq \infty \text{ and } T^-_{t,\boldsymbol{X}}(\eta(t)) < s\big\}.
\end{aligned}
\tag{4.26}
$$

We deduce that

$$
\begin{aligned}
\rho_s \llcorner \{T^+_{s,\boldsymbol{X}} > t\} &= (e_s)_\#\Big(\boldsymbol{\eta} \llcorner \big\{\eta \in C([0,T];\mathring{\mathbb{R}}^d) : \eta(s) \neq \infty \\
&\qquad\qquad \text{and } T^+_{s,\boldsymbol{X}}(\eta(s)) > t\big\}\Big) \\
&= (e_s)_\#\Big(\boldsymbol{\eta} \llcorner \big\{\eta \in C([0,T];\mathring{\mathbb{R}}^d) : \eta(t) \neq \infty \\
&\qquad\qquad \text{and } T^-_{t,\boldsymbol{X}}(\eta(t)) < s\big\}\Big).
\end{aligned}
\tag{4.27}
$$

By (4.25) and (4.27), we proved (4.24).

In order to prove (4.23), we apply the triangular inequality to infer that

$$
\int_{B_r} |\rho_s - \rho_t|\,dx \le \int_{B_r} |\rho_s - \tilde{\rho}^t_s|\,dx + \int_{B_r} |\tilde{\rho}^t_s - \rho_t|\,dx.
$$

The second term in the right-hand side converges to 0 when $s \uparrow t$ by the strong L^1_{loc} continuity of ρ^t_s with respect to s proved in statement (i). To see that also the first term converges to 0, we rewrite it using (4.24), $\rho_t \mathscr{L}^d = (e_t)_\# \boldsymbol{\eta} \llcorner \mathbb{R}^d$, and the fact that $\boldsymbol{\eta}$ is transported by the maximal flow to obtain

$$
\begin{aligned}
\int_{B_r} |\rho_s - \tilde{\rho}^t_s|\,dx &= \int_{B_r} \rho_s 1_{\{T^+_{s,\boldsymbol{X}} \le t\}}\,dx \\
&= \int 1_{B_r \cap \{T^+_{s,\boldsymbol{X}} \le t\}}(\eta(s))\,d\boldsymbol{\eta}(\eta) \\
&= \boldsymbol{\eta}\Big(\big\{\eta : \eta(s) \in B_r \cap \{T^+_{s,\boldsymbol{X}} \le t\} \\
&\qquad\qquad \text{and } \eta(\cdot) = \boldsymbol{X}(\cdot, s, \eta(s)) \text{ in } [s, T^+_{s,\boldsymbol{X}}(\eta(s)))\big\}\Big).
\end{aligned}
$$

Every curve η which belongs to the set in the last line belongs to B_r at time s and blows up in $[s,t]$, since it coincides with the maximal regular flow and $T^+_{s,X}(\eta(s)) \le t$. Hence, for some $s \le s' \le s'' \le t$, it satisfies that $\eta(s') \in B_r$ and $\eta(s'') = \infty$ (we could take $s' = s$, but in order to guarantee the monotonicity with respect to s of the sets below, we prefer to enlarge the set of curves in this way). We obtain that

$$\int_{B_r} |\rho_s - \rho_s^t|\, dx \le \boldsymbol{\eta}\Big(\big\{\eta : \eta(s') \in B_r \text{ and } \eta(s'') = \infty \text{ for some } s', s'' \in [s,t]\big\}\Big).$$

The set in the right-hand side monotonically decreases to the empty set as $s \uparrow t$, therefore its $\boldsymbol{\eta}$-measure converges to 0. This concludes the proof of (4.23). □

Under certain conditions on the generalized flow $\boldsymbol{\eta}$, the most common being

$$\int_0^T \int_{\mathbb{R}^d} \frac{|\boldsymbol{b}_t|(x)}{1+|x|}\, d\mu_t(x)\, dt < \infty, \tag{4.28}$$

where $\mu_t = (e_t)_\# \boldsymbol{\eta} \llcorner \mathbb{R}^d$, one can show that $\boldsymbol{\eta}$ is concentrated on curves that do not blow up. This result is in the same spirit as the no blow-up criterion of Theorem 3.13. We state the result under a more precise assumption than (4.28) (see (4.29)), since this will be important for the application to the Vlasov-Poisson system in Corollaries 8.3 and 8.4.

Proposition 4.7 (No blow-up criterion). *Let $\boldsymbol{b} \in L^1_{loc}([0,T] \times \mathbb{R}^d; \mathbb{R}^d)$ be a Borel vector field, let $\boldsymbol{\eta} \in \mathscr{M}_+\big(C([0,T]; \mathring{\mathbb{R}}^d)\big)$ be a generalized flow of $\boldsymbol{b}$, and for $t \in [0,T]$ let $\mu_t = (e_t)_\# \boldsymbol{\eta} \llcorner \mathbb{R}^d$. Let η_∞ denote the constant curve $\eta \equiv \infty$, and assume that $\boldsymbol{\eta}(\{\eta_\infty\}) = 0$ and*

$$\int_0^T \int_{\mathbb{R}^d} \frac{|\boldsymbol{b}_t|(x)}{(1+|x|)\log(2+|x|)}\, d\mu_t(x)\, dt < \infty. \tag{4.29}$$

Then $\boldsymbol{\eta}$ is concentrated on curves that do not blow up, namely

$$\boldsymbol{\eta}\big(\{\eta \in C([0,T]; \mathring{\mathbb{R}}^d) : \eta(t) = \infty \text{ for some } t \in [0,T]\}\big) = 0.$$

In particular, if we assume that $\mu_t \ll \mathscr{L}^d$ for every $t \in [0,T]$ and that $\boldsymbol{\eta}$ is concentrated on the maximal regular flow $\boldsymbol{X}$ associated to $\boldsymbol{b}$, then $\boldsymbol{X}$ is globally defined on $[0,T]$ for μ_0-a.e. x, namely the trajectories $\boldsymbol{X}(\cdot, x)$ belong to $AC([0,T]; \mathbb{R}^d)$ for μ_0-a.e. $x \in \mathbb{R}^d$.

Proof. Since $\boldsymbol{\eta}(\{\eta_\infty\}) = 0$ we know that $\boldsymbol{\eta}$-a.e. curve is finite at some time. In particular, if we fix a dense set of rational times $\{t_n\}_{n\in\mathbb{N}} \subset [0,T]$, we see that (by continuity of the curves) $\boldsymbol{\eta}$ is concentrated on $\cup_{n\in\mathbb{N}} \Gamma_n$ with

$$\Gamma_n := \{\eta \in C([0,T]; \mathring{\mathbb{R}}^d) : \eta(t_n) \in \mathbb{R}^d\},$$

so it is enough to show that $\boldsymbol{\eta}\llcorner\Gamma_n$ is concentrated on curves that do not blow up.

By applying Theorem 4.3 with $s = t_n$ it follows that $\boldsymbol{\eta}\llcorner\Gamma_n$ is concentrated on curves η that are finite on the time interval $(T^-_{t_n,\boldsymbol{X}}(\eta(t_n)), T^+_{t_n,\boldsymbol{X}}(\eta(t_n))) \subset [0,T]$. Hence, since $(e_t)_\#(\boldsymbol{\eta}\llcorner\Gamma_n) \le \mu_t$, by Fubini theorem and assumption (4.29) we get

$$\begin{aligned}
&\int\int_{T^-_{t_n,\boldsymbol{X}}(\eta(t_n))}^{T^+_{t_n,\boldsymbol{X}}(\eta(t_n))} \left|\frac{d}{dt}\big[\log\log(2+|\eta(t)|)\big]\right| dt\, d[\boldsymbol{\eta}\llcorner\Gamma_n](\eta)\\
&\le \int\int_{T^-_{t_n,\boldsymbol{X}}(\eta(t_n))}^{T^+_{t_n,\boldsymbol{X}}(\eta(t_n))} \frac{|\dot\eta(t)|}{(1+|\eta(t)|)\log(2+|\eta(t)|)}\, dt\, d[\boldsymbol{\eta}\llcorner\Gamma_n](\eta)\\
&= \int\int_{T^-_{t_n,\boldsymbol{X}}(\eta(t_n))}^{T^+_{t_n,\boldsymbol{X}}(\eta(t_n))} \frac{|\boldsymbol{b}_t|(\eta(t))}{(1+|\eta(t)|)\log(2+|\eta(t)|)}\, dt\, d[\boldsymbol{\eta}\llcorner\Gamma_n](\eta)\\
&\le \int_0^T\int_{\mathbb{R}^d} \frac{|\boldsymbol{b}_t|(x)}{(1+|x|)\log(2+|x|)}\, d\mu_t(x)\, dt < \infty.
\end{aligned}$$

This implies that, for $\boldsymbol{\eta}$-a.e. curve $\eta \in \Gamma_n$,

$$\begin{aligned}
&\sup_{T^-_{t_n,\boldsymbol{X}}(\eta(t_n)) \le s < \tau \le T^+_{t_n,\boldsymbol{X}}(\eta(t_n))} \big|\log\log(2+|\eta(s)|) - \log\log(2+|\eta(\tau)|)\big|\\
&\qquad \le \int_{T^-_{t_n,\boldsymbol{X}}(\eta(t_n))}^{T^+_{t_n,\boldsymbol{X}}(\eta(t_n))} \left|\frac{d}{dt}\big[\log\log(2+|\eta(t)|)\big]\right| dt < \infty,
\end{aligned}$$

which in turn says that $T^-_{t_n,\boldsymbol{X}}(\eta(t_n)) = 0$, $T^+_{t_n,\boldsymbol{X}}(\eta(t_n)) = T$, and the curve η cannot blow up in $[0,T]$, as desired.

To show the second part of the statement, le us consider the disintegration of $\boldsymbol{\eta}$ with respect to e_0. By the properties of $\boldsymbol{\eta}$ we have that, for μ_0-a.e. x, the probability measure $\boldsymbol{\eta}_x$ is concentrated on the set

$$\big\{\eta : \eta(0) = x,\ \eta \ne \infty \text{ in } [0,T], \eta = \boldsymbol{X}(\cdot,x) \text{ in } [0, T_{\boldsymbol{X}}(x))\big\}.$$

Since $\boldsymbol{\eta}_x$ is a probability measure it follows that this set is nonempty, that $T_{\boldsymbol{X}}(x) = T$, and this set has to coincide with $\{\boldsymbol{X}(\cdot,x)\}$, thus $\boldsymbol{\eta}_x = \delta_{\boldsymbol{X}(\cdot,x)}$, as desired. In particular, we deduce that for every $t > 0$

$$\mu_t = (e_t)_\#\boldsymbol{\eta}\llcorner\mathbb{R}^d = (e_t)_\#\boldsymbol{\eta} = (e_t)_\#\int_{\mathbb{R}^d}\delta_{\boldsymbol{X}(\cdot,x)}\, d\mu_0(x) = \boldsymbol{X}(t,\cdot)_\#\mu_0. \ \square$$

Remark 4.8. In Proposition 4.7, the assumption that the curve $\eta \equiv \infty$ has $\boldsymbol{\eta}$-measure 0 follows easily from the property

$$|\boldsymbol{\eta}|(C([0,T];\mathring{\mathbb{R}}^d)) \le \sup_{t\in[0,T]} \mu_t(\mathbb{R}^d)$$

(that, as we will see in (4.30) below, is a property of the measures $\boldsymbol{\eta}$ built with the generalized superposition principle, Theorem 4.9).

Assumption (4.29) of Proposition 4.7 could be replaced by

$$\int_0^T \int_{\mathbb{R}^d} \frac{|\boldsymbol{b}_t|(x)}{\omega(|x|)}\, d\mu_t(x)\, dt < \infty,$$

for any nondecreasing function $\omega : [0,\infty) \to [0,\infty)$ with $\omega(0) > 0$ and

$$\int_0^\infty \frac{1}{\omega(r)}\, dr = \infty.$$

4.3. The superposition principle under local integrability bounds on the velocity

In order to represent the solution to the continuity equation by means of a generalized flow, we would like to apply the superposition principle (see Theorem 1.6). However, the lack of global bounds makes this approach very difficult to implement. An analogous of the classical superposition principle is the content of the following theorem.

Theorem 4.9 (Extended superposition principle). *Let $\boldsymbol{b} \in L^1_{\mathrm{loc}}([0,T] \times \mathbb{R}^d; \mathbb{R}^d)$ be a Borel vector field. Let $\rho_t \in L^\infty((0,T); L^1_+(\mathbb{R}^d))$ be a distributional solution of the continuity equation, weakly continuous on* $[0,T]$ *in duality with $C_c(\mathbb{R}^d)$. Assume that:*

(i) *either $|\boldsymbol{b}_t|\rho_t \in L^1_{\mathrm{loc}}([0,T] \times \mathbb{R}^d)$;*
(ii) *or* $\mathrm{div}\, \boldsymbol{b}_t = 0$ *and ρ_t is a renormalized solution.*

Then there exists $\boldsymbol{\eta} \in \mathscr{M}_+\big(C([0,T]; \mathring{\mathbb{R}}^d)\big)$ with

$$|\boldsymbol{\eta}|(C([0,T]; \mathring{\mathbb{R}}^d)) \le \sup_{t\in[0,T]} \|\rho_t\|_{L^1(\mathbb{R}^d)}, \tag{4.30}$$

which is concentrated on the set Γ defined in (4.1) *and satisfies*

$$(e_t)_\# \boldsymbol{\eta} \llcorner \mathbb{R}^d = \rho_t \mathscr{L}^d \qquad \text{for every } t \in [0,T].$$

In addition, if ρ_t belongs also to $L^\infty((0,T); L^\infty_+(\mathbb{R}^d))$ (or ρ_t is renormalized), $\boldsymbol{b}$ is divergence-free and satisfies (a)-(b) *of Section* 3.5*, then $\boldsymbol{\eta}$ is transported by the Maximal Regular Flow of $\boldsymbol{X}$. In particular, ρ_t is a Lagrangian solution.*

Remark 4.10. If, in addition to the last assumptions of the Theorem, we assume that

$$\int_0^T \int_{\mathbb{R}^d} \frac{|\boldsymbol{b}_t|(x)}{1+|x|} \rho_t(x)\, dx\, dt < \infty, \tag{4.31}$$

then ρ_t is transported by the Maximal Flow, namely $T^+_{0,\boldsymbol{X}}(x)=T$, $\boldsymbol{X}(\cdot,0,x)$ belongs to $AC([0,T];\mathbb{R}^d)$ for $\mathscr{L}^d$-a.e. $x \in \{\rho_0 > 0\}$ and $\rho_t \mathscr{L}^d = \boldsymbol{X}(t,\cdot)_\#(\rho_0 \mathscr{L}^d)$.

Indeed, by Theorem 3.13 and (4.31) we know that the Maximal Regular Flow is well defined in $[0,T]$ for $\mathscr{L}^d$-a.e. $x \in \mathbb{R}^d$. Since $\boldsymbol{\eta}$ is transported by $\boldsymbol{X}$, for $\boldsymbol{\eta}$-a.e. η we know that $\eta = \boldsymbol{X}(\cdot,0,\eta(0))$ in $[0,T]$. This implies that for $\mathscr{L}^d$-a.e. $x \in \{\rho_0 > 0\}$ the measure $\boldsymbol{\eta}_x$, obtained through disintegration of $\boldsymbol{\eta}$ with respect to e_0, coincides with $\delta_{\boldsymbol{X}(\cdot,0,x)}$, therefore

$$\begin{aligned}(e_t)_\#\boldsymbol{\eta} &= \int_{\mathbb{R}^d} (e_t)_\#\boldsymbol{\eta}_x \rho_0(x)\, dx = \int_{\mathbb{R}^d} (e_t)_\#\delta_{\boldsymbol{X}(\cdot,0,x)} \rho_0(x)\, dx \\ &= \boldsymbol{X}(\cdot,0,x)_\#\big(\rho_0 \mathscr{L}^d\big),\end{aligned}$$

as desired.

Remark 4.11. Thanks to Theorem 4.9, one can prove that, if $\boldsymbol{b} : (0,T) \times \mathbb{R}^d \to \mathbb{R}^d$ is a locally integrable, divergence-free vector field, then assumption (b) of Section 3.5 is equivalent to

(b′) for any closed interval $I = [a,b] \subset [0,T]$, every bounded distributional solution of the continuity equation

$$\frac{d}{dt}\rho_t + \operatorname{div}(\boldsymbol{b}\rho_t) = 0 \qquad \text{in } (a,b) \times \mathbb{R}^d$$

is renormalized (according to Definition 4.4).

Indeed, if (b′) holds then, given any couple of bounded, compactly supported solutions of the continuity equation in $(a,b) \times \mathbb{R}^d$ with the same initial datum, their difference u is a bounded, compactly supported solution starting from 0. By (b′), it is renormalized, and therefore

$$\frac{d}{dt}\arctan(|u_t|^2) + \operatorname{div}(\boldsymbol{b}\arctan(|u_t|^2)) = 0 \qquad \text{in } (a,b) \times \mathbb{R}^d.$$

Multiplying this equation by a test function $\varphi(x) \in C_c^\infty(\mathbb{R}^d)$ which is 1 on the support of u_t, we find that

$$\int_{\mathbb{R}^d} \arctan(|u_t|^2) = 0 \qquad \forall t \in [0,T]$$

and therefore $u_t \equiv 0$.

On the other hand, if (b) holds and u is a bounded, distributional solution of the continuity equation, by Theorem 4.9 applied to $u + \|u\|_{L^\infty((0,T)\times\mathbb{R}^d)}$ and by Theorem 4.6 we find that $u + \|u\|_{L^\infty((0,T)\times\mathbb{R}^d)}$ is a renormalized solution of the continuity equation, according to Definition 4.4; this implies that the same holds for u.

Let us first briefly explain the idea behind the proof of Theorem 4.9. To overcome the lack of global bounds on $\boldsymbol{b}$ we introduce a kind of "damped" stereographic projection, with damping depending on the growth of $|\boldsymbol{b}|$ at ∞, and we look at the flow of $\boldsymbol{b}$ on the d-dimensional sphere $\mathbb{S}^d$ in such a way that the north pole N of the sphere corresponds to the points at infinity of $\mathbb{R}^d$ (see Figure 4.2).

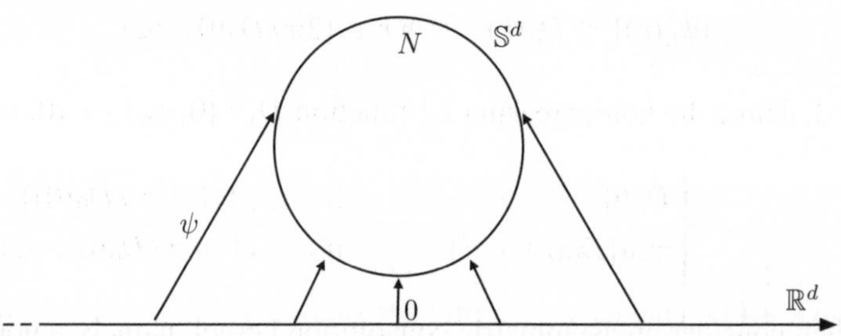

Figure 4.2. The map ψ "wraps" $\mathbb{R}^d$ onto $\mathbb{S}^d \setminus \{N\}$ through a diffeomorphism whose gradient has a controlled growth at ∞, in terms of a prescribed function $D(r) : [0,\infty) \to (0,1]$. This function will be chosen, in turn, in the proof of Theorem 4.9 in terms of the L^1 norms of the vector field $\boldsymbol{b}$ in $(0,T) \times B_r$.

Then we apply the superposition principle in these new variables and eventually, reading this limit in the original variables, we obtain a representation of the solution as a generalized flow. Let us observe that it is crucial for us that the map sending $\mathbb{R}^d$ onto $\mathbb{S}^d$ is chosen a function of $\boldsymbol{b}$: indeed, as we shall see, by shrinking enough distances at infinity we can ensure that the vector field read on the sphere becomes *globally* integrable.

We denote by N be the north pole of the d-dimensional sphere $\mathbb{S}^d$, thought of as a subset of $\mathbb{R}^{d+1}$. For our constructions, we will use a smooth diffeomorphism which maps $\mathbb{R}^d$ onto $\mathbb{S}^d \setminus \{N\}$ and whose derivative has a prescribed decay at ∞.

Lemma 4.12. *Let $D : [0,\infty) \to (0,1]$ be a nonincreasing function. Then there exist $r_0 > 0$ and a smooth diffeomorphism $\psi : \mathbb{R}^d \to \mathbb{S}^d \setminus \{N\} \subset \mathbb{R}^{d+1}$ such that*

$$\psi(x) \to N \quad \text{as } |x| \to \infty, \tag{4.32}$$

$$|\nabla\psi(x)| \le D(0) \qquad \forall x \in \mathbb{R}^d, \tag{4.33}$$

$$|\nabla\psi(x)| \le D(|x|) \qquad \forall x \in \mathbb{R}^d \setminus B_{r_0}. \tag{4.34}$$

Proof. We split the construction in two parts: first we perform a 1-dimensional construction, and then we use this construction to build the desired diffeomorphism.

Step 1: 1-dimensional construction. Let $D_0 : [0, \infty) \to (0, 1]$ be a nonincreasing function. We claim that there exists a smooth diffeomorphism $\psi_0 : [0, \infty) \to [0, \pi)$ such that

$$\lim_{r\to\infty} \psi_0(r) = \pi, \qquad \lim_{r\to\infty} \psi_0'(r) = 0, \tag{4.35}$$

$$\psi_0(r) = c_0 D_0(0) r \ \ \forall r \in [0, \pi/D_0(0)), \text{ for some } c_0 \in (0, 1), \tag{4.36}$$

$$|\psi_0'(r)| \le D_0(0) \qquad \forall r \in [0, \infty), \tag{4.37}$$

$$|\psi_0'(r)| \le D_0(r) \qquad \forall r \in [2\pi/D_0(0), \infty). \tag{4.38}$$

Indeed, define the nonincreasing L^1 function $D_1 : [0, \infty) \to (0, \infty)$ as

$$D_1(r) := \begin{cases} D_0(0) & \text{if } r \in [0, 1 + \pi/D_0(0)] \\ \min\{D_0(r), r^{-2}\} & \text{if } r \in (1 + \pi/D_0(0), \infty). \end{cases}$$

We then consider an asymmetric convolution kernel, namely a nonnegative function $\sigma \in C_c^\infty((0, 1))$ with $\int_{\mathbb{R}} \sigma = 1$, and consider the convolution of $D_1(r)$ with $\sigma(-r)$:

$$\psi_1(r) := \int_0^1 \sigma(r') D_1(r + r')\, dr' \quad \forall r \in [0, \infty).$$

Notice that ψ_1 is smooth on $(0, \infty)$, positive, nonincreasing, and $\psi_1 \le D_1$ in $[0, \infty)$ (in particular $\psi_1 \in L^1(0, \infty)$). Moreover we have that $\psi_1 \equiv D_0(0)$ in $[0, \pi/D_0(0)]$, hence $\|\psi_1\|_{L^1(0,\infty)} \ge \pi$ and $c_0 := \pi \|\psi_1\|^{-1}_{L^1(0,\infty)} \in (0, 1)$. Finally, we define ψ_0 as

$$\psi_0(r) := c_0 \int_0^r \psi_1(s)\, ds \qquad \forall r \in [0, \infty).$$

Since $|\psi_0'(r)| = c_0|\psi_1(r)| \le D_1(r)$, taking into account that $\pi/D_0(0) > 1$ it is easy to check that all the desired properties are satisfied.

Step 2: "radial" diffeomorphism in any dimension. Let $D_0 : [0,\infty) \to (0, 1]$ to be chosen later and consider ψ_0, c_0 as in Step 1. We define $\psi : \mathbb{R}^d \to \mathbb{S}^d \setminus \{N\} \subset \mathbb{R}^{d+1}$ which maps every half-line starting at the origin to an arc of sphere between the south pole and the north pole:

$$\psi(x) := \sin(\psi_0(|x|))\Big(\frac{x}{|x|}, 0\Big) - \cos(\psi_0(|x|))(0, \dots, 0, 1).$$

Thanks to (4.36) and to the fact that the functions $x \mapsto |x|^2$, $t \mapsto \sin(\sqrt{t})/\sqrt{t}$, and $t \mapsto \cos(\sqrt{t})$ are all of class C^∞, we obtain that $\psi \in C^\infty(\mathbb{R}^d; \mathbb{R}^{d+1})$. We also notice that its inverse $\phi : \mathbb{S}^d \setminus \{N\} \to \mathbb{R}^d$ can be explicitly computed:

$$\begin{aligned}\phi(x_1, \dots, x_{d+1}) &= \psi_0^{-1}(\arccos(-x_{d+1}))\frac{(x_1, \dots, x_d)}{|(x_1, \dots, x_d)|}\\ &= \psi_0^{-1}(\arcsin(|(x_1, \dots, x_d)|))\frac{(x_1, \dots, x_d)}{|(x_1, \dots, x_d)|}.\end{aligned}$$

Writing $r = |x|$ and denoting by I_d the identity matrix on the first d components, we compute the gradient of ψ:

$$\begin{aligned}\nabla\psi(x) = &\frac{\cos(\psi_0(r))\psi_0'(r)r - \sin(\psi_0(r))}{r^3}(x, 0) \otimes (x, 0)\\ &+ \frac{\sin(\psi_0(r))}{r}I_d - \frac{\sin(\psi_0(r))\psi_0'(r)}{r}(x, 0) \otimes (0, \dots, 0, 1).\end{aligned}$$

It is immediate to check that $|\nabla\psi(x)| \neq 0$ for all $x \in \mathbb{R}^d$, so it follows by the Inverse Function Theorem that ϕ is smooth as well. Also, we can estimate

$$|\nabla\psi(x)| \leq 2|\psi_0'(r)| + 2\frac{\sin(\psi_0(r))}{r}. \tag{4.39}$$

Using now (4.37) and (4.38), the first term in the right hand side above can be estimated with $2D_0(0)$ for every $x \in \mathbb{R}^d$, and with $2D_0(r)$ for every $x \in \mathbb{R}^d$ such that $r \geq 2\pi/D_0(0)$. As regards the second term, for $r \in [0, \pi/D_0(0)]$ we have that

$$\frac{\sin(\psi_0(r))}{r} = \frac{\sin(c_0 D_0(0)r)}{r} \leq c_0 D_0(0), \tag{4.40}$$

while for $r \in [\pi/D_0(0), \infty)$ we estimate the numerator with 1 to get

$$\frac{\sin(\psi_0(r))}{r} \leq \frac{D_0(0)}{\pi}. \tag{4.41}$$

Therefore, since $c_0 < 1$, by (4.39), (4.40), and (4.41) we get

$$|\nabla\psi(x)| \leq 4D_0(0) \qquad \forall x \in \mathbb{R}^d. \tag{4.42}$$

Now, for $r \in [2\pi/D_0(0), \infty)$, thanks to (4.35) and (4.38) we can estimate

$$\begin{aligned}\frac{\sin(\psi_0(r))}{r} &= \frac{1}{r}\int_r^\infty -\cos(\psi_0(s))\psi_0'(s)\,ds \leq \frac{1}{r}\int_r^\infty |\psi_0'(s)|\,ds\\ &\leq \frac{1}{r}\int_r^\infty D_0(s)\,ds,\end{aligned} \tag{4.43}$$

thus by (4.38), (4.39), and (4.43) we obtain

$$|\nabla\psi(x)| \le 2D_0(r) + \frac{2}{r}\int_r^\infty D_0(s)\,ds \qquad \forall x \in \mathbb{R}^d \setminus B_{2\pi/D_0(0)}. \quad (4.44)$$

So, provided we choose $D_0(r) := \min\{4^{-1}, r^{-2}\}D(r)$ we obtain that (4.42) implies (4.33). Also, by choosing $r_0 := 2\pi/D_0(0) > 2$, from (4.44) and because D is nonincreasing we deduce that

$$|\nabla\psi(x)| \le \frac{D(r)}{2} + \frac{1}{r}\int_r^\infty \frac{D(r)}{s^2}ds \le \frac{D(r)}{2} + \frac{D(r)}{r^2} \le D(r) \;\; \forall x \in \mathbb{R}^d \setminus B_{r_0},$$

proving (4.34) and concluding the proof. □

Proof of Theorem 4.9. We first assume that $|\boldsymbol{b}_t|\rho_t \in L^1_{\mathrm{loc}}([0,T] \times \mathbb{R}^d)$ and we prove the result in this case. This is done in two steps:
- In Step 1, based on Lemma 4.12, we construct a diffeomorphism between $\mathbb{R}^d$ and $\mathbb{S}^d \setminus \{N\}$ with the property that the vector field $\boldsymbol{b}$, read on the sphere, becomes globally integrable.
- In Step 2 we associate a solution of the continuity equation on the sphere to the solution of the continuity equation ρ_t; this is done by adding a time-dependent mass in the north pole. Then, the superposition principle applies on the sphere.

Once the theorem has been proved for $|\boldsymbol{b}_t|\rho_t \in L^1_{\mathrm{loc}}([0,T] \times \mathbb{R}^d)$, we show in Step 3 how to handle the case when ρ_t is a renormalized solution.

Finally, in Step 4 we exploit the results of Section 4.1 to show that ρ_t is transported by the Maximal Regular Flow.

Step 1: construction of a diffeomorphism between $\mathring{\mathbb{R}}^d$ and $\mathbb{S}^d$. We build a diffeomorphism $\psi \in C^\infty(\mathbb{R}^d; \mathbb{S}^d \setminus \{N\})$ such that

$$\lim_{x\to\infty} \psi(x) = N, \qquad (4.45)$$

$$\int_0^T \int_{\mathbb{R}^d} |\nabla\psi(x)||\boldsymbol{b}_t(x)|\rho_t(x)\,dx\,dt < \infty. \qquad (4.46)$$

To this end, we apply Lemma 4.12 with $D(r) = 1$ in $[0, 1)$ and $D(r) = (2^n C_n)^{-1}$ for $r \in [2^{n-1}, 2^n)$, where

$$C_n := 1 + \int_0^T \int_{B_{2^n}} |\boldsymbol{b}_t(x)|\rho_t(x)\,dx\,dt \qquad \text{for every } n \in \mathbb{N}.$$

In this way we obtain a smooth diffeomorphism ψ which maps $\mathbb{R}^d$ onto $\mathbb{S}^d \setminus \{N\}$ such that (4.45) holds, $|\nabla\psi(x)| \leq 1$ on $\mathbb{R}^d$, and

$$|\nabla\psi(x)| \leq \frac{1}{2^n C_n} \qquad \forall x \in B_{2^n} \setminus B_{2^{n-1}},\ n \geq n_0, \tag{4.47}$$

for some $n_0 > 0$. Thanks to these facts we deduce that

$$\begin{aligned} &\int_0^T \int_{\mathbb{R}^d} |\nabla\psi(x)||\boldsymbol{b}_t(x)|\rho_t(x)\,dx\,dt \\ &\leq \int_0^T \int_{B_{2^{n_0}}} |\boldsymbol{b}_t(x)|\rho_t(x)\,dx\,dt \\ &\quad + \sum_{i=n_0+1}^{\infty} \int_0^T \int_{B_{2^i}\setminus B_{2^{i-1}}} |\nabla\psi(x)||\boldsymbol{b}_t(x)|\rho_t(x)\,dx\,dt \\ &\leq \int_0^T \int_{B_{2^{n_0}}} |\boldsymbol{b}_t(x)|\rho_t(x)\,dx\,dt + \sum_{i=n_0+1}^{\infty} \frac{1}{2^i} < \infty, \end{aligned} \tag{4.48}$$

which proves (4.46).

Step 2: superposition principle on the sphere. We build $\boldsymbol{\eta} \in \mathscr{M}_+\big(C([0,T]; \mathring{\mathbb{R}}^d)\big)$ such that $|\boldsymbol{\eta}|(C([0,T]; \mathring{\mathbb{R}}^d)) \leq \sup_{t\in[0,T]} \|\rho_t\|_{L^1(\mathbb{R}^d)}$, $\boldsymbol{\eta}$ is concentrated on curves η which are locally absolutely continuous integral curves of $\boldsymbol{b}$ in $\{\eta \neq \infty\}$, and whose marginal at time t in $\mathbb{R}^d$ is $\rho_t \mathscr{L}^d$.

Without loss of generality, possibly dividing every ρ_t by $\sup_{t\in[0,T]}\|\rho_t\|_{L^1(\mathbb{R}^d)}$, we can assume that $\sup_{t\in[0,T]} \|\rho_t\|_{L^1(\mathbb{R}^d)}$. Define $m_t := \|\rho_t\|_{L^1(\mathbb{R}^d)} \leq 1$,

$$\boldsymbol{c}_t(y) := \begin{cases} \nabla\psi(\phi(y))\boldsymbol{b}_t(\phi(y)) & \text{if } y \in \mathbb{S}^d \setminus \{N\} \\ 0 & \text{if } y = N \end{cases} \tag{4.49}$$

and

$$\mu_t := \psi_\#(\rho_t \mathscr{L}^d) + (1 - m_t)\delta_N \in \mathscr{P}\big(\mathbb{S}^d\big), \qquad t \in [0,T].$$

Since $\boldsymbol{c}_t(N) = 0$ we can neglect the mass at $N = \psi(\infty)$ to get

$$\begin{aligned} \int_0^T \int_{\mathbb{S}^d} |\boldsymbol{c}_t|\,d\mu_t\,dt &= \int_0^T \int_{\mathbb{S}^d\setminus\{N\}} |\nabla\psi|(\phi(y))|\boldsymbol{b}_t|(\phi(y))\,d\mu_t(y)\,dt \\ &= \int_0^T \int_{\mathbb{R}^d} |\nabla\psi|(x)|\boldsymbol{b}_t|(x)\rho_t(x)\,dx\,dt < \infty, \end{aligned}$$

where in the last inequality we used (4.46).

We now show that the probability measure μ_t is a solution to the continuity equation on $\mathbb{S}^d \subset \mathbb{R}^{d+1}$ with vector field $\boldsymbol{c}_t$. To this end we first notice that, by the weak continuity in duality with $C_c(\mathbb{R}^d)$ of ρ_t and by the fact that all the measures μ_t have unit mass, we deduce that μ_t is weakly continuous in time. Indeed, any limit point of μ_s as $s \to t$ is uniquely determined on $\mathbb{S}^d \setminus \{N\}$, and then the mass normalization gives that it is completely determined. We want to prove that the function $t \mapsto \int_{\mathbb{S}^d} \varphi \, d\mu_t$ is absolutely continuous and satisfies

$$\frac{d}{dt} \int_{\mathbb{S}^d} \varphi \, d\mu_t = \int_{\mathbb{S}^d} \boldsymbol{c}_t \cdot \nabla \varphi \, d\mu_t \qquad \mathscr{L}^1\text{-a.e. on } (0, T) \tag{4.50}$$

for every $\varphi \in C^\infty(\mathbb{R}^{d+1})$. We remark that, since ρ_t is a solution to the continuity equation in $\mathbb{R}^d$ with vector field $\boldsymbol{b}_t$, changing variables with the diffeomorphism ψ we obtain that (4.50) holds for every $\varphi \in C_c^\infty(\mathbb{R}^{d+1} \setminus \{N\})$, hence we are left to check that (4.50) holds also when φ is not necessarily 0 in a neighborhood of the north pole.

Let us consider $\varphi \in C_c^\infty(\mathbb{R}^{d+1})$. By $\mu_t(N) = 1 - m_t = 1 - \mu_t(\mathbb{S}^d \setminus \{N\})$, for every $t \in [0, T]$ we have that

$$\begin{aligned} \int_{\mathbb{S}^d} \varphi \, d\mu_t &= \int_{\mathbb{S}^d \setminus \{N\}} \varphi \, d\mu_t + \varphi(N)\mu_t(N) \\ &= \varphi(N) + \int_{\mathbb{S}^d} (\varphi - \varphi(N)) \, d\mu_t . \end{aligned} \tag{4.51}$$

For every $\varepsilon > 0$ let us consider a function $\chi_\varepsilon \in C^\infty(\mathbb{R}^{d+1})$ which is 0 in $B_\varepsilon(N)$, 1 outside $B_{2\varepsilon}(N)$, and whose gradient is bounded by $2/\varepsilon$. Since ρ_t is a solution to the continuity equation in $\mathbb{R}^d$ and since $\chi_\varepsilon(\varphi - \varphi(N))$ is a smooth, compactly supported function in $C_c^\infty(\mathbb{R}^{d+1} \setminus \{N\})$ we deduce that

$$\begin{aligned} \frac{d}{dt} \int_{\mathbb{S}^d} \chi_\varepsilon(\varphi - \varphi(N)) \, d\mu_t &= \int_{\mathbb{S}^d \setminus \{N\}} \boldsymbol{c}_t \cdot \nabla[\chi_\varepsilon(\varphi - \varphi(N))] \, d\mu_t \\ &= \int_{\mathbb{S}^d \setminus \{N\}} (\varphi - \varphi(N)) \boldsymbol{c}_t \cdot \nabla \chi_\varepsilon \, d\mu_t \\ &\quad + \int_{\mathbb{S}^d \setminus \{N\}} \chi_\varepsilon \boldsymbol{c}_t \cdot \nabla \varphi \, d\mu_t . \end{aligned} \tag{4.52}$$

To estimate the first term in the right-hand side of (4.52) we use that $|\varphi - \varphi(N)| \leq \varepsilon \|\nabla \varphi\|_\infty$ in $B_\varepsilon(N)$ and that $|\nabla \chi_\varepsilon| \leq 2/\varepsilon$ to get that

$$\left| \int_{\mathbb{S}^d \setminus \{N\}} \boldsymbol{c}_t \cdot \nabla \chi_\varepsilon (\varphi - \varphi(N)) \, d\mu_t \right| \leq 2 \|\nabla \phi\|_\infty \int_{B_{2\varepsilon}(N) \setminus B_\varepsilon(N)} |\boldsymbol{c}_t| \, d\mu_t ,$$

and notice the latter goes to 0 in $L^1(0,T)$ as $\varepsilon \to 0$ since $|\boldsymbol{c}|$ is integrable with respect to $\mu_t dt$ in space-time thanks to (4.50). Since the second term in the right-hand side of (4.52) converges in $L^1(0,T)$ to $\int_{\mathbb{S}^d\setminus\{N\}} \boldsymbol{c}_t \cdot \nabla\varphi\, d\mu_t$, taking the limit as $\varepsilon \to 0$ in (4.52) we obtain that $t \mapsto \int_{\mathbb{S}^d}(\varphi - \varphi(N))\, d\mu_t$ is absolutely continuous in $[0,T]$ and that for $\mathscr{L}^1$-a.e. $t \in (0,T)$ one has

$$\frac{d}{dt}\int_{\mathbb{S}^d}(\varphi - \varphi(N))\, d\mu_t = \int_{\mathbb{S}^d} \boldsymbol{c}_t \cdot \nabla\varphi\, d\mu_t.$$

Using the identity (4.51), this formula can be rewritten in the form (4.50), as desired.

Since μ_t is a weakly continuous solution of the continuity equation and the integrability condition (4.50) holds, we can apply the superposition principle (see Theorem 1.6) to deduce the existence of a measure $\boldsymbol{\sigma} \in \mathscr{P}\big(C([0,T];\mathbb{S}^d)\big)$ which is concentrated on integral curves of $\boldsymbol{c}$ and such that $(e_t)_\#\boldsymbol{\sigma} = \mu_t$ for all $t \in [0,T]$.

We then consider $\phi : \mathbb{S}^d \to \mathring{\mathbb{R}}^d$ to be the inverse of ψ extended to N as $\phi(N) = \infty$, and define $\Phi : C([0,T];\mathbb{S}^d) \to C([0,T];\mathring{\mathbb{R}}^d)$ as $\Phi(\eta) := \phi \circ \eta$. Then the measure

$$\boldsymbol{\eta} := \Phi_\#\boldsymbol{\sigma} \in \mathscr{P}\big(C([0,T];\mathring{\mathbb{R}}^d)\big)$$

is concentrated on locally absolutely continuous integral curves of $\boldsymbol{b}$ in the sense stated in (4.1), and

$$(e_t)_\#\boldsymbol{\eta} \llcorner \mathbb{R}^d = \phi_\#(e_t)_\#\boldsymbol{\sigma} \llcorner \mathbb{R}^d = \phi_\#\mu_t \llcorner \mathbb{R}^d = \rho_t \mathscr{L}^d.$$

Step 3: the case of renormalized solutions. We now show how to prove the result when $\operatorname{div} \boldsymbol{b}_t = 0$ and ρ_t is a renormalized solution. Notice that in this case we have no local integrability information on $|\boldsymbol{b}_t|\rho_t$, so the argument above does not apply. However, exploiting the fact that ρ_t is renormalized we can easily reduce to that case.

More precisely, we begin by observing that, by a simple approximation argument, the renormalization property (see Definition 4.4) is still true when β is a bounded Lipschitz function. Thanks to this observation we consider, for $k \geq 0$, the functions

$$\beta_k(s) := \begin{cases} 0 & \text{if } s \leq k, \\ s - k & \text{if } k \leq s \leq k+1, \\ 1 & \text{if } s \geq k+1. \end{cases}$$

Since ρ_t is renormalized, $\beta_k(\rho_t)$ is a bounded distributional solution of the continuity equation, hence by Steps 1-2 above there exists a measure $\boldsymbol{\eta}_k \in \mathscr{M}_+\big(C([0,T];\mathring{\mathbb{R}}^d)\big)$ with

$$|\boldsymbol{\eta}_k|(C([0,T];\mathring{\mathbb{R}}^d)) \leq \sup_{t\in[0,T]} \|\beta_k(\rho_t)\|_{L^1(\mathbb{R}^d)},$$

which is concentrated on the set defined in (4.1) and satisfies

$$(e_t)_\# \boldsymbol{\eta}_k \llcorner \mathbb{R}^d = \beta_k(\rho_t)\mathscr{L}^d \qquad \text{for every } t \in [0,T].$$

Since $\sum_{k\geq 0} \beta_k(s) = s$, we immediately deduce that the measure $\boldsymbol{\eta} := \sum_{k\geq 0} \boldsymbol{\eta}_k$ satisfies all the desired properties.

Step 4: representation via the Maximal Regular Flow. If we assume in addition that $\boldsymbol{b}$ is divergence-free and satisfies (a)-(b) of Section 3.5 and that $\rho_t \in L^\infty((0,T)\times\mathbb{R}^d)$ (respectively that ρ_t is renormalized), then $\boldsymbol{\eta}$ (respectively every $\boldsymbol{\eta}_k$) is a regular generalized flow and by Theorem 4.3 it is transported by the Maximal Regular Flow.

□

Chapter 5
The continuity equation with an integrable damping term

In this chapter we consider the Cauchy problem for the continuity equation with a linear source term, namely

$$\begin{cases} \partial_t u_t(x) + \nabla \cdot (\boldsymbol{b}_t(x) u_t(x)) = c_t(x) u_t(x) \\ u_0(x) = \bar{u}(x) \end{cases} \tag{5.1}$$

where $(t, x) \in (0, T) \times \mathbb{R}^d$, $u_t(x) \in \mathbb{R}$, $\boldsymbol{b}_t(x) \in \mathbb{R}^d$ and $c_t(x) \in \mathbb{R}$. This kind of equation appears in many nonlinear systems of PDEs and, in analogy with fluid dynamics, we call *damping* the coefficient c. As it happens in the case $c \equiv 0$ (see Section 1.1), the continuity equation (5.1) is strictly related to the ordinary differential equation

$$\begin{cases} \partial_t \boldsymbol{X}(t, x) = \boldsymbol{b}_t(\boldsymbol{X}(t, x)) & \forall t \in (0, T) \\ \boldsymbol{X}(0, x) = x \end{cases} \tag{5.2}$$

for $x \in \mathbb{R}^d$. Indeed, assuming that $\bar{u}$, $\boldsymbol{b}$ and c are smooth and compactly supported and denoting by $\boldsymbol{X} : [0, T] \times \mathbb{R}^d \to \mathbb{R}^d$ the flow of $\boldsymbol{b}$, the map $\boldsymbol{X}(t, \cdot)$ is a diffeomorphism. We denote by $\boldsymbol{X}^{-1}(t, \cdot)$ its inverse and we set $J\boldsymbol{X}(t, x) := \det(\nabla_x \boldsymbol{X}(t, x)) \neq 0$. A solution of (5.1) is then given in term of the flow $\boldsymbol{X}$ by the following explicit formula

$$u_t(x) = \frac{\bar{u}(\boldsymbol{X}^{-1}(t, \cdot)(x))}{J\boldsymbol{X}(t, \boldsymbol{X}^{-1}(t, \cdot)(x))} \exp\left(\int_0^t c_\tau(\boldsymbol{X}(\tau, \boldsymbol{X}^{-1}(t, \cdot)(x)))\, d\tau \right). \tag{5.3}$$

Moreover, (5.3) can be equivalently rewritten as

$$u_t \mathscr{L}^d = \boldsymbol{X}(t, \cdot)_\sharp \left(\bar{u} \exp\left(\int_0^t c_\tau(\boldsymbol{X}(\tau, \cdot))\, d\tau \right) \mathscr{L}^d \right). \tag{5.4}$$

If $c \in L^\infty((0,T) \times \mathbb{R}^d)$, under suitable (regularity and growth) assumptions on the velocity field ensuring the existence and uniqueness of a Lagrangian flow, DiPerna and Lions [81] showed that (5.4) is the unique distributional solution of (5.1) with initial datum $\bar{u}$. At a very formal level, their strategy to prove uniqueness consists in considering the difference u between two solutions with the same initial datum, which by linearity solves (5.1) with initial datum 0, and multiplying the equation by $2u$. They obtain

$$\begin{aligned} \frac{d}{dt} \int_{\mathbb{R}^d} u_t(x)^2\,dx &= \int_{\mathbb{R}^d} (2c_t(x) - \operatorname{div} \boldsymbol{b}_t(x)) u_t(x)^2\,dx \\ &\leq (2\|c_t\|_{L^\infty(\mathbb{R}^d)} + \|\operatorname{div} \boldsymbol{b}_t\|_{L^\infty(\mathbb{R}^d)}) \int_{\mathbb{R}^d} u_t(x)^2\,dx. \end{aligned} \tag{5.5}$$

They conclude thanks to Gronwall lemma that $\int_{\mathbb{R}^d} u_t(x)^2\,dx = 0$ for every $t \in [0,T]$, which implies uniqueness.

If $c \in L^1((0,T) \times \mathbb{R}^d)$ then (5.4) does not make sense as distributional solution even in the simplest autonomous cases. For instance, let $\boldsymbol{b}_t(x) = 0$, $\bar{u} = 1_{[0,1]^d}$, and $c \in L^1(\mathbb{R}^d)$. A solution of (5.1) is given by $u_t(x) = \bar{u}(x)e^{tc(x)}$; however u_t may not belong to $L^1_{\mathrm{loc}}(\mathbb{R}^d)$ due to the low integrability of c. In this case (5.3) is not a distributional solution of (5.1).

We notice however that, if we assume $c \in L^1((0,T) \times \mathbb{R}^d)$, the function u defined in (5.3) is almost everywhere pointwise defined since the flow is assumed to preserve the Lebesgue measure, up to a multiplicative constant, and hence

$$\int_{\mathbb{R}^d} \int_0^T c_\tau(\boldsymbol{X}(\tau, x))\,d\tau\,dx \leq C \int_0^T \int_{\mathbb{R}^d} c_\tau(x)\,d\tau\,dx < \infty. \tag{5.6}$$

In the following, we introduce a natural notion of renormalized solution of (5.1) (see Definition 5.3) following [81] and we prove that the function defined in (5.3) is a renormalized solution of (5.10). Then we move to the more delicate problem of uniqueness with this weak notion of solution. Here a different estimate with respect to (5.5) is needed, since already the formal computation (5.5) fails if we assume lower summability than L^∞ for the damping c. In analogy with the logarithmic estimates introduced by Ambrosio, Lecumberry and Maniglia [22], Crippa and De Lellis [67] for solutions to the ODE (5.2), we perform a logarithmic estimate for solutions of the PDE (5.1). As in the computation (5.5), we consider the difference u of two solutions with the same initial datum and we multiply

(5.1) by $u/(\delta + u^2)$, where $\delta > 0$ is fixed, to obtain

$$\begin{aligned}\frac{d}{dt}\int_{\mathbb{R}^d} \log\Big(1+\frac{u_t(x)^2}{\delta}\Big)\,dx &= \int_{\mathbb{R}^d} \operatorname{div} \boldsymbol{b}_t(x) \log\Big(1+\frac{u_t(x)^2}{\delta}\Big)\,dx \\ &\quad + \int_{\mathbb{R}^d} (c_t(x) - \operatorname{div} \boldsymbol{b}_t(x)) \frac{u_t(x)^2}{\delta + u_t(x)^2}\,dx \\ &\le \|\operatorname{div} \boldsymbol{b}_t\|_{L^\infty(\mathbb{R}^d)} \int_{\mathbb{R}^d} \log\Big(1+\frac{u_t(x)^2}{\delta}\Big)\,dx \\ &\quad + 2\int_{\mathbb{R}^d} |c_t(x)| + |\operatorname{div} \boldsymbol{b}_t(x)|\,dx. \end{aligned} \tag{5.7}$$

By Gronwall lemma we deduce that for every $t \in [0, T]$

$$\begin{aligned}\int_{\mathbb{R}^d} \log\Big(1+\frac{u_t(x)^2}{\delta}\Big)\,dx &\le \exp\Big(\int_0^T \|\operatorname{div} \boldsymbol{b}_t\|_{L^\infty(\mathbb{R}^d)}\,dt\Big) \\ &\quad \cdot \int_0^T \int_{\mathbb{R}^d} 2(|c_t(x)| + |\nabla \cdot \boldsymbol{b}_t(x)|)\,dx\,dt;\end{aligned}$$

letting finally δ go to 0, since the right-hand side is independent on δ we obtain that $u_t = 0$. A justification of the estimate (5.7) in a non-smooth setting requires some work, as the one performed in [81] to justify (5.5). First, one needs to prove that the difference of renormalized solutions is still renormalized, which is not an automatic consequence of the linearity of the equation and of the theory of renormalized solutions. Moreover, to allow general growth conditions on $\boldsymbol{b}$, one would like to localize the estimate. In [81], general growth conditions were considered by means of a cutoff function and by a duality argument. Instead, we refine the estimate (5.7) by means of a decaying function.

The plan of the paper is the following. In Section 5.1 we introduce the notions of regular Lagrangian flow and of renormalized solution; then we state our existence and uniqueness result. Sections 5.2 and 5.3 are devoted to the proof of the main theorem.

5.1. Existence and uniqueness of renormalized solutions

We denote by $B_r(x) \subseteq \mathbb{R}^d$ the open ball of centre $x \in \mathbb{R}^d$ and radius $r > 0$, shortened to B_r if $x = 0$. In the case of a smooth, divergence free vector field $\boldsymbol{b}$, the solution to the equation (5.1), given by the explicit formula (5.3) with $J\boldsymbol{X}(t, x) = 1$, is obtained by transporting the initial datum $\bar{u}$ along the flow of the vector field $\boldsymbol{b}$, together with a correction due to the damping term c. To obtain a similar statement in the non-smooth setting, we consider the regular Lagrangian flow of $\boldsymbol{b}$ (see Definition 1.4) and we point out some useful properties in the remark below.

Remark 5.1. Under the assumptions of Theorem 1.5, which guarantee the existence and uniqueness of a regular lagrangian flow, if we further assume a two-sided bound on div $\boldsymbol{b}$, namely div $\boldsymbol{b} \in L^1((0,T);L^\infty(\mathbb{R}^d))$, the map $\boldsymbol{X}(t,\cdot)$ is almost everywhere invertible for every $t \in [0,T]$. We denote by $\boldsymbol{X}^{-1}(t,\cdot)$ the inverse map, which satisfies for every $t \in [0,T]$

$$\boldsymbol{X}(t,\boldsymbol{X}^{-1}(t,x)) = x \ \text{ and } \ \boldsymbol{X}^{-1}(t,\boldsymbol{X}(t,x)) = x \qquad \text{for } \mathscr{L}^d\text{-a.e. } x \in \mathbb{R}^d. \tag{5.8}$$

Moreover, we recall that the compressibility constant $C(\boldsymbol{X})$ in Definition 1.4 (ii) can be chosen as $\exp\big(\int_0^T \|\operatorname{div} \boldsymbol{b}_t\|_{L^\infty(\mathbb{R}^d)}\,dt\big)$.

When the vector field $\boldsymbol{b}$ is divergence-free, the Jacobian of the flow is equal to 1 in the explicit solution (5.3) of (5.1). Instead, when the vector field $\boldsymbol{b}$ is not divergence-free, the Jacobian of the flow appears in (5.3). In the smooth setting, the Jacobian is defined as $J\boldsymbol{X}(t,x) = \det(\nabla_x \boldsymbol{X}(t,x))$, and satisfies the differential equation

$$\partial_t J\boldsymbol{X}(t,x) = J\boldsymbol{X}(t,x)\operatorname{div} \boldsymbol{b}_t(\boldsymbol{X}(t,x)) \qquad \forall (t,x) \in (0,T)\times\mathbb{R}^d.$$

In the non-smooth setting, we define the Jacobian through an explicit formula; we will see in Lemma 5.7 that this object satisfies a change of variable formula.

Definition 5.2. Let $T > 0$, let $\boldsymbol{b}$ be a Borel, locally integrable vector field, and let $\boldsymbol{X}$ as in Definition 1.4. Assume moreover that div $\boldsymbol{b} \in L^1((0,T);L^1_{\mathrm{loc}}(\mathbb{R}^d))$. We define the Jacobian of $\boldsymbol{X}$ as the measurable function $J\boldsymbol{X} : (0,T)\times\mathbb{R}^d \to \mathbb{R}^d$ given by

$$J\boldsymbol{X}(t,x) = \exp\Big(\int_0^t \operatorname{div} \boldsymbol{b}_s(\boldsymbol{X}(s,x))\,ds\Big).$$

Thanks to the compressibility condition (ii) in the definition of regular Lagrangian flow and to the local integrability of div $\boldsymbol{b}$, a computation like (5.6) shows that $J\boldsymbol{X}$ is well defined and absolutely continuous in $[0,T]$ for $\mathscr{L}^d$-a.e. $x \in \mathbb{R}^d$.

We present now a notion of solution of (5.1) which does not even require local integrability of u and was first introduced in [81]. This notion adapts Definition 1.3 to our context by imposing more constraints on the renormalization function β (which, in turn, guarantee the correct integrability of every term). In the sequel of this chapter all the functions involved will be defined up to a set of Lebesgue measure zero.

Definition 5.3. Let $\bar{u} : \mathbb{R}^d \to \mathbb{R}$ be a measurable function, let $\boldsymbol{b} \in L^1_{\text{loc}}((0,T)\times\mathbb{R}^d;\mathbb{R}^d)$ be a vector field such that $\operatorname{div}\boldsymbol{b} \in L^1_{\text{loc}}((0,T)\times\mathbb{R}^d)$ and let $c \in L^1_{\text{loc}}((0,T)\times\mathbb{R}^d)$. A measurable function $u : [0,T]\times\mathbb{R}^d \to \mathbb{R}$ is a renormalized solution of (5.1) if for every function $\beta : \mathbb{R}\to\mathbb{R}$ satisfying

$$\beta \in C^1 \cap L^\infty(\mathbb{R}), \qquad \beta'(z)z \in L^\infty(\mathbb{R}), \qquad \beta(0) = 0 \tag{5.9}$$

we have that

$$\partial_t\beta(u) + \nabla\cdot(\boldsymbol{b}\beta(u)) + \operatorname{div}\boldsymbol{b}\big(u\beta'(u) - \beta(u)\big) = cu\beta'(u) \tag{5.10}$$

in the sense of distributions, namely for every $\phi \in C^\infty_c([0,T)\times\mathbb{R}^d)$

$$\begin{aligned}\int_{\mathbb{R}^d}\phi(0,x)\beta(\bar{u})\,dx &+ \int_0^T\int_{\mathbb{R}^d}[\partial_t\phi + \nabla\phi\cdot\boldsymbol{b}]\beta(u)\,dx\,dt\\ &+ \int_0^T\int_{\mathbb{R}^d}\phi\Big[\operatorname{div}\boldsymbol{b}\big(\beta(u) - u\beta'(u)\big) + cu\beta'(u)\Big]\,dx\,dt = 0.\end{aligned} \tag{5.11}$$

The second assumption in (5.9) is exploited to give a distributional meaning to the right-hand side of (5.10), which becomes locally integrable despite the lack of integrability of u.

Remark 5.4. As precised in a similar context in Remark 4.5, in Definition 5.3, we can equivalently test equation (5.10) with compactly supported space functions φ; in other words, (5.11) holds if and only if for every $\varphi \in C^\infty_c(\mathbb{R}^d)$ the function $\int_{\mathbb{R}^d}\varphi(x)\beta(u_t(x))\,dx$ coincides $\mathscr{L}^1$-a.e. in $(0,T)$ with an absolutely continuous function $\Gamma(t)$ such that $\Gamma(0) = \int_{\mathbb{R}^d}\varphi(x)\beta(\bar{u}(x))\,dx$ and for $\mathscr{L}^1$-a.e. $t \in [0,T]$

$$\begin{aligned}\frac{d}{dt}\Gamma(t) = &\int_{\mathbb{R}^d}\nabla\varphi\cdot\boldsymbol{b}_t\beta(u_t)\,dx\\ &+ \int_{\mathbb{R}^d}\varphi\Big[\operatorname{div}\boldsymbol{b}_t\big(\beta(u_t) - u_t\beta'(u_t)\big) + c_t u_t\beta'(u_t)\Big]\,dx.\end{aligned} \tag{5.12}$$

This follows by the choice $\phi(t,x) = \varphi(x)\eta(t)$ in (5.11) with $\eta \in C^\infty_c([0,T))$; by the density of the linear span of these functions in $C^\infty_c([0,T)\times\mathbb{R}^d)$, it is possibile to deduce the equivalence (see for instance [19, Section 8.1]). Notice moreover that, with a standard approximation argument, we are allowed to use every Lipschitz, compactly supported test function $\varphi : \mathbb{R}^d \to \mathbb{R}$ as a test function for the computation (5.12).

We now state the main result of this chapter, namely the existence and uniqueness of renormalized solutions to the continuity equation with integrable, unbounded damping.

Theorem 5.5. *Let $\boldsymbol{b} \in L^1((0,T); BV_{\mathrm{loc}}(\mathbb{R}^d;\mathbb{R}^d))$ be a vector field that satisfies a bound on the divergence* $\operatorname{div}\boldsymbol{b} \in L^1((0,T); L^\infty(\mathbb{R}^d))$ *and the growth condition*

$$\frac{|\boldsymbol{b}_t(x)|}{1+|x|} \in L^1((0,T); L^1(\mathbb{R}^d)) + L^1((0,T); L^\infty(\mathbb{R}^d)). \tag{5.13}$$

Let

$$c \in L^1((0,T)\times\mathbb{R}^d)$$

and let $\bar u : \mathbb{R}^d \to \mathbb{R}$ be a measurable function. Then there exists a unique renormalized solution $u : [0,T]\times\mathbb{R}^d \to \mathbb{R}$ of (5.1) *starting from $\bar u$ and it is given by the formula*

$$u_t(x) = \frac{\bar u(\boldsymbol{X}^{-1}(t,\cdot)(x))}{J\boldsymbol{X}(t,\boldsymbol{X}^{-1}(t,\cdot)(x))} \exp\left(\int_0^t c_\tau(\boldsymbol{X}(\tau,\boldsymbol{X}^{-1}(t,\cdot)(x)))\,d\tau\right). \tag{5.14}$$

Remark 5.6. The same statement holds for vector fields $\boldsymbol{b}$ satisfying other local regularity assumptions than BV; more precisely, Theorem 5.5 holds for every $\boldsymbol{b}$ such that every bounded, distributional solution of the continuity equation is renormalized. In turn, this property is needed both for the existence and uniqueness of the regular lagrangian flow (since, as it is shown in Remark 4.11, it implies property (b-$\mathbb{R}^d$) of Section 1.3, and therefore we can apply Remark 1.8 and Theorem 1.5) and for Lemma 5.9. Moreover, as shown in Remark 4.11 when the vector field is divergence-free, the property that every bounded, distributional solution of the continuity equation is renormalized is equivalent to (b-$\mathbb{R}^d$). Hence, to see some classes of vector fields other than BV which satisfy this assumption, we refer to Remark 1.9 (for the sake of completeness, we also mention that in many of these explicit examples the property that every bounded, distributional solution of the continuity equation is renormalized is proven directly through commutator estimates).

5.2. Proof of existence

To prove existence in Theorem 5.5, we show by explicit computation that (5.3) provides a renormalized solution to (5.1). In the case of a divergence-free vector field, the flow $\boldsymbol{X}(t,\cdot)$ is measure preserving and (5.14) can be rewritten as

$$u_t(x) = \bar u(\boldsymbol{X}^{-1}(t,\cdot)(x)) \exp\left(\int_0^t c_\tau(\boldsymbol{X}(\tau,\boldsymbol{X}^{-1}(t,\cdot)(x)))\,d\tau\right).$$

An easy computation shows that this function is a renormalized solution of (5.1):

$$\begin{aligned}
\frac{d}{dt}\int_{\mathbb{R}^d} \varphi\beta(u_t)\,dx &= \frac{d}{dt}\int_{\mathbb{R}^d} \varphi(\boldsymbol{X})\beta(u_t(\boldsymbol{X}))\,dx \\
&= \int_{\mathbb{R}^d} \Big[\nabla\varphi(\boldsymbol{X})\cdot \boldsymbol{b}_t(\boldsymbol{X})\beta(u_t(\boldsymbol{X})) + \varphi(\boldsymbol{X})\beta'(u_t(\boldsymbol{X}))u_t(\boldsymbol{X})c_t(\boldsymbol{X})\Big]\,dx \\
&= \int_{\mathbb{R}^d} \Big[\nabla\varphi\cdot \boldsymbol{b}_t\beta(u) + \varphi\beta'(u_t)u_t c_t\Big]\,dx,
\end{aligned}$$

(compare with (5.12)). Note that in the above calculation it has been used that when the representation formula is considered along the flow it holds that

$$\frac{d}{dt}(u_t(\boldsymbol{X})) = u_t(\boldsymbol{X})c_t(\boldsymbol{X}).$$

The computation can be made rigorous thanks to the absolute continuity of $\boldsymbol{X}(\cdot, x)$.

The following lemma, regarding a change of variable formula and time regularity of the Jacobian of regular Lagrangian flows, is useful in the proof when $\boldsymbol{b}$ is not divergence-free.

Lemma 5.7 (Properties of the Jacobian). *Let $\boldsymbol{b}$ as in Theorem 5.5 and let $\boldsymbol{X}$ be the regular Lagrangian flow of $\boldsymbol{b}$. Then, the function $J\boldsymbol{X}$ in Definition 5.2 is in $L^1((0,T); L^\infty(\mathbb{R}^d))$ and for every $t > 0$ and every $\phi \in L^1(\mathbb{R}^d)$ satisfies the following change of variable formula:*

$$\int_{\mathbb{R}^d} \phi(\boldsymbol{X}(t,x))J\boldsymbol{X}(t,x)\,dx = \int_{\mathbb{R}^d} \phi(x)\,dx. \tag{5.15}$$

Moreover, $e^{-L} \le J\boldsymbol{X} \le e^{L}$ with $L = \int_0^T \|\operatorname{div}\boldsymbol{b}_t\|_{L^\infty(\mathbb{R}^d)}\,dt$, $J\boldsymbol{X}(\cdot, x)$ and $J\boldsymbol{X}^{-1}(\cdot, x)$ are absolutely continuous in $[0,T]$ and satisfy

$$\partial_t J\boldsymbol{X}(t,x) = J\boldsymbol{X}(t,x)\operatorname{div}\boldsymbol{b}_t(\boldsymbol{X}(t,x)) \quad \textit{for } \mathscr{L}^1\textit{-a.e. } t \in (0,T), \tag{5.16}$$

$$\partial_t\left[\frac{1}{J\boldsymbol{X}(t,x)}\right] = -\left(\frac{1}{J\boldsymbol{X}(t,x)}\right)\operatorname{div}\boldsymbol{b}_t(\boldsymbol{X}(t,x)) \ \textit{for } \mathscr{L}^1\textit{-a.e. } t\in (0,T) \tag{5.17}$$

for $\mathscr{L}^d$-a.e. $x \in \mathbb{R}^d$.

Proof. **Step 1: approximation with smooth vector fields.** Let us approximate the vector field $\boldsymbol{b}$ by convolution. In particular let $\boldsymbol{b}^\varepsilon$ be the convolution between $\boldsymbol{b}$, extended to 0 in $(\mathbb{R} \setminus [0,T]) \times \mathbb{R}^d$, and a kernel of the form $\varepsilon^{-d-1}\rho_1(t/\varepsilon)\rho_2(x/\varepsilon)$, where $\rho_1 \in C_c^\infty(\mathbb{R})$ and $\rho_2 \in C_c^\infty(\mathbb{R}^d)$ are standard convolution kernels, so that

$$\| \operatorname{div} \boldsymbol{b}_t^\varepsilon \|_{L^\infty(\mathbb{R}^d)} \leq \int_{\mathbb{R}} \rho_1(t') \| \operatorname{div} \boldsymbol{b}_{t-\varepsilon t'} \|_{L^\infty(\mathbb{R}^d)} \, dt'. \tag{5.18}$$

Let $\boldsymbol{X}^\varepsilon \in C^\infty([0,T] \times \mathbb{R}^d; \mathbb{R}^d)$ be the flow of $\boldsymbol{b}^\varepsilon$; for every $t > 0$ the function $\boldsymbol{X}^\varepsilon(t,\cdot)$ is a diffeomorphism of $\mathbb{R}^d$ and, setting $J\boldsymbol{X}^\varepsilon(t,x) = \det \nabla_x \boldsymbol{X}^\varepsilon(t,x)$, we have the change of variable formula

$$\int_{\mathbb{R}^d} \phi(\boldsymbol{X}^\varepsilon(t,x)) J\boldsymbol{X}^\varepsilon(t,x)\, dx = \int_{\mathbb{R}^d} \phi(x)\, dx \qquad \forall \phi \in C_c(\mathbb{R}^d). \tag{5.19}$$

Moreover for every $x \in \mathbb{R}^d$ the function $J\boldsymbol{X}^\varepsilon(\cdot, x)$ solves the ODE

$$\begin{cases} \partial_t J\boldsymbol{X}^\varepsilon(t,x) = J\boldsymbol{X}^\varepsilon(t,x) \operatorname{div} \boldsymbol{b}_t^\varepsilon(\boldsymbol{X}^\varepsilon(t,x)) & \text{for any } t \in (0,T) \\ J\boldsymbol{X}^\varepsilon(0,x) = x, \end{cases}$$

hence it is given by the expression

$$J\boldsymbol{X}^\varepsilon(t,x) = \exp\Big(\int_0^t \operatorname{div} \boldsymbol{b}_s^\varepsilon(\boldsymbol{X}^\varepsilon(s,x))\, ds \Big) \qquad \forall (t,x) \in [0,T] \times \mathbb{R}^d.$$

Integrating (5.18) in $(0,T)$, we find that

$$e^{-L} \leq J\boldsymbol{X}^\varepsilon \leq e^L \;\; \forall (t,x) \in [0,T] \times \mathbb{R}^d \;\text{ with } L = \int_0^T \| \operatorname{div} \boldsymbol{b}_t \|_{L^\infty(\mathbb{R}^d)}\, dt. \tag{5.20}$$

Step 2: pointwise convergence of Jacobians. We show that, up to a subsequence (not relabeled) in ε, for $\mathscr{L}^d$-a.e. $x \in \mathbb{R}^d$

$$\lim_{\varepsilon \to 0} J\boldsymbol{X}^\varepsilon(t,x) = J\boldsymbol{X}(t,x) \qquad \text{for every } t \in (0,T), \tag{5.21}$$

where $J\boldsymbol{X}$ is defined in Definition 5.2.

To this end, let us first prove that, up to a subsequence (not relabeled),

$$\lim_{\varepsilon \to 0} \operatorname{div} \boldsymbol{b}_t^\varepsilon(\boldsymbol{X}^\varepsilon(t,x)) = \operatorname{div} \boldsymbol{b}_t(\boldsymbol{X}(t,x)) \qquad \text{in } L^1_{\text{loc}}([0,T] \times \mathbb{R}^d). \tag{5.22}$$

By the stability of regular Lagrangian flows (see [5, 81, 67] or [13, Section 5] or Theorem 3.2 noticing that assumption (5.13) prevents finite-time blow up of trajectories thanks to Theorem 3.13), for every $t \in [0,T]$ we have that, up to a subsequence (not relabelled)

$$\lim_{\varepsilon \to 0} \boldsymbol{X}^\varepsilon(t,x) = \boldsymbol{X}(t,x) \quad \text{pointwise for } \mathscr{L}^{d+1}\text{-a.e. } (t,x) \in [0,T] \times \mathbb{R}^d. \tag{5.23}$$

Let us consider $r > 0$ and let us prove the convergence in (5.22) in $[0,T] \times B_r$. Let $R > 0$ and $\eta > 0$ to be chosen later. The estimate on superlevels in [67, Proposition 3.2], which depends on the growth assumptions (5.20) and on the compressibility of the flows, implies that

$$\mathscr{L}^d(\{x \in B_r : \boldsymbol{X}^\varepsilon(t,x) \in \mathbb{R}^d \setminus B_R\}) \le g(R,r), \tag{5.24}$$

for a function $g(R,r)$ which converges to 0 as $R \to \infty$ for every $r > 0$ (and it is independent on ε and t). The analogous of (5.24) holds also with $\boldsymbol{X}$ in place of $\boldsymbol{X}^\varepsilon$.

By Egorov theorem, there exists a measurable set $E \subseteq [0,T] \times B_R$ of small measure $\mathscr{L}^{d+1}(E) \le \eta$ such that

$$\lim_{\varepsilon \to 0} \operatorname{div} \boldsymbol{b}^\varepsilon = \operatorname{div} \boldsymbol{b} \qquad \text{uniformly in } \big([0,T] \times B_R\big) \setminus E. \tag{5.25}$$

As a consequence, $\operatorname{div} \boldsymbol{b}_t$ is continuous on $\big([0,T] \times B_R\big) \setminus E$. Let us consider E^t to be the intersection of E with $\{t\} \times \mathbb{R}^d$. Letting

$$\begin{aligned} E^t_{\varepsilon,R} = \{x \in B_r : \boldsymbol{X}^\varepsilon(t,x) \in E^t \\ \cup (\mathbb{R}^d \setminus B_R)\} \cup \{x \in B_r : \boldsymbol{X}(t,x) \in E^t \cup (\mathbb{R}^d \setminus B_R)\}, \end{aligned}$$

we have that

$$\begin{aligned} &\int_0^T \int_{B_r} |\operatorname{div} \boldsymbol{b}^\varepsilon_t(\boldsymbol{X}^\varepsilon) - \operatorname{div} \boldsymbol{b}_t(\boldsymbol{X})| \, dx \, dt \\ &\le \int_0^T \int_{B_r \setminus E^t_{\varepsilon,R}} |\operatorname{div} \boldsymbol{b}^\varepsilon_t(\boldsymbol{X}^\varepsilon) - \operatorname{div} \boldsymbol{b}_t(\boldsymbol{X})| \, dx \, dt \\ &\quad + \int_0^T \mathscr{L}^d(E^t_{\varepsilon,R}) \| \operatorname{div} \boldsymbol{b}^\varepsilon_t \|_{L^\infty(\mathbb{R}^d)} \, dt \\ &\quad + \int_0^T \mathscr{L}^d(E^t_{\varepsilon,R}) + \| \operatorname{div} \boldsymbol{b}_t \|_{L^\infty(\mathbb{R}^d)}) \, dt \end{aligned} \tag{5.26}$$

The second and the third term in the right-hand side of (5.26) can be estimated uniformly in ε thanks to the compressibility of $\boldsymbol{X}^\varepsilon(t,\cdot)$ and $\boldsymbol{X}(t,\cdot)$, which is less or equal, in both cases, than e^L thanks to (5.20) and Remark 5.1. More precisely

$$\begin{aligned} \mathscr{L}^d(\{\boldsymbol{X}^\varepsilon(t,\cdot) \in E^t \cup (\mathbb{R}^d \setminus B_R)\}) &\le \mathscr{L}^d(\{\boldsymbol{X}^\varepsilon(t,\cdot) \in E^t\}) \\ &\quad + \mathscr{L}^d(\{\boldsymbol{X}^\varepsilon(t,\cdot) \in \mathbb{R}^d \setminus B_R\}) \\ &\le e^L \mathscr{L}^d(E^t) + g(R,r) \end{aligned}$$

and a similar computation holds for the set $\{X(t,\cdot) \in E^t \cup (\mathbb{R}^d \setminus B_R)\}$, so that overall

$$\mathscr{L}^d(E^t_{\varepsilon,R}) \le 2e^L \mathscr{L}^d(E^t) + 2g(R,r).$$

Thanks to (5.26), it implies that

$$\begin{aligned}
&\int_0^T \int_{B_r} |\operatorname{div} \boldsymbol{b}^\varepsilon_t(\boldsymbol{X}^\varepsilon) - \operatorname{div} \boldsymbol{b}_t(\boldsymbol{X})|\, dx\, dt \\
&\le \int_0^T \int_{B_r \setminus E^t_{\varepsilon,R}} |\operatorname{div} \boldsymbol{b}^\varepsilon_t(\boldsymbol{X}^\varepsilon) - \operatorname{div} \boldsymbol{b}_t(\boldsymbol{X})|\, dx\, dt \\
&\quad + 2e^L \int_0^T \mathscr{L}^d(E_t) \|\operatorname{div} \boldsymbol{b}_t\|_{L^\infty(\mathbb{R}^d)}\, dt \\
&\quad + 2e^L \int_0^T \mathscr{L}^d(E_t) \|\operatorname{div} \boldsymbol{b}^\varepsilon_t\|_{L^\infty(\mathbb{R}^d)}\, dt \\
&\quad + 4g(R,r) \int_0^T \|\operatorname{div} \boldsymbol{b}_t\|_{L^\infty(\mathbb{R}^d)}\, dt
\end{aligned}$$

which can be written as follows:

$$\begin{aligned}
&\int_0^T \int_{B_r} |\operatorname{div} \boldsymbol{b}^\varepsilon_t(\boldsymbol{X}^\varepsilon) - \operatorname{div} \boldsymbol{b}_t(\boldsymbol{X})|\, dx\, dt \\
&\le \int_0^T \int_{B_r \setminus E^t_{\varepsilon,R}} |\operatorname{div} \boldsymbol{b}^\varepsilon_t(\boldsymbol{X}^\varepsilon) - \operatorname{div} \boldsymbol{b}_t(\boldsymbol{X})|\, dx\, dt \\
&\quad + 4g(R,r) \int_0^T \|\operatorname{div} \boldsymbol{b}_t\|_{L^\infty(\mathbb{R}^d)}\, dt \\
&\quad + 2e^L \int_E \|\operatorname{div} \boldsymbol{b}_t\|_{L^\infty(\mathbb{R}^d)}\, dx\, dt \\
&\quad + 2e^L \int_E \|\operatorname{div} \boldsymbol{b}^\varepsilon_t\|_{L^\infty(\mathbb{R}^d)}\, dx\, dt.
\end{aligned} \tag{5.27}$$

The first term in (5.27) converges to 0 as $\varepsilon \to 0$ because $\operatorname{div} \boldsymbol{b}^\varepsilon_t(\boldsymbol{X}^\varepsilon)$ converges pointwise to $\operatorname{div} \boldsymbol{b}_t(\boldsymbol{X})$ in $B_r \setminus E^t_{\varepsilon,R}$ and $\operatorname{div} \boldsymbol{b}$ is continuous on E^t:

$$\begin{aligned}
|\operatorname{div} \boldsymbol{b}^\varepsilon_t(\boldsymbol{X}^\varepsilon) - \operatorname{div} \boldsymbol{b}_t(\boldsymbol{X})| &\le |\operatorname{div} \boldsymbol{b}^\varepsilon_t(\boldsymbol{X}^\varepsilon) - \operatorname{div} \boldsymbol{b}_t(\boldsymbol{X}^\varepsilon)| \\
&\quad + |\operatorname{div} \boldsymbol{b}_t(\boldsymbol{X}^\varepsilon) - \operatorname{div} \boldsymbol{b}_t(\boldsymbol{X})| \\
&\le \|\operatorname{div} \boldsymbol{b}^\varepsilon_t - \operatorname{div} \boldsymbol{b}_t\|_{L^\infty(B_R \setminus E^t)} \\
&\quad + |\operatorname{div} \boldsymbol{b}_t(\boldsymbol{X}^\varepsilon) - \operatorname{div} \boldsymbol{b}_t(\boldsymbol{X})|.
\end{aligned} \tag{5.28}$$

The second term goes to 0 because $g(R,r) \to 0$ for $R \to \infty$. The last terms, in turn, converge to 0 as $\eta \to 0$, where η has been chosen in (5.25) and is independent on ε, by the absolute continuity of the Lebesgue integral. Indeed, each function is dominated by

$$t \mapsto \|\operatorname{div} \boldsymbol{b}_t^\varepsilon\|_{L^\infty(\mathbb{R}^d)} + \|\operatorname{div} \boldsymbol{b}_t\|_{L^\infty(\mathbb{R}^d)} \le (\varepsilon^{-1}\rho_1(\cdot/\varepsilon)) * \|\operatorname{div} \boldsymbol{b}_t\|_{L^\infty(\mathbb{R}^d)} + \|\operatorname{div} \boldsymbol{b}_t\|_{L^\infty(\mathbb{R}^d)}$$

and the last function converges in $L^1([0,T])$ to $2\|\operatorname{div} \boldsymbol{b}_t\|_{L^\infty(\mathbb{R}^d)}$, so that we can take the limit in the right-hand side of (5.27) by the absolute continuity of the Lebesgue integral. Finally, choosing first R and η small enough, and then letting ε go to 0 in (5.27), we find (5.22). By (5.22), up to a subsequence, for $\mathscr{L}^d$-a.e. $x \in \mathbb{R}^d$, $\operatorname{div} \boldsymbol{b}_t^\varepsilon(\boldsymbol{X}^\varepsilon(t,x))$ converges to $\operatorname{div} \boldsymbol{b}_t(\boldsymbol{X}(t,x))$ in $L^1([0,T])$. Hence for $\mathscr{L}^d$-a.e. x we deduce (5.21).

Step 3: conclusion. Let us fix $t > 0$ and $\phi \in C_c(B_R)$ with $R > 0$. We take the limit as ε goes to 0 in (5.19) to get (5.15). More precisely, to show that the limit of (5.19) is (5.15), we estimate the difference of the two terms by adding and subtracting $\phi(\boldsymbol{X})J\boldsymbol{X}^\varepsilon$ and using the bound on $J\boldsymbol{X}^\varepsilon$ given by (5.20)

$$\left|\int_{\mathbb{R}^d} (\phi(\boldsymbol{X}^\varepsilon)J\boldsymbol{X}^\varepsilon - \phi(\boldsymbol{X})J\boldsymbol{X})\,dx\right| \le \int_{\mathbb{R}^d} \big(|\phi(\boldsymbol{X})||J\boldsymbol{X}^\varepsilon - J\boldsymbol{X}| + e^L|\phi(\boldsymbol{X}^\varepsilon) - \phi(\boldsymbol{X})|\big)\,dx.$$

The first term goes to 0 as $\varepsilon \to 0$ by (5.21) and the dominated convergence theorem, since the functions are nonzero only on the set $\{x : \boldsymbol{X}(t,x) \in B_R\}$ and this set has finite measure.

Regarding the second term, for every $\tilde{R} > 0$ we have

$$\int_{\mathbb{R}^d} |\phi(\boldsymbol{X}^\varepsilon) - \phi(\boldsymbol{X})|\,dx \le 2\|\phi\|_{L^\infty}\mathscr{L}^d(\{x \notin B_{\tilde{R}} : \boldsymbol{X}(t,x) \in B_R \text{ or } \boldsymbol{X}^\varepsilon(t,x) \in B_R\}) + \int_{B_{\tilde{R}}} |\phi(\boldsymbol{X}^\varepsilon) - \phi(\boldsymbol{X})|\,dx.$$

By choosing $\tilde{R}$ sufficiently big, the first term can be made as small as we want independently on ε thanks to the estimate on superlevels in [67, Proposition 3.2] (see also (5.24)). Finally, letting $\varepsilon \to 0$ in the second term with $\tilde{R}$ fixed, we obtain that it converges to 0 by dominated convergence. Hence, (5.15) holds true for every $\phi \in C_c(\mathbb{R}^d)$. Then we approximate every $\phi \in L^1(\mathbb{R}^d)$ with compactly supported, continuous functions $\{\phi_n\}_{n\in\mathbb{N}}$ and we take the limit in (5.15) applied to ϕ_n. The left-hand side converges thanks to the bound on the Jacobian and to the bounded compressibility of $\boldsymbol{X}$:

$$\left| \int_{\mathbb{R}^d} (\phi_n(\boldsymbol{X}) - \phi(\boldsymbol{X})) J\boldsymbol{X}\, dx \right| \le e^L \int_{\mathbb{R}^d} |\phi_n(\boldsymbol{X}) - \phi(\boldsymbol{X})|\, dx$$
$$\le C e^L \int_{\mathbb{R}^d} |\phi_n - \phi|\, dx,$$

hence we obtain (5.15) with ϕ.

Finally, (5.16) and (5.17) are easily checked by direct computation and using the fact that $J\boldsymbol{X}$ is absolutely continuous in the time variable. □

Proof of Theorem 5.5, *Existence*. Let $\beta : \mathbb{R} \to \mathbb{R}$ be a function satisfying (5.9). From the expression (5.3) we compute an equation involving $\beta(u_t(x))$. Let $\varphi \in C_c^\infty$ be a test function. By the change of variable formula (5.15) applied with $\phi(x) = \varphi(x)\beta(u_t(x))$ we have that

$$\int_{\mathbb{R}^d} \varphi(x)\beta(u_t(x))\, dx = \int_{\mathbb{R}^d} \varphi(\boldsymbol{X}(t,x))\beta(u_t(\boldsymbol{X}(t,x))) J\boldsymbol{X}(t,x)\, dx.$$

Thanks to the absolute continuity of $\boldsymbol{X}(\cdot, x)$, of $J\boldsymbol{X}(\cdot, x)$, and of $1/J\boldsymbol{X}(\cdot, x)$ and since the set of bounded, absolutely continuous functions is an algebra, for every $x \in \mathbb{R}^d$ the functions $t \to u_t(\boldsymbol{X}(t,x))$ and $t \to \varphi(\boldsymbol{X}(t,x))\beta(u_t(\boldsymbol{X}(t,x))) J\boldsymbol{X}(t,x)$ are absolutely continuous. Their derivative can be computed by the explicit formula for u given in (5.14) thanks to (5.16) and (5.17): for $\mathscr{L}^1$-a.e. $s \in [0,T]$

$$\begin{aligned}
\partial_s \left[u_s(\boldsymbol{X})\right] &= \partial_s \left[\frac{\bar u}{J\boldsymbol{X}} \exp\left(\int_0^t c_\tau(\boldsymbol{X}(\tau))\, d\tau \right) \right] \\
&= \frac{\bar u}{J\boldsymbol{X}} \exp\left(\int_0^t c_\tau(\boldsymbol{X}(\tau))\, d\tau \right) c_s(\boldsymbol{X}) \\
&\quad + \partial_s \left[\frac{1}{J\boldsymbol{X}} \right] \bar u \exp\left(\int_0^t c_\tau(\boldsymbol{X}(\tau))\, d\tau \right) \\
&= u_s(\boldsymbol{X}) c_s(\boldsymbol{X}) + u_s(\boldsymbol{X}) \partial_s \left[\frac{1}{J\boldsymbol{X}} \right] J\boldsymbol{X} \\
&= u_s(\boldsymbol{X}) c_s(\boldsymbol{X}) - u_s(\boldsymbol{X}) \operatorname{div} \boldsymbol{b}_s(\boldsymbol{X}) \frac{1}{J\boldsymbol{X}}
\end{aligned}$$

and therefore

$$
\begin{aligned}
\partial_s\big[\varphi(\boldsymbol{X}(s,x))\beta(u_s(\boldsymbol{X}(s,x)))J\boldsymbol{X}(s,x)\big] = {} & \nabla\varphi(\boldsymbol{X})\cdot \boldsymbol{b}_s(\boldsymbol{X})\beta(u_s(\boldsymbol{X}))J\boldsymbol{X} \\
& + \varphi(\boldsymbol{X})\beta'(u_s(\boldsymbol{X}))\partial_s\big[u_s(\boldsymbol{X})\big]J\boldsymbol{X} + \varphi(\boldsymbol{X})\beta(u_s(\boldsymbol{X}))\partial_s J\boldsymbol{X} \\
= {} & \big[\nabla\varphi(\boldsymbol{X})\cdot \boldsymbol{b}_s(\boldsymbol{X})\beta(u_s(\boldsymbol{X})) + \varphi(\boldsymbol{X})\beta'(u_s(\boldsymbol{X}))u_s(\boldsymbol{X})c_s(\boldsymbol{X}) \\
& - \varphi(\boldsymbol{X})\beta'(u_s(\boldsymbol{X}))u_s(\boldsymbol{X})\operatorname{div}\boldsymbol{b}_s(\boldsymbol{X}) + \varphi(\boldsymbol{X})\beta(u_s(\boldsymbol{X}))\operatorname{div}\boldsymbol{b}_s(\boldsymbol{X})\big]J\boldsymbol{X}
\end{aligned}
$$

(for the sake of brevity we sometimes write $\boldsymbol{X}$ in place of $\boldsymbol{X}(s,x)$ and $J\boldsymbol{X}$ in place of $J\boldsymbol{X}(s,x)$). Hence, by Fubini theorem and by the change of variable (5.15), we have that

$$
\begin{aligned}
& \int_{\mathbb{R}^d}\varphi(x)\beta(u_t(x))\,dx - \int_{\mathbb{R}^d}\varphi(x)\beta(\bar{u}(x))\,dx \\
& = \int_{\mathbb{R}^d}\int_0^t \partial_s\big[\varphi(\boldsymbol{X}(s,x))\beta(u_s(\boldsymbol{X}(s,x)))J\boldsymbol{X}(s,x)\big]\,ds\,dx \\
& = \int_0^t\int_{\mathbb{R}^d}\Big[\nabla\varphi(\boldsymbol{X})\cdot \boldsymbol{b}_s(\boldsymbol{X})\beta(u_s(\boldsymbol{X})) + \varphi(\boldsymbol{X})\beta'(u_s(\boldsymbol{X}))u_s(\boldsymbol{X})c_s(\boldsymbol{X}) \\
& -\varphi(\boldsymbol{X})\beta'(u_s(\boldsymbol{X}))u_s(\boldsymbol{X})\operatorname{div}\boldsymbol{b}_s(\boldsymbol{X}) + \varphi(\boldsymbol{X})\beta(u_s(\boldsymbol{X}))\operatorname{div}\boldsymbol{b}_s(\boldsymbol{X})\Big]J\boldsymbol{X}\,dx\,ds \\
& = \int_0^t\int_{\mathbb{R}^d}\Big[\nabla\varphi(x)\cdot \boldsymbol{b}_s(x)\beta(u_s(x)) + \varphi(x)c_s(x)u_s(x)\beta'(u_s(x)) \\
& + \varphi(x)\operatorname{div}\boldsymbol{b}_s(x)\big(-u_s(x)\beta'(u_s(x)) + \beta(u_s(x))\big)\Big]\,dx\,ds.
\end{aligned}
$$

Notice that the integrand in the right-hand side is in $L^1((0,T)\times\mathbb{R}^d))$ thanks to the properties of β and since φ is compactly supported. We have therefore verified that the function $t \to \int_{\mathbb{R}^d}\varphi(x)\beta(u_t(x))\,dx$ is absolutely continuous in $[0,T]$ and that (5.12) holds; we conclude that u is a renormalized solution thanks to Remark 5.4. □

5.3. Proof of uniqueness

In this section we are going to prove the uniqueness part of Theorem 5.5. In Lemma 5.9 we prove that under our assumptions the difference of renormalized solutions is still a renormalized solution following the lines of [81, Lemma II.2]. Therefore, to prove uniqueness in Theorem 5.5 it is enough to show that every renormalized solution starting from $\bar{u} = 0$ is identically 0. The following simple lemma states the property of the particular renormalization function which allows to pass to the limit in the damping term.

Lemma 5.8. *Let* $\beta(r) = \arctan(r) : \mathbb{R} \to (-\pi/2, \pi/2)$ *and, for every* $M > 0$*, let* $\beta_M(r) = M\beta(r/M)$*. Then we have that*

$$|r_1\beta_M'(r_1) - r_2\beta_M'(r_2)| \leq |\beta_M(r_1) - \beta_M(r_2)| \qquad \forall r_1, r_2 \in \mathbb{R}. \quad (5.29)$$

Proof. First we prove the inequality for $M = 1$, namely

$$\left|\frac{r_1}{1+r_1^2} - \frac{r_2}{1+r_2^2}\right| \leq |\arctan(r_1) - \arctan(r_2)| \qquad \forall r_1, r_2 \in \mathbb{R}. \quad (5.30)$$

Setting $t_i = \arctan(r_i)$, $i = 1, 2$, the inequality is equivalent to

$$\left|\frac{\tan(t_1)}{1+\tan^2(t_1)} - \frac{\tan(t_2)}{1+\tan^2(t_2)}\right| \leq |t_1 - t_2| \qquad \forall t_1, t_2 \in \left(-\frac{\pi}{2}, \frac{\pi}{2}\right).$$

Since the left-hand side can be rewritten as $|\sin(2t_1)/2 - \sin(2t_2)/2|$ and the function $\sin(2t)/2$ is Lipschitz with constant 1, the previous inequality is satisfied. To prove (5.29) with $M > 0$, we apply (5.30) at r_1/M and r_2/M to obtain

$$\left|\frac{Mr_1}{M^2+r_1^2} - \frac{Mr_2}{M^2+r_2^2}\right| \leq \left|\arctan\left(\frac{r_1}{M}\right) - \arctan\left(\frac{r_2}{M}\right)\right| \qquad \forall r_1, r_2 \in \mathbb{R}.$$

Multiplying both sides by M we obtain (5.29). □

Although the continuity equation is linear, it does not follow from its definition that the class of renormalized solutions is linear. However, thanks to the regularity of the vector field and to a particular choice of renormalization functions, we prove that the difference of renormalized solutions is still a renormalized solution.

Lemma 5.9. *Let us consider a vector field* $\boldsymbol{b} \in L^1((0,T); BV_{\mathrm{loc}}(\mathbb{R}^d; \mathbb{R}^d))$ *with* $\operatorname{div} \boldsymbol{b} \in L^1((0,T); L^1_{\mathrm{loc}}(\mathbb{R}^d))$*, a damping* $c \in L^1_{\mathrm{loc}}((0,T) \times \mathbb{R}^d)$*, and a measurable initial datum* $\bar{u} : \mathbb{R}^d \to \mathbb{R}$*. Let* u_1 *and* u_2 *be renormalized solutions of* (5.1) *with initial datum* $\bar{u}$*.*

Then $u := u_1 - u_2$ *is a renormalized solution with initial datum* 0.

Proof. Let $M > 0$ and $\beta_M(r) = M\arctan(r/M)$ for every $r \in \mathbb{R}$. Notice that β_M satisfies (5.9), so that in the sense of distributions

$$\partial_t\beta_M(u_i) + \nabla\cdot(\boldsymbol{b}\beta_M(u_i)) + \operatorname{div}\boldsymbol{b}\big(u_i\beta_M'(u_i) - \beta_M(u_i)\big) = cu_i\beta_M'(u_i) \quad i = 1, 2.$$

Taking the difference between these equations and setting $v_M = \beta_M(u_1) - \beta_M(u_2)$ we obtain that v_M solves in the sense of distributions

$$\partial_t v_M + \nabla\cdot(\boldsymbol{b}v_M) = (c - \operatorname{div}\boldsymbol{b})[u_1\beta_M'(u_1) - u_2\beta_M'(u_2)] + \operatorname{div}\boldsymbol{b}\, v_M.$$

Thanks to the assumptions on $\boldsymbol{b}$, since the right hand side of the previous equation is locally integrable, and since $v_M \in L^\infty((0,T)\times\mathbb{R}^d)$, it follows by [5] (see also [13, Theorem 35] and the discussion of Remark 5.6) that v_M is also a renormalized solution, namely for every γ which satisfies (5.9) we have

$$\partial_t\gamma(v_M) + \nabla\cdot(\boldsymbol{b}\gamma(v_M)) = (c - \operatorname{div}\boldsymbol{b})\gamma'(v_M)v_M\frac{u_1\beta_M'(u_1) - u_2\beta_M'(u_2)}{\beta_M(u_1)-\beta_M(u_2)} + \operatorname{div}\boldsymbol{b}\,\gamma(v_M).$$

This means that, since $v_M(0,\cdot) = 0$, for every $\phi \in C_c^\infty([0,T)\times\mathbb{R}^d)$ we have

$$\begin{aligned}&-\int_0^T\int_{\mathbb{R}^d}[\partial_t\phi+\nabla\phi\cdot\boldsymbol{b}]\gamma(v_M)dxdt\\ &=\int_0^T\int_{\mathbb{R}^d}\phi\Big[(c-\operatorname{div}\boldsymbol{b})\gamma'(v_M)v_M\frac{u_1\beta_M'(u_1)-u_2\beta_M'(u_2)}{\beta_M(u_1)-\beta_M(u_2)}+\operatorname{div}\boldsymbol{b}\gamma(v_M)\Big]dxdt.\end{aligned} \tag{5.31}$$

Then, we let M go to ∞ in the previous equation. First, we note that since $\beta_M(r)\to r$ as $M\to\infty$ it follows that v_M converges to u_1-u_2 pointwise as $M\to\infty$. As regards the left-hand side of (5.31), $\gamma(v_M)$ converges pointwise to $\gamma(u_1-u_2)$ and these functions are bounded by $\|\gamma\|_\infty$. The right-hand side of (5.31) converges pointwise to the right-hand side of (5.32) below and by Lemma 5.8 it is bounded by the L^1_{loc} function $(|c| + 2|\operatorname{div}\boldsymbol{b}|)\|z\gamma'(z)\|_{L^\infty(\mathbb{R}^d)}$. Hence by dominated convergence we get

$$\begin{aligned}&-\int_0^T\int_{\mathbb{R}^d}[\partial_t\phi+\nabla\phi\cdot\boldsymbol{b}]\gamma(u)\,dx\,dt\\ &=\int_0^T\int_{\mathbb{R}^d}\phi\Big[(c-\operatorname{div}\boldsymbol{b})u\gamma'(u)+\operatorname{div}\boldsymbol{b}\gamma(u)\Big]\,dx\,dt\end{aligned} \tag{5.32}$$

for every γ which satisfies (5.9). □

In the following lemma we enlarge the class of admissible test functions in (5.11). As it will be clear from the proof of Theorem 5.5, a particular Lipschitz, decaying test function will play an important role. In particular in the proof of the uniqueness in Theorem 5.5 the estimate (5.43) fails when only compactly supported smooth test functions are considered.

Lemma 5.10. *Let $C > 0$ and let* $\mathbf{b}$ *and* $\bar{u}$ *be as in Theorem* 5.5. *Let* u *be a renormalized solution of* (5.1) *and let* $\varphi \in W^{1,\infty}(\mathbb{R}^d)$ *be a function with the following decay*

$$|\varphi(x)| \leq \frac{C}{(1+|x|)^{d+1}}, \quad |\nabla\varphi(x)| \leq \frac{C}{(1+|x|)^{d+2}} \text{ for } \mathscr{L}^d\text{-a.e. } x \in \mathbb{R}^d. \tag{5.33}$$

Then the function $\int_{\mathbb{R}^d} \varphi(x)\beta(u_t(x))\,dx$ *coincides* $\mathscr{L}^1$*-a.e. with an absolutely continuous function* $\Gamma(t)$ *such that* $\Gamma(0) = \int_{\mathbb{R}^d} \varphi(x)\beta(\bar{u}(x))\,dx$ *and for* $\mathscr{L}^1$*-a.e.* $t \in [0, T]$

$$\begin{aligned} \frac{d}{dt}\Gamma(t) = & \int_{\mathbb{R}^d} \nabla\varphi \cdot \boldsymbol{b}_t \beta(u_t)\,dx \\ & + \int_{\mathbb{R}^d} \varphi\Big[\operatorname{div} \boldsymbol{b}_t\big(\beta(u_t) - u_t\beta'(u_t)\big) + c_t u_t \beta'(u_t)\Big]\,dx. \end{aligned} \tag{5.34}$$

Proof. Although the proof is a standard argument via approximation, we sketch it for the sake of completeness. We approximate the function φ by means of smooth, compactly supported functions φ_n satisfying the same decay (5.33) with C independent on n. By Remark 5.4, the function $t \to \int_{\mathbb{R}^d} \varphi_n(x)\beta(u_t(x))\,dx$ coincides for $\mathscr{L}^1$-a.e. $t \in [0, T]$ with an absolutely continuous function $\Gamma_n(t)$ which satisfies (5.12) and $\Gamma_n(0) = \int_{\mathbb{R}^d} \varphi_n(x)\beta(\bar{u}(x))\,dx$. Thanks to (5.33), to the growth assumptions on $\boldsymbol{b}$, and to the integrability of c, by dominated convergence we get that

$$\begin{aligned} \lim_{n\to\infty} \frac{d}{dt}\Gamma_n(t) = & \lim_{n\to\infty} \int_{\mathbb{R}^d} \nabla\varphi_n \cdot \boldsymbol{b}_t \beta(u_t)\,dx \\ & + \int_{\mathbb{R}^d} \varphi_n\Big[\operatorname{div} \boldsymbol{b}_t\big(\beta(u_t) - u_t\beta'(u_t)\big) + c u_t \beta'(u_t)\Big]\,dx \\ = & \int_{\mathbb{R}^d} \nabla\varphi \cdot \boldsymbol{b}_t \beta(u_t)\,dx \\ & + \int_{\mathbb{R}^d} \varphi\Big[\operatorname{div} \boldsymbol{b}_t\big(\beta(u_t) - u_t\beta'(u_t)\big) + c_t u_t \beta'(u_t)\Big]\,dx \end{aligned} \tag{5.35}$$

in $L^1(0, T)$. Moreover by dominated convergence we have

$$\lim_{n\to\infty} \Gamma_n(0) = \int_{\mathbb{R}^d} \varphi(x)\beta(\bar{u}(x))\,dx$$

and for $\mathscr{L}^1$-a.e. $t \in [0, T]$

$$\begin{aligned} \Gamma(t) = \lim_{n\to\infty} \Gamma_n(t) = & \lim_{n\to\infty} \int_{\mathbb{R}^d} \varphi_n(x)\beta(u_t(x))\,dx \\ = & \int_{\mathbb{R}^d} \varphi(x)\beta(u_t(x))\,dx. \end{aligned} \tag{5.36}$$

Hence the functions Γ_n pointwise converge to an absolutely continuous function $\Gamma: [0,T] \to \mathbb{R}$ such that (5.34) holds, $\Gamma(0) = \int_{\mathbb{R}^d} \varphi(x)\beta(\bar{u}(x))dx$, and

$$\Gamma(t) = \int_{\mathbb{R}^d} \varphi(x)\beta(u_t(x))\,dx$$

for $\mathscr{L}^1$-a.e. $t \in [0, T]$. □

Proof of Theorem 5.5, *Uniqueness*. Up to taking the difference of two renormalized solutions, which is still a renormalized solution with initial datum 0 by Lemma 5.9, it is enough to show that if u is a renormalized solution with initial datum 0 then $u = 0$ in $[0, T] \times \mathbb{R}^d$.

Let $\delta > 0$. We consider the positive function

$$\beta_\delta(r) = \log\left(1 + \frac{[\arctan(r)]^2}{\delta}\right) \qquad \forall r \in \mathbb{R}, \tag{5.37}$$

which satisfies (5.9) and in particular, thanks to (5.29) applied with $M = 1, r_1 = r, r_2 = 0$

$$|r\beta_\delta'(r)| = \left|\frac{\arctan(r)}{\delta + [\arctan(r)]^2} r \arctan'(r)\right| \leq 1 \qquad \forall r \in \mathbb{R}. \tag{5.38}$$

For every $R > 0$ consider

$$\varphi_R(x) = \begin{cases} \dfrac{1}{2^{d+1}} & x \in \mathbb{R}^d,\ |x| < R \\ \dfrac{R^{d+1}}{(R + |x|)^{d+1}} & x \in \mathbb{R}^d,\ |x| > R. \end{cases} \tag{5.39}$$

We use β_δ to renormalize the solution u and φ_R as a test function. Notice that $\varphi_R \in L^1 \cap W^{1,\infty}(\mathbb{R}^d)$ with $0 \leq \varphi_R \leq 1$ and by Lemma 5.10 the function φ_R is an admissible test function in (5.34). Hence there exists an absolutely continuous function $\Gamma_{\delta,R} : [0, T] \to \mathbb{R}$ such that $\Gamma_{\delta,R}(0) = 0$ and for $\mathscr{L}^1$-a.e. $t \in [0, T]$

$$\Gamma_{\delta,R}(t) = \int_{\mathbb{R}^d} \varphi_R(x)\beta_\delta(u_t(x))\,dx,$$

$$\begin{aligned} \frac{d}{dt}\Gamma_{\delta,R}(t) = &\int_{\mathbb{R}^d} \nabla\varphi_R \cdot \boldsymbol{b}_t\beta_\delta(u_t)\,dx \\ &+ \int_{\mathbb{R}^d} \varphi_R(c_t - \operatorname{div} \boldsymbol{b}_t)u_t\beta_\delta'(u_t)\,dx \\ &+ \int_{\mathbb{R}^d} \varphi_R \operatorname{div} \boldsymbol{b}_t\beta_\delta(u_t)\,dx \end{aligned} \tag{5.40}$$

(here and in the following we omit the dependence of $\boldsymbol{b}$, c, u on (t, x) and of φ_R on x). We estimate each term in the right-hand side of (5.40).

The third term can be estimated thanks to the condition on the divergence of $\boldsymbol{b}$

$$\int_{\mathbb{R}^d} \varphi_R \operatorname{div} \boldsymbol{b}_t \beta_\delta(u_t)\, dx \leq \| \operatorname{div} \boldsymbol{b}_t \|_{L^\infty(\mathbb{R}^d)} \int_{\mathbb{R}^d} \varphi_R \beta_\delta(u_t)\, dx. \tag{5.41}$$

As regards the second term, we use (5.38) to deduce

$$\begin{aligned}\int_{\mathbb{R}^d} \varphi_R(c_t - \operatorname{div} \boldsymbol{b}_t) u_t \beta'_\delta(u_t)\, dx &\leq \int_{\mathbb{R}^d} \varphi_R(|c_t| + |\operatorname{div} \boldsymbol{b}_t|)\, dx \\ &\leq \int_{\mathbb{R}^d} |c_t|\, dx + \| \operatorname{div} \boldsymbol{b}_t \|_{L^\infty(\mathbb{R}^d)} \int_{\mathbb{R}^d} \varphi_R\, dx.\end{aligned} \tag{5.42}$$

To estimate the first term, we take into account the growth condition (5.13) on $\boldsymbol{b}$. Let $\boldsymbol{b}_1$ and $\boldsymbol{b}_2$ two nonnegative functions such that

$$\frac{|\boldsymbol{b}_t(x)|}{1+|x|} \leq \boldsymbol{b}_{1t}(x) + \boldsymbol{b}_{2t}, \qquad \boldsymbol{b}_1 \in L^1((0,T)\times\mathbb{R}^d), \qquad \boldsymbol{b}_2 \in L^1((0,T)).$$

Notice that $\nabla\varphi_R(x)$ can be explicitly computed; for every $x \in \mathbb{R}^d$ with $|x| < R$ it is 0 and if $|x| > R$ we have that $|\nabla\varphi_R(x)| \leq (d+1)\varphi_R(x)(R+|x|)^{-1}$. If $R > 1$, we have

$$\begin{aligned}&\int_{\mathbb{R}^d} \nabla\varphi_R \cdot \boldsymbol{b}_t \beta_\delta(u_t)\, dx \\ &\leq (d+1) \int_{\mathbb{R}^d\setminus B_R} \frac{\varphi_R}{R+|x|}(1+|x|)(\boldsymbol{b}_{1t} + \boldsymbol{b}_{2t})\beta_\delta(u_t)\, dx \\ &\leq (d+1) \int_{\mathbb{R}^d\setminus B_R} \varphi_R(\boldsymbol{b}_{1t} + \boldsymbol{b}_{2t})\beta_\delta(u_t)\, dx \\ &\leq (d+1) \log\left(1 + \frac{\pi^2}{4\delta}\right) \int_{\mathbb{R}^d\setminus B_R} \boldsymbol{b}_{1t}\, dx \\ &+ (d+1)\boldsymbol{b}_{2t} \int_{\mathbb{R}^d} \varphi_R \beta_\delta(u_t)\, dx.\end{aligned} \tag{5.43}$$

Setting for every $t \in [0,T]$ the L^1 functions:

$$a(t) = \| \operatorname{div} \boldsymbol{b}_t \|_{L^\infty(\mathbb{R}^d)} + (d+1)\boldsymbol{b}_{2t},$$

$$b_R(t) = \|c_t\|_{L^1(\mathbb{R}^d)} + \| \operatorname{div} \boldsymbol{b}_t \|_{L^\infty(\mathbb{R}^d)} \|\varphi_R\|_{L^1(\mathbb{R}^d)},$$

$$c_R(t) = (d+1)\|\boldsymbol{b}_{1t}\|_{L^1(\mathbb{R}^d\setminus B_R)},$$

from (5.40), (5.41), (5.42), and (5.43) we deduce that for $\mathscr{L}^1$-a.e. $t \in [0, T]$

$$\frac{d}{dt}\Gamma_{\delta,R}(t) \leq a(t)\Gamma_{\delta,R}(t) + b_R(t) + c_R(t)\log\left(1 + \frac{\pi^2}{4\delta}\right).$$

Since $\Gamma_{\delta,R}(0) = 0$, by Gronwall lemma we obtain that for every $t \in [0, T]$

$$\begin{aligned}\Gamma_{\delta,R}(t) &\leq \exp\left(\int_0^T a(s)ds\right)\left(\int_0^T b_R(s)ds + \log\left(1+\frac{\pi^2}{4\delta}\right)\int_0^T c_R(s)ds\right)\\ &= \exp(A)\left(B_R + \log\left(1 + \frac{\pi^2}{4\delta}\right)C_R\right).\end{aligned} \tag{5.44}$$

Notice that by definition

$$\lim_{R\to\infty} C_R = (d+1)\lim_{R\to\infty}\int_0^T\int_{\mathbb{R}^d\setminus B_R} \boldsymbol{b}_{1s}(x)\,dx\,ds = 0. \tag{5.45}$$

We conclude finding a contradiction as in [67, 36]. Let us assume that u_t is not identically 0 for some $t \in [0, T]$; then $\arctan u_t$ is not identically 0 and there exists $R_0 > 0$ and $\gamma > 0$ such that $\mathscr{L}^d(\{x \in B_{R_0} : [\arctan u_t(x)]^2 > \gamma\}) > 0$. Dividing (5.44) by $\log(1 + \gamma/\delta)$ we obtain that for every $R \geq R_0$

$$\begin{aligned}0 &< \frac{\mathscr{L}^d(\{x \in B_{R_0} : [\arctan u_t(x)]^2 > \gamma\})}{2^{d+1}} \leq \left(\log\left(1 + \frac{\gamma}{\delta}\right)\right)^{-1}\Gamma_{\delta,R}(t)\\ &\leq \exp(A)\left(\log\left(1+\frac{\gamma}{\delta}\right)\right)^{-1}\left(B_R + \log\left(1 + \frac{\pi^2}{4\delta}\right)C_R\right).\end{aligned}$$

Letting δ go to 0 we find

$$0 < \frac{\mathscr{L}^d(\{x \in B_{R_0} : [\arctan u_t(x)]^2 > \gamma\})}{2^{d+1}} \leq \exp(A)C_R,$$

which is a contradiction thanks to (5.45) provided that R is chosen big enough. □

Chapter 6
Regularity results for very degenerate elliptic equations

In the following informal discussion, we describe the connections between some traffic models, involving in their formulation different tools such as the Lagrangian point of view, the variational minimization and some degenerate elliptic equations. We refer to the lecture notes of Santambrogio [113] for a wider presentation of the topic.

A Lagrangian problem. Let $\Omega \subseteq \mathbb{R}^d$ be an open domain; in the application, it may represent an urban area. Let μ, ν be two probability measures on Ω which may describe the initial and final distribution of workers, commuting from their houses to the work offices. In a continuous setting, we describe the transport pattern with a probability measure $\boldsymbol{\eta}$ on the set of absolutely continuous paths $AC([0,1];\Omega)$, where each path represents the choice of a traveler. We associate to $\boldsymbol{\eta}$ the traffic intensity $i_\eta \in \mathscr{M}_+(\Omega)$ defined by duality through

$$\int_\Omega \varphi(x)\, di_\eta(x) = \int \int_0^1 \varphi(\eta(t))|\eta'(t)|\, dt\, d\boldsymbol{\eta}(\eta)$$

for every $\varphi \in C(\Omega)$. Intuitively, in the smooth setting $i_\eta(x)$ represents at any point $x \in \Omega$ the total traffic that flows through x (in any direction). In order to prescribe the initial and final distributions $\mu, \nu \in \mathscr{P}(\Omega)$, we consider a given convex closed subset Γ of the set of transport plans between μ and ν

$$\Pi(\mu,\nu) = \{\gamma \in \mathscr{P}(\Omega \times \Omega) : (\pi_1)_\sharp\gamma = \mu,\ (\pi_2)_\sharp\gamma = \nu\}.$$

For a measure $\boldsymbol{\eta}$ to be admissible, we require $(e_0, e_1)_\#\boldsymbol{\eta} \in \Gamma$; the two most natural choices for Γ are either $\Gamma = \{\gamma_0\}$ for some $\gamma_0 \in \Pi(\mu,\nu)$ (corresponding to the case when each traveler chooses his initial position and final destination), or $\Gamma = \Pi(\mu,\nu)$ (this second condition is natural in long-term city planning). In the following discussion, we make always the second choice.

In order to describe the congestion effects, we consider a given non-decreasing function $g(i) : \mathbb{R}^+ \to \mathbb{R}^+$. The case $g \equiv 1$ would correspond to the case where we don't consider congestion and it leads to a formulation of the classical Kantorovich problem with the cost function $c(x, y) = |x - y|$; instead, here g is chosen to have $g(0) > 0$ (so that, as we will see below, empty streets have nonzero cost) and to be unbounded (in order to penalize congestion); our model function is $g(i) = 1 + pi^{p-1}$ for some $p > 1$. One may allow g to depend on the point x as well and require the monotonicity only in i variable, but for simplicity we avoid this analysis. Given a transport pattern $\boldsymbol{\eta}$, we associate to every curve η its weighted length (or traveling time) as

$$L_{\boldsymbol{\eta}}(\eta) = \int_0^1 g(i_{\boldsymbol{\eta}}(\eta(t)))|\eta'(t)|\, dt$$

and we define the distance

$$d_{\boldsymbol{\eta}}(x, y) = \inf\{L_{\boldsymbol{\eta}}(\eta) : \eta \in AC([0, 1]; \Omega),\ \eta(0) = x, \eta(1) = y\}.$$

We call *geodesics* the curves that minimize this distance between given points, namely such that $d_{\boldsymbol{\eta}}(\eta(0), \eta(1)) = L_{\boldsymbol{\eta}}(\eta)$.

A plan $\boldsymbol{\eta}$ satisfies a *Wardrop equilibrium condition* if no traveler wants to change his path, provided that all other travelers keep the same strategy. In other words, a Wardrop equilibrium $\boldsymbol{\eta}$ satisfies the property to be concentrated on geodesics in the metric induced by $\boldsymbol{\eta}$ itself

$$\boldsymbol{\eta}\left(\left\{\eta \in AC([0, 1]; \Omega) : L_{\boldsymbol{\eta}}(\eta) = d_{\boldsymbol{\eta}}(\eta(0), \eta(1))\right\}\right) = 0.$$

Under some technical assumptions, Carlier, Jimenez and Santambrogio [49] show that Wardrop equilibria can be found as minimizers of the variational problem

$$\min\left\{\int_\Omega G(i_{\boldsymbol{\eta}}(x))\, dx : (e_0, e_1)_\# \boldsymbol{\eta} \in \Pi(\mu, \nu)\right\}, \tag{6.1}$$

where G is the primitive of g, namely $G(0) = 0$ and $G' = g$.

Beckmann's minimal flow problem and its dual formulation. The problem (6.1), in turn, can be also reformulated in terms of a minimization problem over measurable vector field $\boldsymbol{w} : \Omega \to \mathbb{R}^d$. In other words, we consider the *Beckmann's minimal flow problem*

$$\min\left\{\int_\Omega \mathcal{G}(\boldsymbol{w}(x))\, dx : \operatorname{div} \boldsymbol{w} = \mu - \nu,\ \boldsymbol{w} \cdot \nu_{\partial\Omega} = 0\right\}, \tag{6.2}$$

where $\mathcal{G}(x) = H(|x|)$ for every $x \in \mathbb{R}^d$. It is clear that the infimum in (6.1) is less or equal than the infimum in (6.2). Indeed, given an admissible $\boldsymbol{\eta}$ for (6.1), we can associate a natural flow $\boldsymbol{w}_\eta$ defined by duality

$$\int_\Omega \varphi(x) \cdot d\boldsymbol{w}_\eta(x) = \int \int_0^1 \varphi(\eta(t)) \cdot \eta'(t)\, dt\, d\boldsymbol{\eta}(\eta)$$

for every $\varphi \in C(\Omega; \mathbb{R}^d)$. It can be easily checked that, with this definition, $\boldsymbol{w}_\eta$ is admissible in (6.2) and, since $|\boldsymbol{w}_\eta| \leq i_\eta$, one sees that

$$\int_\Omega \mathcal{G}(\boldsymbol{w}_\eta(x))\, dx \leq \int_\Omega G(i_\eta(x))\, dx.$$

Actually, the equality between the two problems in (6.1) and (6.2) holds, but this requires more work to be seen.

In turn, problem (6.2) can be rewritten, formally, by means of some convex analysis tools, allowing to exchange max and min

$$\begin{aligned}
&\min_{\boldsymbol{w}} \left\{ \int_\Omega \mathcal{G}(\boldsymbol{w})\, dx : \operatorname{div} \boldsymbol{w} = \rho_0 - \rho_1,\ \ \boldsymbol{w} \cdot \nu_{\partial\Omega} = 0 \right\} \\
&= \min_{\boldsymbol{w}} \left\{ \int_\Omega \mathcal{G}(\boldsymbol{w})\, dx - \min_u \left\{ \int_\Omega \big[(\rho_0 - \rho_1)u + \boldsymbol{w} \cdot \nabla u\big]\, dx \right\} \right\} \\
&= - \max_{\boldsymbol{w}} \min_u \left\{ \int_\Omega \big[- \mathcal{G}(\boldsymbol{w}) + (\rho_0 - \rho_1)u + \boldsymbol{w} \cdot \nabla u \big]\, dx \right\} \qquad (6.3) \\
&= - \min_u \max_{\boldsymbol{w}} \left\{ \int_\Omega \big[- \mathcal{G}(\boldsymbol{w}) + \boldsymbol{w} \cdot \nabla u + (\rho_0 - \rho_1)u \big]\, dx \right\} \\
&= - \min_u \left\{ \int_\Omega \big[\mathcal{G}^*(\nabla u) + (\rho_0 - \rho_1)u\big]\, dx \right\},
\end{aligned}$$

where $\mathcal{G}^*$ denotes the convex conjugate of $\mathcal{G}$ and ρ_0 and ρ_1 denote the (smooth) densities of μ and ν with respect to the Lebesgue measure. We notice that the minimizer $\boldsymbol{w}$ in (6.2) can be obtained from the minimizer u in (6.3) through the formula $\boldsymbol{w} = \nabla\mathcal{G}(\nabla u)$ and that u solves the very degenerate elliptic equation

$$\operatorname{div}\big(\nabla\mathcal{G}(\nabla u)\big) = \rho_0 - \rho_1$$

with Neumann boundary conditions.

Sobolev regularity of $\boldsymbol{w}$ has been proven in [38] and it allows to associate to $\boldsymbol{w}$ a regular lagrangian flow. In this chapter, we study the continuity properties of $\boldsymbol{w}$, which are, in turn, crucial to rigorously justify the previous formal discussion (for instance, the equivalence between problem (6.1) and (6.2)) and to formulate the geodesic problem presented above in a relatively nice Riemannian setting.

6.1. Degenerate elliptic equations

Given a bounded open subset Ω of $\mathbb{R}^d$, a convex function $\mathcal{F} : \mathbb{R}^d \to \mathbb{R}$, and an integrable function $f : \Omega \to \mathbb{R}$, we consider a function $u : \Omega \to \mathbb{R}$ which locally minimizes the functional

$$\int_\Omega \mathcal{F}(\nabla u) + fu. \tag{6.4}$$

When $\nabla^2 \mathcal{F}$ is uniformly elliptic, namely there exist $\lambda, \Lambda > 0$ such that

$$\lambda \, \mathrm{Id} \leq \nabla^2 \mathcal{F} \leq \Lambda \, \mathrm{Id},$$

the regularity results of u in terms of $\mathcal{F}$ and f rely on De Giorgi theorem and Schauder estimates (see Theorems 1.19 and 1.20). If $\mathcal{F}$ degenerates at only one point, then several results are still available. For instance, in the case of the p-Laplace equation, the $C^{1,\alpha}$ regularity of u has been stated in Theorem 1.22.

More in general, one can consider functions whose degeneracy set is a convex set: for example, for $p > 1$ one may consider

$$\mathcal{F}(v) = \frac{1}{p}(|v| - 1)_+^p \qquad \forall\, v \in \mathbb{R}^d, \tag{6.5}$$

so that the degeneracy set is the entire unit ball. There are many Lipschitz results on u in this context [87, 84, 37]: they are based on the observation that the equation solved by each partial derivative $\partial_e u$ is elliptic where the gradient is large and this allows to build suitable subsolutions of an elliptic equation starting from $\partial_e u$; in turn these subsolutions are bounded by standard elliptic theory (see Theorem 1.21). Instead, in general no more regularity than L^∞ can be expected on ∇u. Indeed, when $\mathcal{F}$ is given by (6.5) and f is identically 0, every 1-Lipschitz function solves the equation. However, as proved in [114] in dimension 2, something more can be said about the regularity of $\nabla \mathcal{F}(\nabla u)$, since either it vanishes or we are in the region where the equation is more elliptic.

In this chapter we prove that, if $\mathcal{F}$ vanishes on some convex set E and is elliptic outside such a set, and if u is a local minimizer of (6.4)[1] then $\mathcal{H}(\nabla u)$ is continuous for any continuous function $\mathcal{H} : \mathbb{R}^d \to \mathbb{R}$

[1] Recall that a function $u \in W^{1,1}_{\mathrm{loc}}(\Omega)$ is said a local minimizer of a function of the form (6.4) (with $f \in L^n_{\mathrm{loc}}(\Omega)$) if, for every $\Omega' \Subset \Omega$, we have

$$\int_{\Omega'} \mathcal{F}(\nabla u + \nabla \phi) + f(u + \phi) \geq \int_{\Omega'} \mathcal{F}(\nabla u) + fu \qquad \forall\, \phi \in W^{1,1}_0(\Omega').$$

which vanishes on E. In particular, by applying this result with $\mathcal{H} = \partial_i \mathcal{F}$ ($i = 1, \dots, d$) where $\mathcal{F}$ is as in (6.5), our continuity result implies that $\nabla\mathcal{F}(\nabla u)$ (the minimizer of (6.2)) is continuous in the interior of Ω.

Since we want to allow any bounded convex set as degeneracy set for $\mathcal{F}$, before stating the result we introduce the notion of norm associated to a convex set, which is used throughout the chapter to identify the nondegenerate region. Given a bounded closed convex set $E \subseteq \mathbb{R}^d$ such that 0 belongs to $\mathrm{Int}(E)$ (the interior of E), and denoting by tE the dilation of E by a factor t with respect to the origin, we define $|\cdot|_E$ as

$$|e|_E := \inf\{t > 0 : e \in tE\}. \tag{6.6}$$

Notice that $|\cdot|_E$ is a convex positively 1-homogeneous function. However $|\cdot|_E$ is not symmetric unless E is symmetric with respect to the origin.

The main result of this chapter proves that, in the context introduced before, $\nabla\mathcal{F}(\nabla u)$ is continuous.

Theorem 6.1. *Let d be a positive integer, $0 < \lambda \le \Lambda$, Ω a bounded open subset of $\mathbb{R}^d$, $f \in L^q(\Omega)$ for some $q > d$. Let E be a bounded, convex set with $0 \in \mathrm{Int}(E)$. Let $\mathcal{F} : \mathbb{R}^d \to \mathbb{R}$ be a convex nonnegative function such that $\mathcal{F} \in C^2(\mathbb{R}^d \setminus \overline{E})$. Let us assume that for every $\delta > 0$ there exist $\lambda_\delta, \Lambda_\delta > 0$ such that*

$$\lambda_\delta I \le \nabla^2\mathcal{F}(x) \le \Lambda_\delta I \quad \text{for a.e. } x \text{ such that } 1 + \delta \le |x|_E \le 1/\delta. \tag{6.7}$$

Let $u \in W^{1,\infty}_{\mathrm{loc}}(\Omega)$ be a local minimizer of the functional

$$\int_\Omega \mathcal{F}(\nabla u) + fu.$$

Then, for any continuous function $\mathcal{H} : \mathbb{R}^d \to \mathbb{R}$ such that $\mathcal{H} = 0$ on E, we have

$$\mathcal{H}(\nabla u) \in C^0(\Omega). \tag{6.8}$$

More precisely, for every open set $\Omega' \Subset \Omega$ there exists a modulus of continuity $\omega : [0, \infty) \to [0, \infty)$ for $\mathcal{H}(\nabla u)$ on Ω', which depends only on the modulus of continuity of $\mathcal{H}$, on the modulus of continuity of $\nabla^2\mathcal{F}$, on the functions $\delta \to \lambda_\delta, \delta \to \Lambda_\delta$, and on $\|\nabla u\|_\infty$ in a neighborhood of Ω', such that

$$\omega(0) = 0 \text{ and } \big|\mathcal{H}(\nabla u(x)) - \mathcal{H}(\nabla u(y))\big| \le \omega(|x - y|) \text{ for any } x, y \in \Omega'. \tag{6.9}$$

In particular, if $\mathcal{F} \in C^1(\mathbb{R}^d)$ then $\nabla\mathcal{F}(\nabla u) \in C^0(\Omega)$.

Remark 6.2. In the hypothesis of Theorem 6.1 the Lipschitz regularity of u is always satisfied under mild assumptions on $\mathcal{F}$. For instance, if $\mathcal{F}$ is uniformly elliptic outside a fixed ball, then $u \in W^{1,\infty}_{\text{loc}}(\Omega)$. In [37] many other cases are studied. For example, the Lipschitz regularity of u holds true for our model case $(|x|-1)^p_+$ for every $p > 1$.

Remark 6.3. The regularity result of Theorem 6.1 is optimal without any further conditions about the degeneracy of $\mathcal{F}$ near E. More precisely, there exist functions $\mathcal{F}$ satisfying our assumptions and $\mathcal{H}$ Lipschitz such that $\mathcal{H}(\nabla u)$ is not Hölder continuous for any exponent. Indeed, let us consider the minimizer of the functional (6.4) with $f = d$. The minimizer can be explicitly computed from the Euler equation and turns out to be $\mathcal{F}^*$, where $\mathcal{F}^*$ is the convex conjugate of $\mathcal{F}$. We consider a radial function $\mathcal{F}$. Let ω be a modulus of strict convexity for $\mathcal{F}$ outside E, i.e.,

$$\begin{aligned}\big(\nabla\mathcal{F}(x)-\nabla\mathcal{F}(y)\big)\cdot(x-y)\geq\omega(|x-y|)|x-y|\\ \forall x,y\in\mathbb{R}^d\setminus B_1,\ x=ty,\ t>0.\end{aligned}\tag{6.10}$$

Then the function ω^{-1} is a modulus of continuity of $\nabla\mathcal{F}^*$. Hence it suffices to choose $\mathcal{F}$ so that ω^{-1} is not Hölder continuous.

For simplicity, we construct an explicit example in dimension 1, although it can be easily generalized to any dimension considering a radial function $\mathcal{F}$. Let

$$G(t) := \begin{cases} e^{-1/(|t|-1)^2} & \text{if } |t| > 1, \\ 0 & \text{if } |t| \leq 1, \end{cases}$$

and let $F \in C^\infty(\mathbb{R})$ be a convex function which coincides with G in a $(-1-\varepsilon, 1+\varepsilon)$ for some $\varepsilon > 0$ (see Figure 6.1). Then the function $u : \mathbb{R} \to \mathbb{R}$ defined as

$$u(x) := \int_0^{|x|} [F']^{-1}(s)\, ds$$

solves the Euler-Lagrange equation $\big(F'(u'(x))\big)' = 1$ (note that the function $F' : \mathbb{R}\setminus[-1,1] \to \mathbb{R}\setminus\{0\}$ is invertible, so u is well defined), and it is easy to check that, given $\mathcal{H}(x) := (|x|-1)_+$, the function $\mathcal{H}(u') = \big([F']^{-1}-1\big)_+$ is not Hölder continuous at 0.

Theorem 6.1 has been proved in dimension 2 with $E = B_1(0)$ by Santambrogio and Vespri in [114]. Their proof is based on a method by Di Benedetto and Vespri [77], which is very specific to the two dimensional case: using the equation they prove that either the oscillation of the solution is reduced by a constant factor when passing from a ball $B_r(0)$ to a

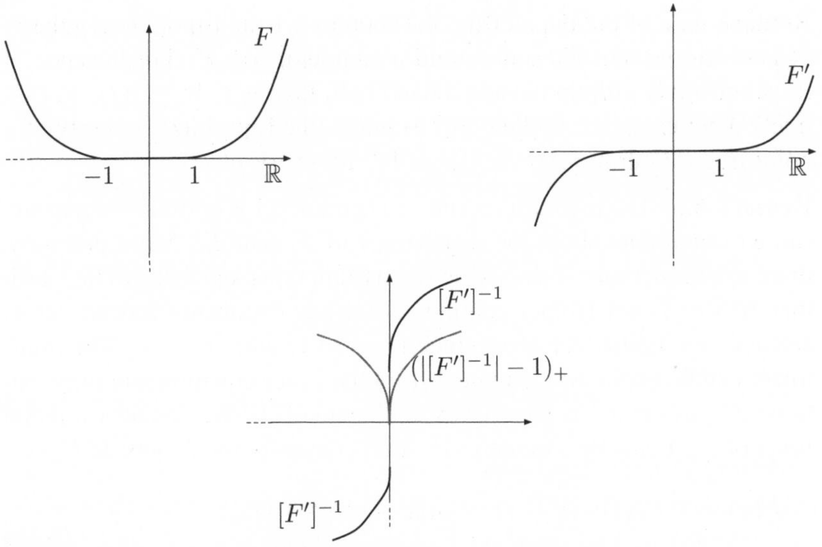

Figure 6.1. The figure shows the functions F, F', $[F']^{-1}$, and $(u' - 1)_+ = \left([F']^{-1} - 1\right)_+$ of the 1-dimensional counterexample of Remark 6.3; since, for every $\alpha \in (0, 1)$, the function F' is smaller than $|x - 1|^{1/\alpha}$ in a neighborhood of 1, the inverse F' is not α-Hölder continuous in a right-neighborhood of 0.

smaller ball $B_{\varepsilon r}(0)$, or the Dirichlet energy in the annulus $B_r(0) \setminus B_{\varepsilon r}(0)$ is at least a certain value, which is scale invariant in dimension 2. Since the Dirichlet energy is assumed to be finite in the whole domain, this proves a decay for the oscillation.

In this chapter we present a generalization of the result to dimension d and with a general convex set of degeneracy, using a different method and following some ideas of a paper by Wang [122] in the case of the p-laplacian. We divide regions where the gradient is degenerate from non-degeneracy regions. The rough idea is the following: if no partial derivative of u is close to $|\nabla u|$ in a set of positive measure inside a ball, then $|\nabla u|$ is smaller (by a universal factor) in a smaller ball. If u has a nondegenerate partial derivative in a set of large measure, then its slope in the center of the ball is nondegenerate and the ellipticity of the equation provides regularity of u, through an improvement of flatness lemma, which requires in turn a compactness result for degenerate equations presented in Section 6.2. An alternative approach of variational nature to handle the case when u has a nondegenerate partial derivative is described in Chapter 7 (see Corollary 7.3 and Theorem 7.6, which generalizes Theorem 6.1 by weakening the regularity assumptions on the integrand); this time, the proof is based on an excess decay result at nondegenerate points.

Theorem 6.1 is obtained from the following result through an approximation argument, which allows us to deal with smooth functions.

Theorem 6.4. *Let E be a bounded, strictly convex set with* $0 \in \mathrm{Int}(E)$. *Let $f \in C^0(B_2(0))$ and let $q > d$. Let $\mathcal{F} \in C^\infty(\mathbb{R}^d)$ be a convex function, fix $\delta > 0$, and assume that there exist constants $\lambda, \Lambda > 0$ such that*

$$\lambda I \le \nabla^2 \mathcal{F}(x) \le \Lambda I \qquad \text{for every } x \text{ such that } 1 + \frac{\delta}{2} \le |x|_E. \quad (6.11)$$

Let $u \in C^2(B_2(0))$ be a solution of

$$\nabla \cdot (\nabla \mathcal{F}(\nabla u)) = f \qquad \text{in } B_2(0). \quad (6.12)$$

satisfying $\|\nabla u\|_{L^\infty(B_2(0))} \le M$.

Then there exist $C > 0$ and $\alpha \in (0,1)$, depending only on the modulus of continuity of $\nabla^2 \mathcal{F}$, and on $E, \delta, M, q, \|f\|_{L^q(B_2(0))}, \lambda$, and Λ, such that

$$\|(|\nabla u|_E - (1+\delta))_+\|_{C^{0,\alpha}(B_1(0))} \le C. \quad (6.13)$$

The chapter is structured as follows: in Section 6.2 we prove a compactness result for a class of elliptic equations which are nondegenerate only in a small neighborhood of the origin. Then, in Section 6.3, we provide a way of separating degeneracy points from nondegeneracy points, and in Section 6.4 we prove $C^{1,\alpha}$ regularity of u at any point where the equation is nondegenerate. Finally, Section 6.5 is devoted to the proof of Theorems 6.4 and 6.1.

6.2. Compactness result for a degenerate equation

In this section we prove a regularity result for a class of degenerate fully nonlinear elliptic equations. The argument follows the lines of [115, Corollary 3.3], although there are some main differences: First, in [115, Corollary 3.3] regularity is proved in the class of fully nonlinear equations with a degeneracy depending on the hessian of the solution, whereas in our case the degeneracy is in the gradient. Moreover only right hand sides in L^∞ are considered there, while in our context we are allowed to take them in L^d. Allowing f to be in L^d introduce several additional difficulties, in particular in the proof of Lemma 6.8. In addition, we would like to notice that the proofs of Lemmas 6.7 and 6.8 do not seem to easily adapt to the case $f \in L^d$ if in addition we allow a degeneracy in the hessian as in [115] (more precisely, in this latter case neither (6.22) nor (6.32) would allow to deduce that the equation is uniformly elliptic at the contact points).

We also notice that, with respect to [115], we prove a slightly weaker statement which is however enough for our purposes: instead of showing the L^∞ norm of u decays geometrically, we only prove that its oscillation decays. The reason for this is just that the proof of this latter result is slightly simpler. However, by using the whole argument in the proof of [115, Theorem 1.1] one could replace $\operatorname{osc} u$ with $\|u\|_\infty$ in the statements of Proposition 6.6 and Theorem 6.5.

We keep the notation as similar as possible to the one of [115]. We assume for simplicity that $u \in C^2$ and f continuous, but these regularity assumptions are not needed (though verified for our application) and the same proof could be carried out in the context of viscosity solutions (as done in [115]).

Let $\mathcal{S} \subseteq \mathbb{R}^{d\times d}$ be the space of symmetric matrices in $\mathbb{R}^d$, $F : B_1(0) \times \mathbb{R} \times \mathbb{R}^d \times \mathcal{S} \to \mathbb{R}$ be a measurable function, and consider the fully nonlinear equation

$$F(x, u(x), \nabla u(x), \nabla^2 u(x)) = f(x). \tag{6.14}$$

Let $\delta > 0$. We consider the following assumptions on F.

(H1) F is elliptic, namely for every $x \in B_1(0), z \in \mathbb{R}, v \in \mathbb{R}^d, M, N \in \mathcal{S}$ with $N \geq 0$

$$F(x, z, v, M + N) \geq F(x, z, v, M).$$

(H2) F is uniformly elliptic in a neighborhood of $\nabla u = 0$ with ellipticity constants $0 < \lambda \leq \Lambda$: namely, for every $x \in B_1(0), z \in \mathbb{R}$, $v \in B_\delta(0), M, N \in \mathcal{S}$ with $N \geq 0$

$$\Lambda\|N\| \geq F(x, z, v, M + N) - F(x, z, v, M) \geq \lambda\|N\|.$$

(H3) Small planes are solutions of (6.14), namely for every $x \in B_1(0)$, $z \in \mathbb{R}, v \in B_\delta(0)$,

$$F(x, z, v, 0) = 0.$$

Given $M \in \mathcal{S}$, let M^+ and M^- denote its positive and negative part, respectively, so that $M = M^+ - M^-$ and $M^+, M^- \geq 0$. Applying (H2) twice and using (H3), we have

$$\Lambda\|M^+\| - \lambda\|M^-\| \geq F(x, z, p, M) \geq \lambda\|M^+\| - \Lambda\|M^-\| \tag{6.15}$$

for every $x \in B_1(0), z \in \mathbb{R}, v \in B_\delta(0), M \in \mathcal{S}$.

In this section we will call *universal* any positive constant which depends only on d, λ, Λ.

Theorem 6.5. *Let $\delta > 0$, $F : B_1(0) \times \mathbb{R} \times \mathbb{R}^d \times \mathcal{S} \to \mathbb{R}$ a measurable function which satisfies* (H1), (H2), *and* (H3), $f \in C^0(B_1(0))$, *and assume that $u \in C^2(B_1(0))$ solves* (6.14). *Then there exist universal constants $\nu, \varepsilon, \kappa, \rho \in (0,1)$ such that if $\delta' > 0$ and $k \in \mathbb{N}$ satisfy*

$$\operatorname*{osc}_{B_1(0)} u \le \delta' \le \rho^{-k}\kappa\delta, \qquad \|f\|_{L^d(B_1(0))} \le \varepsilon\delta', \tag{6.16}$$

then

$$\operatorname*{osc}_{B_{\rho^s}(0)} u \le (1-\nu)^s\delta' \qquad \forall s = 0, ..., k+1. \tag{6.17}$$

As we will show at the end of this section, Theorem 6.5 follows by an analogous result at scale 1 (stated in the following proposition) and a scaling argument.

Proposition 6.6. *Let $\delta > 0$, $F : B_1(0) \times \mathbb{R} \times \mathbb{R}^d \times \mathcal{S} \to \mathbb{R}$ a measurable function which satisfies* (H1), (H2), *and* (H3), $f \in C^0(B_1(0))$, *and assume that $u \in C^2(B_1(0))$ solves* (6.14).

Then there exist universal constants $\nu, \varepsilon, \kappa, \rho \in (0,1)$ such that if δ' satisfies

$$\operatorname*{osc}_{B_1(0)} u \le \delta' \le \kappa\delta, \qquad \|f\|_{L^d(B_1(0))} \le \varepsilon\delta', \tag{6.18}$$

then

$$\operatorname*{osc}_{B_\rho(0)} u \le (1-\nu)\delta'.$$

Before proving this result, we state and prove three basic lemmas. The first lemma gives an estimate on the contact set of a family of paraboloids with fixed opening in terms of the measure of the set of vertices. The proof is a simple variant of the one of [115, Lemma 2.1].

Lemma 6.7. *Let $\delta > 0$, F, λ, Λ, f, and u be as in Proposition 6.6. Fix $a \in (0, \delta/2)$, let $K \subseteq B_1(0)$ be a compact set, and define $A \subseteq \overline{B_1(0)}$ to be the set of contact point of paraboloids with vertices in K and opening $-a$, namely the set of points $x \in \overline{B_1(0)}$ such that there exists $y \in K$ which satisfies*

$$\inf_{z \in B_1(0)} \left\{ \frac{a}{2}|y-z|^2 + u(z) \right\} = \frac{a}{2}|y-x|^2 + u(x). \tag{6.19}$$

Assume that $A \subset B_1(0)$.

Then there exists a universal constant $c_0 > 0$, such that

$$c_0|K| \le |A| + \int_A \frac{|f(x)|^d}{a^d}\,dx. \tag{6.20}$$

Proof. Since by assumption $A \subset B_1(0)$, for every $x \in A$, given $y \in K$ which satisfies (6.19), we have that

$$\nabla u(x) = -a(x-y). \tag{6.21}$$

Let $T : A \to K$ be the map which associates to every contact point x the vertex of the paraboloid, namely

$$T(x) := \frac{\nabla u(x)}{a} + x.$$

Notice that $T \in C^1(\overline{A})$ and $K = T(A)$. From (6.21) we have that, at each contact point $x \in A$,

$$|\nabla u(x)| = a|x-y| \leq 2a \leq \delta,$$

hence from (H2) the equation is uniformly elliptic at x. Moreover we have that $-a\,\mathrm{Id} \leq \nabla^2 u(x)$, so it follows by (6.15) that

$$-a\,\mathrm{Id} \leq \nabla^2 u(x) \leq \frac{\Lambda a + |f(x)|}{\lambda}\,\mathrm{Id} \qquad \forall x \in A. \tag{6.22}$$

In addition, from the change of variable formula we have that

$$|K| = |T(A)| \leq \int_A \det \nabla T(x)\,dx = \int_A \det\left(\frac{\nabla^2 u(x)}{a} + \mathrm{Id}\right) dx \tag{6.23}$$

Since each eigenvalue of the matrix $\nabla u(x)/a + \mathrm{Id}$ lies in the interval $[0, (1+\Lambda/\lambda) + |f(x)|/(\lambda a)]$ (see (6.22)), we get

$$\det\left(\frac{\nabla^2 u(x)}{a} + I\right) \leq C_0\left[1 + \frac{|f(x)|^d}{a^d}\right]$$

for some universal constant C_0. Hence, it follows from (6.23) that

$$|K| \leq C_0|A| + C_0 \int_A \frac{|f(x)|^d}{a^d}\,dx,$$

which proves (6.20) with $c_0 = 1/C_0$. □

Before stating the next lemma we introduce some notation.

Given u as before, for every $b > 0$ we define A_b be the set of $x \in B_1(0)$ such that $u(x) \leq b$ and the function u can be touched from below at x with a paraboloid of opening $-b$, namely there exists $y \in \overline{B_1(0)}$ such that

$$\inf_{z \in B_1(0)} \left\{\frac{b}{2}|y-z|^2 + u(z)\right\} = \frac{b}{2}|y-x|^2 + u(x). \tag{6.24}$$

In addition, given $g \in L^1(B_1(0))$, we denote by $M[g]$ the maximal function associated to g, namely

$$M[g](x) := \sup \left\{ \fint_{B_r(z)} g(y)\, dy : B_r(z) \subseteq B_1(0), x \in B_r(z) \right\}.$$

Maximal functions enjoy weak-L^1 estimates (see for instance [117]): there exists a constant C_d depending only on the dimension such that

$$|\{x : M[g](x) > t\}| \leq \frac{C_d \|g\|_{L^1(B_1(0))}}{t} \qquad \forall t > 0, \ \forall g \in L^1(B_1(0)). \tag{6.25}$$

Given f as before, for every $b > 0$ we denote by M_b the set

$$M_b := \{x \in B_1(0) : M[|f|^n](x) \leq b^n\}.$$

Lemma 6.8. *Let $\delta > 0$, F, λ, Λ, f and u be as in Proposition 6.6. Let $a > 0$, $\overline{B_{4r}(x_0)} \subset B_1(0)$.*

Then there exist universal constants $\tilde{C} \geq 2$ and $\tilde{c}$, $\mu > 0$, such that if $a \leq \delta/\tilde{C}$, and

$$B_r(x_0) \cap A_a \cap M_{\mu a} \neq \emptyset$$

then

$$|B_{r/8}(x_0) \cap A_{\tilde{C}a}| \geq \tilde{c}|B_r(x_0)|. \tag{6.26}$$

Proof. Let $x_1 \in B_r(x_0) \cap A_a \cap M_{\mu a}$ and $y_1 \in B_1(0)$ be the vertex of the paraboloid which satisfies (6.24) with x_1. Let $P_{y_1}(x)$ be the tangent paraboloid, namely

$$P_{y_1}(x) = u(x_1) + \frac{a}{2}|x_1 - y_1|^2 - \frac{a}{2}|x - y_1|^2.$$

Step 1. We prove that there exist universal constants $C_0, C_1 > 0$ such that if $a \leq \delta/C_0$, then there is $z \in B_{r/16}(x_0)$ such that

$$u(z) \leq P_{y_1}(z) + C_1 a r^2. \tag{6.27}$$

Let $\alpha > 0$ be a large universal constant which we choose later, and define $\varphi : \mathbb{R}^d \to \mathbb{R}$ as

$$\varphi(x) := \begin{cases} \alpha^{-1}(32^\alpha - 1) & \text{if } |x| < 32^{-1} \\ \alpha^{-1}(|x|^{-\alpha} - 1) & \text{if } 32^{-1} \leq |x| \leq 1 \\ 0 & \text{if } 1 < |x|. \end{cases} \tag{6.28}$$

Given $x_3 \in B_r(x_1) \cap B_{r/32}(x_0)$ we consider the function $\psi : \mathbb{R}^d \to \mathbb{R}$ given by

$$\psi(x) := P_{y_1}(x) + ar^2\varphi\left(\frac{x - x_3}{r}\right) \qquad \forall x \in \mathbb{R}^d.$$

We slide the function ψ from below until it touches the function u. Let x_4 be the contact point. Since the function φ is radial and decreasing in the radial direction, from

$$\begin{aligned} -ar^2\varphi\left(\frac{x_4 - x_3}{r}\right) &\leq u(x_4) - \psi(x_4) \leq \min_{x\in B_1(0)} \{u(x) - \psi(x)\} \\ &\leq -ar^2\varphi\left(\frac{x_1 - x_3}{r}\right) \end{aligned} \tag{6.29}$$

we deduce that $|x_4 - x_3| \leq |x_1 - x_3| \leq r$. In particular since $|x_4 - x_0| \leq |x_4 - x_3| + |x_3 - x_0| \leq 2r$ and $\overline{B_{2r}(x_0)} \subset B_1(0)$ (by assumption), the contact point is inside $B_1(0)$. We now distinguish two cases:

– *Case* 1: *There exists $x_3 \in B_r(x_1) \cap B_{r/32}(x_0)$ such that the contact point x_4 lies inside $B_{r/32}(x_3)$.*

In this case we have $|x_4 - x_0| \leq |x_4 - x_3| + |x_3 - x_0| \leq r/16$. In addition, the last two inequalities in (6.29) give that $u(x_4) - \psi(x_4) \leq 0$. Hence

$$u(x_4) \leq \psi(x_4) = P_{y_1}(x_4) + ar^2\varphi\left(\frac{x_4 - x_3}{r}\right) \leq P_{y_1}(x_4) + ar^2\|\varphi\|_{L^\infty(\mathbb{R}^d)},$$

which proves that $z = x_4$ satisfies (6.27) with $C_1 := \|\varphi\|_{L^\infty(\mathbb{R}^d)}$ (without any restriction on a).

– *Case* 2: *For every $x_3 \in B_r(x_1) \cap B_{r/32}(x_0)$ the contact point x_4 satisfies $1/32 < |x_4 - x_3| < 1$.*

At the contact point we have that

$$\nabla u(x_4) = \nabla\psi(x_4) = -a(x_4 - y_1) + ar\nabla\varphi\left(\frac{x_4 - x_3}{r}\right). \tag{6.30}$$

Hence, if we choose C_0 such that $C_0 \geq 2 + \|\varphi\|_{L^\infty(\mathbb{R}^d)}$ we get

$$|\nabla u(x_4)| \leq a|x_4 - y_1| + ar\|\varphi\|_{L^\infty(\mathbb{R}^d)} < a(2 + \|\varphi\|_{L^\infty(\mathbb{R}^d)}) \leq C_0 a \leq \delta,$$

which shows that the equation (6.14) is uniformly elliptic at x_4 thanks to our assumptions on F.

Computing the second derivatives of ψ at x_4 we get

$$\begin{aligned}\nabla^2\psi(x_4) &= -aI + a\nabla^2\varphi\left(\frac{x_4-x_3}{r}\right)\\ &= a\left(-I - \left(\frac{r}{|x_4-x_3|}\right)^{2+\alpha} I\right.\\ &\qquad\left. +(2+\alpha)\frac{(x_4-x_3)\otimes(x_4-x_3)}{r^2}\left(\frac{r}{|x_4-x_3|}\right)^{4+\alpha}\right),\end{aligned}$$

hence from (H1) and (6.15) applied with $M = \nabla^2\psi(x_4)$ we obtain (since ψ touches u from below at x_4, we have $\nabla^2 u(x_4) \geq \nabla^2\psi(x_4)$)

$$\begin{aligned} f(x_4) &= F(x_4, u(x_4), \nabla u(x_4), \nabla^2 u(x_4))\\ &\geq F(x_4, u(x_4), \nabla u(x_4), \nabla^2 \psi(x_4))\\ &\geq a\left(-\Lambda - \Lambda\left(\frac{r}{|x_4-x_3|}\right)^{2+\alpha} + (2+\alpha)\lambda\left(\frac{r}{|x_4-x_3|}\right)^{2+\alpha}\right)\\ &= a\left(-\Lambda + ((2+\alpha)\lambda - \Lambda)\left(\frac{r}{|x_4-x_3|}\right)^{2+\alpha}\right).\end{aligned}$$

Choosing α big enough so that $(2+\alpha)\lambda - \Lambda \geq \Lambda + 1$, and using that $|x_4 - x_3| \leq r$, we obtain

$$\frac{f(x_4)}{a} \geq -\Lambda + (\Lambda+1)\left(\frac{r}{|x_4-x_3|}\right)^{2+\alpha} \geq 1. \tag{6.31}$$

In addition,

$$\begin{aligned}\nabla^2 u(x_4) \geq \nabla^2\psi(x_4) &= -aI + a\nabla^2\varphi\left(\frac{x_4-x_3}{r}\right)\\ &\geq a\left(-1 - \left(\frac{r}{|x_4-x_3|}\right)^{2+\alpha}\right) I \geq -(1+32^{2+\alpha})a\,\mathrm{Id},\end{aligned}$$

so by applying the second inequality in (6.15) to $M = \nabla^2 u(x_4)$, we get

$$\begin{aligned}\lambda\|\nabla^2 u(x_4)^+\| &\leq F(x_4, u(x_4), \nabla u(x_4), \nabla^2 u(x_4)) + \Lambda\|\nabla^2 u(x_4)^-\|\\ &\leq |f(x_4)| + \Lambda(1+32^{2+\alpha})a,\end{aligned}$$

that is

$$\frac{\nabla^2 u(x_4)}{a} \leq C_2\left(1 + \frac{|f(x_4)|}{a}\right)\mathrm{Id}, \tag{6.32}$$

for some $C_2 > 0$ universal.

Let us consider K the set of contact points x_4 as x_3 varies in $B_{r/32}(x_0)$ (as we observed before, $K \subseteq B_{2r}(x_0)$), and let $T : K \to \mathbb{R}^d$ be the map which associates to every contact point x_4 the corresponding x_3, which is given by (see (6.30))

$$T(x) = x - r(\nabla\varphi)^{-1}\left(\frac{\nabla u(x) + a(x - y_1)}{ar}\right)$$

(note that $\nabla\varphi$ is an invertible function in the annulus $1/32 < |x| < 1$ and $(\nabla\varphi)^{-1}$ can be explicitly computed). Since $T(K) = B_r(x_1) \cap B_{r/32}(x_0)$, we deduce that there exists a constant c_d, depending only on the dimension, such that $c_d r^d \leq |B_r(x_1) \cap B_{r/32}(x_0)| = |T(K)|$. Therefore, from the area formula,

$$c_d r^d \leq \int_K |\det \nabla T(x)|\, dx \tag{6.33}$$

We now observe that

$$\nabla T(x) = \mathrm{Id} - \left(\nabla^2\varphi \circ (\nabla\varphi)^{-1}\left(\frac{\nabla^2 u(x) + a(x - y_1)}{ar}\right)\right)^{-1} \frac{\nabla u(x) + aI}{a},$$

so from (6.32) and (6.31) we get

$$\begin{aligned}\|\nabla T(x)\| &\leq 1 + \|(\nabla^2\varphi)^{-1}\|_{L^\infty(B_1\setminus B_{1/32})}\left(1 + C_2 + C_2\frac{|f(x)|}{a}\right)\\ &\leq \left(1 + \|(\nabla^2\varphi)^{-1}\|_{L^\infty(B_1\setminus B_{1/32})}(1 + 2C_2)\right)\frac{|f(x)|}{a}.\end{aligned}$$

Hence, combining this bound with (6.33) we get

$$c_d r^d \leq C_3 \int_K \frac{|f(x)|^d}{a^d}\, dx \leq C_3 \int_{B_{2r}(x_0)} \frac{|f(x)|^d}{a^d}\, dx,$$

where $C_3 > 0$ is universal. Since $B_{2r}(x_0) \subseteq B_{3r}(x_1)$ and $B_{3r}(x_1) \subset B_1(0)$ (note $B_{3r}(x_1)$ is included in $B_{4r}(x_0)$, which is contained inside $B_1(0)$ by assumption), we conclude

$$c_d r^d \leq C_3 \int_{B_{3r}(x_1)} \frac{|f(x)|^d}{a^d}\, dx \leq C_3 M[|f|^d](x_1)\frac{|B_{3r}(x_1)|}{a^d}. \tag{6.34}$$

Recalling that by assumption $M(|f|^n)(x_1) \leq \mu^n a^n$, choosing μ small enough so that $\mu^d < c_d/(C_3|B_3(0)|2^d)$, we obtain

$$C_3 M(|f|^d)(x_1)\frac{|B_1(0)|2^d r^d}{a^d} \leq C_3\mu^d |B_1(0)|2^d r^d < c_d r^d,$$

which contradicts (6.34).

Step 2. We conclude the proof. From now on, we assume that $a \le \delta/C_0$, so that the conclusion of Step 1 holds.

Let $C_4 > 0$ be a universal constant which will be fixed later, and for every $y \in B_{r/64}(z)$ we consider the paraboloid

$$Q_y(x) := P_{y_1}(x) - C_4\frac{a}{2}|x-y|^2.$$

It can be easily seen that for every y the function $Q_y(x)$ is a paraboloid with opening $-(C_4+1)a$ and vertex

$$\frac{y_1 + C_4 y}{1+C_4}. \tag{6.35}$$

Let slide Q_y from below until it touches the graph of u. We claim that the contact point $\bar{x}$ lies inside $B_{r/16}(z) \subset B_{r/8}(x_0)$.

Indeed if $|\bar{x}-z| \ge r/16$ we have that

$$|\bar{x}-y| \ge |\bar{x}-z| - |z-y| \ge \frac{r}{16} - \frac{r}{64} \ge \frac{r}{32},$$

so, thanks to (6.27),

$$\begin{aligned}\min_{x\in B_1(0)} \left\{u(x) - P_{y_1}(x) + C_4\frac{a}{2}|x-y|^2\right\} &\le u(z) - P_{y_1}(z) + C_4\frac{a}{2}|z-y|^2 \\ &\le C_1 a r^2 + C_4\frac{a}{2}\left(\frac{r}{64}\right)^2.\end{aligned} \tag{6.36}$$

On the other hand, since $u \ge P_{y_1}$ we have

$$u(\bar{x}) - P_{y_1}(\bar{x}) + C_4\frac{a}{2}|\bar{x}-y|^2 \ge C_4\frac{a}{2}\left(\frac{r}{32}\right)^2,$$

which contradicts (6.36) if we choose C_4 sufficiently large. This proves in particular that

$$\bar{x} \in B_{r/16}(z) \subset B_{r/8}(x_0). \tag{6.37}$$

We now show that the contact points satisfy $u(\bar{x}) \le C_4 a$. Indeed, since by assumption $P_{y_1}(x_1) = u(x_1) \le a$ and all points lie inside $B_1(0)$, we have

$$P_{y_1}(\bar{x}) = u(x_1) + \frac{a}{2}|x_1-y_1|^2 - \frac{a}{2}|\bar{x}-y_1|^2 \le a + 4a = 5a,$$

so from (6.36) we obtain

$$\begin{aligned}u(\bar{x}) &\le P_{y_1}(\bar{x}) - C_4\frac{a}{2}|\bar{x}-y|^2 + C_1 a r^2 + C_4\frac{a}{2}\left(\frac{r}{64}\right)^2 \\ &\le 5a + C_1 a r^2 + C_4\frac{a}{2}\left(\frac{r}{64}\right)^2,\end{aligned}$$

which is less than $C_4 a$ provided that C_4 is chosen sufficiently large.

We now observe that, as y varies in $B_{r/64}(z)$, the set of vertices of the paraboloids is a ball around $\frac{y_1+C_4 z}{1+C_4}$ of radius $\frac{C_4 r}{64(1+C_4)}$ (see (6.35)). Hence, recalling (6.37) and that $u \leq C_4 a$ at the contact points, it follows from Lemma 6.7 that

$$c\left(\frac{C_4 r}{64(1+C_4)}\right)^d |B_1(0)| \leq |B_{r/8}(x_0) \cap A_{C_4 a}| + \int_{B_{r/8}(x_0)} \frac{|f(x)|^d}{a^d}\, dx.$$

Since the last integral can be estimated with

$$\int_{B_{2r}(x_1)} \frac{|f(x)|^d}{a^d}\, dx \leq M[|f|^d](x_1)\frac{|B_{2r}(x_1)|}{a^d} \leq \mu^d r^d |B_2(0)|,$$

we conclude that (6.26) holds with $\tilde{C} := \max\{C_0, C_4\}$, provided μ is sufficiently small. □

The following measure covering lemma is proved by Savin in [115, Lemma 2.3] in a slightly different version.

Lemma 6.9. *Let $\sigma, r_0 \in (0,1)$, and let D_0, D_1 be two closed sets satisfying*

$$\emptyset \neq D_0 \subseteq D_1 \subseteq \overline{B_{r_0}(0)}.$$

Assume that whenever $x \in B_{r_0}(0)$ and $r > 0$ satisfy

$$B_{4r}(x) \subseteq B_1(0), \qquad B_{r/8}(x) \subseteq B_{r_0}(0), \qquad \overline{B_r(x)} \cap D_0 \neq \emptyset$$

then

$$|B_{r/8}(x) \cap D_1| \geq \sigma |B_r(x)|.$$

Then, if $r_0 > 0$ is sufficiently small we get

$$|B_{r_0}(0) \setminus D_1| \leq (1-\sigma)|B_{r_0}(0) \setminus D_0|. \tag{6.38}$$

Although the proof is a minor variant of the argument of Savin in [115, Lemma 2.3], we give the argument for completeness. As we will see from the proof, a possible choice for r_0 is $1/13$.

Proof. Given $x_0 \in B_{r_0}(0) \setminus D_0$, set $\bar{r} := \operatorname{dist}(x_0, D_0) \leq 2r_0$, and define

$$x_1 := x_0 - \frac{\bar{r}}{7}\frac{x_0}{|x_0|}, \qquad r := \frac{8}{7}\bar{r}.$$

Then it is easy to check that

$$B_{r/8}(x_1) \subset B_{r/4}(x_0) \cap B_{r_0}(0), \qquad \overline{B_r(x_1)} \cap D_0 = \emptyset.$$

In addition, since $r \leq 3r_0$ and $|x_1| < r_0$,

$$B_{4r}(x_1) \subset B_{13r_0}(0) \subseteq B_1(0) \qquad \text{provided } r_0 \leq 1/13.$$

Hence, using our assumptions we get

$$\begin{aligned}|B_{r/4}(x_0) \cap B_{r_0}(0) \cap D_1| &\geq |B_{r/8}(x_1) \cap D_1| \geq \sigma |B_r(x_1)| \\ &= \sigma |B_r(x_0)| \geq \sigma |B_{r_0}(0) \cap B_r(x_0)|.\end{aligned}$$

Now, for every $x \in B_{r_0}(0) \setminus D_0$ we consider the ball centered at x and radius $r := \mathrm{dist}(x, D_0)$, and we apply Vitali covering's Lemma to this family to extract a subfamily $\{B_{r_i}(x_i)\}$ such that the balls $B_{r_i/3}(x_i)$ (and so in particular also the balls $B_{r_i/4}(x_i)$) are disjoint. Hence

$$\begin{aligned}\sigma |B_{r_0}(0) \setminus D_0| &\leq \sigma \sum_i \left|\left(B_{r_i}(x_i) \cap B_{r_0}\right) \setminus D_0\right| \\ &\leq \sum_i |B_{r_i/4}(x_i) \cap B_{r_0}(0) \cap (D_1 \setminus D_0)| \\ &\leq |B_{r_0} \cap (D_1 \setminus D_0)|,\end{aligned}$$

from which the result follows easily. □

Proof of Proposition 6.6. Let c_0 be the constant from Lemma 6.7, and $\tilde{C}$, $\tilde{c}$, μ the constants given by Lemma 6.8. Also, we fix $r_0 > 0$ sufficiently small so that Lemma 6.9 applies, and we define $r_1 := r_0/8$.

Let $\nu < 1/2$ and N be universal constants (to be chosen later) satisfying $N\nu \ll 1$, set $a := N\nu\delta'$, $m := \inf_{B_1(0)} u$ and assume by contradiction that there exists $x_0 \in B_{r_0/2}(0)$ such that

$$u(x_0) - m < \nu\delta', \tag{6.39}$$

and in addition

$$\sup_{B_{r_1}(0)} u - m > \delta'/2. \tag{6.40}$$

(Note that if either (6.39) or (6.40) fails, then $\mathrm{osc}_{B_{r_1}(0)} u \leq (1-\nu)\delta'$, so the statement is true with $\rho = r_1$).

We define the sets A_a as before but replacing u with the nonnegative function $u - m$, that is A_a is the set of points where $u - m$ is bounded by a and can be touched from below with a paraboloid of opening $-a$.

Step 1. The following holds:

$$|B_{r_0}(0) \cap A_a| \geq \frac{c_0 |B_{r_1}(0)|}{2}, \qquad |M_{\mu a}| > |B_1| - \frac{c_0 |B_{r_1}(0)|}{2}. \tag{6.41}$$

To prove this, for every $y \in B_{r_1}(0)$ we consider the paraboloid

$$P_y(x) := \frac{a}{2}\left((r_0 - r_1)^2 - |x - y|^2\right).$$

We observe that

$$P_y \le 0 \qquad \text{for } |x| \ge r_0$$

(because $|x-y| \ge |x|-|y| \ge r_0-r_1$), while $|x-y| \le |x|+|y| \le r_0/2+r_1$ for $x \in B_{r_0/2}(0)$, which implies (recall that $a = N\nu\delta'$)

$$\begin{aligned} P_y(x) &\ge \frac{a}{2}\left((r_0 - r_1)^2 - \left(\frac{r_0}{2} + r_1\right)^2\right) > \nu\delta' \\ &\ge u(x_0) - m \qquad \forall x \in B_{r_0/2}(0) \end{aligned} \tag{6.42}$$

provided N is sufficiently large. Moreover $P_y(x) \le a$ for every $x, y \in B_1(0)$.

Hence, let us slide the paraboloids P_y from below until they touch the function $u - m$. Let A be the contact set as y varies inside $B_{r_1}(0)$. By what said before it follows that the contact points are contained inside $B_{r_0}(0)$. In addition, thanks to (6.39) and (6.42), at any contact point x we have

$$\begin{aligned} 0 &> u(x_0) - m - \nu\delta' \ge \min_{z \in B_1(0)} \{u(z) - m - P_y(z)\} \\ &= u(x) - m - P_y(x) \ge u(x) - m - a, \end{aligned}$$

which proves that $A \subset B_{r_0}(0) \cap A_a$. From Lemma 6.7 applied to $K = B_{r_1}(0)$ we obtain

$$\begin{aligned} |B_{r_0}(0) \cap A_a| \ge |A| &\ge c_0|B_{r_1}(0)| - \int_A \frac{|f(x)|^d}{a^d}\,dx \\ &\ge c_0|B_{r_1}(0)| - \int_{B_1(0)} \frac{|f(x)|^d}{a^d}\,dx \\ &\ge c_0|B_{r_1}(0)| - \frac{\varepsilon^d}{N^d\nu^d}, \end{aligned}$$

while the maximal estimate (6.25) gives

$$|B_1(0) \setminus M_{\mu a}| \le \frac{C_d\|f\|^d_{L^d(B_1(0))}}{(\mu a)^d} \le \frac{C_d\varepsilon^d}{\mu^d N^d \nu^d},$$

hence (6.41) is satisfied provided ε is sufficiently small.

Step 2. There exists a constant $\bar{C} > 0$, depending only on the dimension, such that

$$|B_{r_0}(0) \setminus A_{\tilde{C}^k a}| \le \bar{C}(1-\tilde{c})^k \qquad \text{provided } \tilde{C}^{k+1} a \le \delta. \tag{6.43}$$

From (6.41) it follows that

$$B_{r_0}(0) \cap A_a \cap M_{\mu a} \ne \emptyset.$$

Since the sets A_a and M_a are increasing with respect to k, this implies that

$$B_{r_0}(0) \cap A_{\tilde{C}^k a} \cap M_{\mu \tilde{C}^k a} \ne \emptyset \qquad \forall k \in \mathbb{N}, \tag{6.44}$$

where $\tilde{C} \ge 2$ is as in Lemma 6.8.

Now, for every $k \in \mathbb{N}$ such that $\tilde{C}^{k+1} a \le \delta$ we apply Lemma 6.9 to the closed sets

$$D_0 := \overline{B_{r_0}(0)} \cap A_{\tilde{C}^k a} \cap M_{\mu \tilde{C}^k a}, \qquad D_1 := \overline{B_{r_0}(0)} \cap A_{\tilde{C}^{k+1} a}.$$

Since D_0 is nonempty (see (6.44)), Lemma 6.8 applied with $\tilde{C}^k a$ instead of a proves that assumption of Lemma 6.9 are satisfied with $\sigma = \tilde{c} > 0$. Therefore

$$\begin{aligned} |B_{r_0}(0) \setminus A_{\tilde{C}^{k+1} a}| &\le (1-\tilde{c}) |B_{r_0}(0) \setminus (A_{\tilde{C}^k a} \cap M_{\mu \tilde{C}^k a})| \\ &\le (1-\tilde{c}) \left(|B_{r_0}(0) \setminus A_{\tilde{C}^k a}| + |B_{r_0}(0) \setminus M_{\mu \tilde{C}^k a}| \right). \end{aligned} \tag{6.45}$$

Applying (6.45) inductively for every positive integer k such that $\tilde{C}^{k+1} a \le \delta$ and using the maximal estimate (6.25), we obtain

$$\begin{aligned} |B_{r_0}(0) \setminus A_{\tilde{C}^k a}| &\le (1-\tilde{c})^k |B_{r_0}(0) \setminus A_a| + \sum_{i=1}^{k} (1-\tilde{c})^i |B_{r_0}(0) \setminus M_{\tilde{C}^{k-i} a}| \\ &\le (1-\tilde{c})^k |B_{r_0}(0)| + \sum_{i=1}^{k} (1-\tilde{c})^i \frac{C_d \|f\|^d_{L^d(B_1(0))}}{\mu^d \tilde{C}^{d(k-i)} a^d}, \end{aligned}$$

so by (6.18) we get (recall that $a = N\nu\delta'$)

$$\begin{aligned} |B_{r_0}(0) \setminus A_{\tilde{C}^k a}| &\le (1-\tilde{c})^k \left[|B_{r_0}(0)| + \frac{C_d \varepsilon^d}{\mu^d N^d \nu^d} \sum_{i=1}^{k} \frac{1}{((1-\tilde{c})\tilde{C}^d)^{k-i}} \right] \\ &\le (1-\tilde{c})^k \left[|B_{r_0}(0)| + \frac{C_d \varepsilon^d}{\mu^d N^d \nu^d} \sum_{i=0}^{\infty} \frac{1}{((1-\tilde{c})\tilde{C}^d)^{i}} \right]. \end{aligned} \tag{6.46}$$

Assuming without loss of generality that $\tilde{c} \leq 1/2$, $\tilde{C} \geq 3$, and $\varepsilon \leq \mu N \nu / C_d^{-1/d}$ we have

$$|B_{r_0}(0) \setminus A_{\tilde{C}^k a}| \leq (1-\tilde{c})^k \left[|B_{r_0}(0)| + \sum_{i=0}^{\infty} \left(\frac{2}{3^d} \right)^i \right],$$

which proves (6.43).

Step 3. Let $E := \{x \in B_{r_0}(0) : u(x) - m \geq \delta'/4\}$. Then

$$|E| \geq \frac{c_0 |B_{r_1}(0)|}{2}. \tag{6.47}$$

For every $y \in B_{r_1}(0)$ we consider the paraboloid

$$Q_y(x) := \frac{\delta'}{(r_0 - r_1)^2}|x-y|^2 + \frac{\delta'}{4},$$

and we slide it from above (in Step 1 we slided paraboloids from below) until it touches the graph of $u - m$ inside $B_1(0)$. It is easy to check that, since $|x - y| \geq |x| - |y|$, we have

$$Q_y(x) > \delta' \geq u(x) - m \qquad \text{for } |x| \geq r_0$$

(recall that $y \in B_{r_1}(0)$ and $u - m \leq \delta'$ inside $B_1(0)$), while by (6.40)

$$\sup_{B_{r_1}(0)} Q_y \leq \delta'/2 < \sup_{B_{r_1}(0)} u - m \tag{6.48}$$

(recall that $r_0 = 8r_1$), so the contact point lies inside $B_{r_0}(0)$. If we denote by A' the contact set as y varies inside $B_{r_1}(0)$ applying Lemma 6.7 "from above" (namely to the function $-u(x) + m$ touched from below by the paraboloids $-Q_y(x)$) with $a = 2\delta'/(r_0 - r_1)^2$ (notice that $\delta' \leq \kappa\delta$, so $a \leq \delta/2$ if κ is sufficiently small) we obtain

$$|A'| \geq c_0 |B_{r_1}(0)| - \int_{A'} \frac{|f(x)|^d}{a^d} dx \geq c_0 |B_{r_1}(0)| - \frac{\varepsilon^d}{N^d \nu^d}. \tag{6.49}$$

Moreover, it follows by (6.48) tha$u - m \geq \delta'/4$ at every contact point. This implies that the contact set A' is contained in E, so the desired estimate follows by (6.49).

Step 4. Conclusion. Let $k_0 \in \mathbb{N}$ be the largest number such that $\tilde{C}^{k_0+1} a \leq \delta'/4$. Since $\delta' \leq \delta$, by Step 2 we get

$$|B_{r_0}(0) \setminus A_{\tilde{C}^{k_0} a}| \leq \bar{C}(1 - \tilde{c})^{k_0}.$$

On the other hand, since

$$E \subset \left\{x \in B_{r_0}(0) : u(x) - m > \tilde{C}^{k_0} a\right\} \subset B_{r_0}(0) \setminus A_{\tilde{C}^{k_0} a},$$

it follows by Step 3 that

$$\frac{c_0 |B_{r_1}(0)|}{2} \leq \bar{C}(1 - \tilde{c})^{k_0}.$$

Since $k_0 \sim |\log_{\tilde{C}}(N\nu)|$ (recall that $a = N\nu\delta'$), we get a contradiction by first fixing N large enough (so that all the previous arguments apply) and then choosing ν sufficiently small. □

Proof of Theorem 6.5. Let $\nu, \varepsilon, \kappa, \rho \in (0, 1)$ be the constants of Proposition 6.6. Without loss of generality we assume that $\nu, \rho \leq 1/2$. We prove (6.17) by induction on s. For $s = 0$ the result is true by assumption. We prove the result for $s + 1$ given the one for s. Let $\widetilde{F} : B_1(0) \times \mathbb{R} \times \mathbb{R}^d \times \mathcal{S} \to \mathbb{R}$ be

$$\widetilde{F}(x, z, p, M) := \rho^s F(x, \rho^s z, p, \rho^{-s} M),$$

and consider the function

$$v(x) := \rho^{-s} u(\rho^s x) \qquad \forall x \in B_1(0).$$

Then $\widetilde{F}$ satisfies the same assumptions (H1), (H2), and (H3) which are satisfied by F with the same ellipticity constants λ and Λ, and v solves the fully nonlinear equation

$$\widetilde{F}(x, v(x), \nabla v(x), \nabla^2 v(x)) = \rho^s f(\rho^s x).$$

By inductive hypothesis

$$\begin{aligned} \|v\|_{L^\infty(B_{\rho^s}(0))} &= \rho^{-s} \|u\|_{L^\infty(B_{\rho^s}(0))} \leq \rho^{-s}(1 - \nu)^s \delta' \\ &\leq \rho^{-s}\delta' \leq \rho^{k-s}\kappa\delta \leq \kappa\delta. \end{aligned} \tag{6.50}$$

Also, by (6.16),

$$\begin{aligned} \|\rho^s f(\rho^s x)\|_{L^d(B_1(0))} &= \|f\|_{L^d(B_{\rho^s}(0))} \leq \|f\|_{L^d(B_1(0))} \\ &\leq \varepsilon\delta' \leq \varepsilon\rho^{-s}(1 - \nu)^s \delta'. \end{aligned}$$

Hence, we apply Proposition 6.6 to v with $\rho^{-s}(1 - \nu)^s\delta'$ instead of δ', to obtain

$$\rho^{-s} \|u\|_{L^\infty(B_{\rho^{s+1}}(0))} = \|v\|_{L^\infty(B_\rho(0))} \leq \rho^{-s}(1 - \nu)^{s+1}\delta',$$

which proves the inductive step. □

6.3. Separation between degenerancy and nondegeneracy

First, we introduce some notation regarding the norm induced by a convex set E (see (6.6)).

We denote by E^* the ball in the dual norm

$$E^* := \{e^* \in \mathbb{R}^d : e^* \cdot e \leq 1 \quad \forall e \in E\}. \tag{6.51}$$

It can be easily seen that with this definition

$$|e|_E = \sup\{e^* \cdot e : e^* \in E^*\} \qquad \forall e \in \mathbb{R}^d.$$

We denote by d_E (and d_{E^*}, respectively) the smallest radius such that $E \subseteq B_{d_E}(0), (E^* \subseteq B_{d_{E^*}}(0)$, respectively). Notice that

$$d_E = \max\{|e| : |e|_E = 1\} \tag{6.52}$$

Similarly, we denote by $\widetilde{d}_E$ the biggest radius such that $B_{\widetilde{d}_E}(0) \subseteq E$. It satisfies

$$|e|_E \leq |e|/\widetilde{d}_E \qquad \forall e \in \mathbb{R}^d. \tag{6.53}$$

Moreover, if E is strictly convex, then we can define map $\ell : \partial E^* \to \partial E$, where $\ell_{e^*} := \ell(e^*)$ is the unique element of ∂E such that $|\ell_{e^*}|_E = e^* \cdot \ell_{e^*}$ (in other terms, $\{x \cdot e^* = 1\}$ is a supporting hyperplane for E at ℓ_{e^*}). In addition, again by the strict convexity of E, ℓ is continuous in the following sense: for every $\varepsilon_0 > 0$ there exists $\eta(\varepsilon_0) > 0$ such that

$$e \in \overline{E}, \quad e^* \in \partial E^*, \quad 1 - \eta(\varepsilon_0) \leq e^* \cdot e \leq 1 \Rightarrow |e - \ell_{e^*}| \leq \varepsilon_0. \tag{6.54}$$

In the following lemma we prove that, at every scale, if none of the partial derivatives of u is close to the L^∞ norm of $|\nabla u|_E$ in a set of large measure, then $|\nabla u|_E$ decays by a fixed amount on a smaller ball. As we will see in the next section, if this case does not occur, then the equation is nondegenerate and we can prove that u is $C^{1,\alpha}$ there.

As we will see below, a key observation being the proof of the next result is the fact that the function $v_{e^*}(x) := (\partial_{e^*} u(x) - (1+\delta))_+$ solves

$$\partial_i[\partial_{ij}\mathcal{F}(\nabla u(x))\partial_j v_{e^*}(x)] \geq \partial_{e^*} f(x) 1_{\{1/2 - v_{e^*}(x) > 0\}}, \tag{6.55}$$

and the equation might be assumed to be uniformly elliptic, since the values of the coefficients $\partial_{ij}\mathcal{F}(\nabla u(x))$ are not relevant when $|\nabla u(x)| \leq 1+\delta$ (since at that points $v_{e^*} = 0$). In the previous observation, the convexity of E plays a fundamental role. Indeed, the function v_{e^*} vanishes, for any $e^* \in \mathbb{R}^d \setminus \{0\}$, when ∇u belongs to an half-space (namely, the set

$\{x : x \cdot e^* \le 1 + \delta\}$); in order for the equation to be uniformly elliptic, we need to consider only the vectors e^* for which the half-space contains E. On the other hand, convex sets are the only ones that can be written as intersections of half-spaces.

Lemma 6.10. *Fix $\eta > 0$, and let δ, $\mathcal{F}$, E, λ, Λ, M, f, and u be as in Theorem 6.4. For every $i \in \mathbb{N}$ set*

$$d_i := \sup\{(|\nabla u(x)|_E - (1+\delta))_+ : x \in B_{2^{-i}(0)}\},$$

and assume that there exists $k \in \mathbb{N}$ such that for every $i = 0, ..., k$

$$\sup_{e^* \in \partial E^*} |\{x \in B_{2^{-2i-1}}(0) : (\partial_{e^*} u(x) - (1+\delta))_+ \ge (1-\eta) d_{2i}\}| \le (1-\eta)|B_{2^{-2i-1}}(0)|. \tag{6.56}$$

Then there exists $\alpha \in (0,1)$ and $C_0 > 0$, depending only on η, M, q, $\|f\|_{L^q(B_1(0))}$, d_{E^*}, $\widetilde{d}_E$, δ, λ, and Λ, such that

$$d_{2i} \le C_0 2^{-2i\alpha} \qquad \forall i = 0, ..., k+1. \tag{6.57}$$

Proof. Given $e^* \in \partial E^*$, we differentiate (6.12) in the direction of e^* to obtain

$$\partial_i[\partial_{ij}\mathcal{F}(\nabla u(x))\partial_j(\partial_{e^*} u(x))] = \partial_{e^*} f(x).$$

Since the function $t \mapsto (t - (1+\delta))_+$ is convex, it follows that the function $v_{e^*}(x) := (\partial_{e^*} u(x) - (1+\delta))_+$ is a subsolution of the above equation, that is (6.55) holds.

Note that, since $v_{e^*}(x)$ is constant where $|\nabla u|_E \le 1+\delta$ and $\mathcal{F}$ is uniformly elliptic on the set $\{|\nabla u|_E \ge 1 + \delta/2\}$ (see (6.11)), we can change the coefficients outside this region to ensure that the equation is uniformly elliptic everywhere, with constants λ and Λ. We apply the weak Harnack inequality of Theorem 1.21 to the function $d_{2i} - v_{e^*}(x)$ (which is a nonnegative supersolution inside $B_{2^{-2i}}(0)$); notice that the right-hand side of the equation solved by this function is not exactly a divergence, but the proof of the weak Harnack inequality works also in this case. We obtain that there exists a constant $c_0 := c_0(d, \lambda, \Lambda) > 0$ such that

$$\inf\{d_{2i} - v_{e^*}(x) : x \in B_{2^{-2i-2}}(0)\} \ge c_0 2^{2id} \int_{B_{2^{-2i-1}}(0)} (d_{2i} - v_{e^*}(x))\,dx \; - 2^{-2i(1-d/q)}\|fe^*\|_{L^q(B_{2^{-2i}}(0))}$$

We estimate the integral in the right hand side considering only the set

$$\{x \in B_{2^{-2i-1}}(0) : v_{e^*}(x) \le (1-\eta)d_{2i}\}.$$

There, the integrand is greater than ηd_{2i} and the measure of the set is greater than $\eta|B_{2^{-2i-1}}(0)|$ (by (6.56)), hence

$$\begin{aligned}
&\inf\{d_{2i} - v_{e^*}(x) : x \in B_{2^{-2i-2}}(0)\}\\
&\geq c_0 2^{2id}\eta^2 d_{2i}|B_{2^{-2i-1}}(0)| - 2^{-2i(1-d/q)}\|fe^*\|_{L^q(B_1(0))} \qquad (6.58)\\
&\geq c_0\eta^2 d_{2i}|B_{1/2}(0)| - 2^{-2i(1-d/q))}\|f\|_{L^q(B_1(0))}d_{E^*}.
\end{aligned}$$

We now distinguish two cases, depending whether

$$\frac{c_0|B_{1/2}(0)|\eta^2}{2}d_{2i} \geq d_{E^*}2^{-2i(1-d/q))}\|f\|_{L^q(B_1)} \tag{6.59}$$

holds or not.

– *Case* 1*:* (6.59) *holds*. In this case we obtain from (6.58) that

$$v_{e^*}(x) \leq \left(1 - \frac{c_0|B_{1/2}(0)|\eta^2}{2}\right)d_{2i} \qquad \forall x \in B_{2^{-2i-2}}(0).$$

Since $e^* \in \partial E^*$ is arbitrary and

$$\begin{aligned}
\sup_{e^*\in\partial E^*} v_{e^*}(x) &= \left(\sup_{e^*\in\partial E^*} \partial_{e^*}u(x) - (1+\delta)\right)_+\\
&= (|\nabla u(x)|_E - (1+\delta))_+ \qquad \forall x \in B_1(0),
\end{aligned}$$

we get

$$(|\nabla u(x)|_E - (1+\delta))_+ \leq \left(1 - \frac{c_0|B_{1/2}(0)|\eta^2}{2}\right)d_{2i} \quad \forall x \in B_{2^{-2i-2}}(0),$$

that is

$$d_{2(i+1)} \leq \left(1 - \frac{c_0|B_{1/2}(0)|\eta^2}{2}\right)d_{2i}. \tag{6.60}$$

– *Case* 2*:* (6.59) *fails*. In this case we get

$$d_{2(i+1)} \leq d_{2i} \leq C'2^{-2i(1-d/q)}. \tag{6.61}$$

for some constant C' depending only on $\eta, d, \lambda, \Lambda, d_{E^*}$, and $\|f\|_{L^q(B_1(0))}$.

Let us choose $\alpha \in (0,1)$ such that

$$\alpha \leq 1 - d/q, \qquad 1 - \frac{c_0|B_{1/2}(0)|\eta^2}{2} \leq 2^{-2\alpha},$$

and $C_0 := \max\{M/\widetilde{d}_E, 4C'\}$ (recall that M is an upper bound for $|\nabla u|$ inside $B_2(0)$). We prove the result by induction over i.

Since $|\nabla u(x)|_E \leq |\nabla u(x)|/\widetilde{d}_E \leq M/\widetilde{d}_E$ (see (6.53)), we have that $d_0 \leq M/\widetilde{d}_E$, so the statement is true for $i = 0$.

Assuming the result for i, if (6.59) holds, then from (6.60) and the inductive hypothesis we obtain

$$d_{2(i+1)} \leq \left(1 - \frac{c_0|B_{1/2}(0)|\eta^2}{2}\right) d_{2i} \leq 2^{-2\alpha} \cdot C_0 2^{-2i\alpha},$$

while if (6.59) fails then (6.61) gives

$$d_{2(i+1)} \leq C' 2^{-2i(1-d/q)} \leq C' 2^{-2i\alpha} \leq 4C' \cdot 2^{-2(i+1)\alpha} \leq C_0 2^{-2(i+1)\alpha}.$$

This proves the inductive step on $d_{2(i+1)}$, and concludes the proof. □

6.4. Regularity at nondegenerate points

In the following lemma we prove that in a neighborhood of a nondegenerate point the function u is close to a linear function with a nondegenerate slope. In Proposition 6.13 we prove that this implies $C^{1,\alpha}$ regularity of u at the nondegenerate point. The proof is based on an approximation argument with solutions of a smooth elliptic operator, which is stated in Lemma 6.12 and whose proof is based on the compactness result of Section 6.2.

We recall that E^* denotes the dual of a convex set E, and $|\cdot|_E$ the norm associated to E (see (6.51) and (6.6)).

Lemma 6.11. *Let $\delta, \eta, \zeta > 0$, and let E be a strictly convex set.*

Let $u : B_1(0) \to \mathbb{R}$ with $u(0) = 0$ and $|\nabla u(x)|_E \leq \zeta + \delta + 1$ for every $x \in B_1(0)$. Let us assume that there exists $e^* \in \partial E^*$ such that

$$\left|\left\{x \in B_1 : (\partial_{e^*} u(x) - (1+\delta))_+ \geq (1-\eta)\zeta\right\}\right| \geq (1-\eta)|B_1(0)|. \quad (6.62)$$

Then for every $\varepsilon > 0$ there exists η depending only on E and d, and constants $A \in \mathbb{R}^d$ and $b \in \mathbb{R}$, such that

$$|u(x) - A \cdot x - b| \leq \varepsilon(\zeta + \delta + 1) \qquad \forall x \in B_1(0). \quad (6.63)$$

In addition $|A|_E = \zeta + \delta + 1$ and $|b| \leq C(\zeta + \delta + 1)$, where C depends only on E.

Proof. First of all, by standard Sobolev inequalities, there exists a constant C_0 such that for every $u \in W^{1,2d}(B_1(0))$

$$\left|u(x) - \fint_{B_1(0)} u(y)dy\right| \leq C_0 \left(\fint_{B_1(0)} |\nabla u(y)|^{2d} dy\right)^{1/(2d)} \qquad \forall x \in B_1(0). \quad (6.64)$$

Recalling that $\ell : \partial E^* \to \partial E$ denotes the duality map, we apply (6.64) to the function $u(x) - (\zeta+\delta+1)\ell_{e^*} \cdot x$. Thus, setting m to be the average of $u(x)/(\zeta+\delta+1)$ inside $B_1(0)$, we obtain

$$\begin{aligned}&\left|u(x) - (\zeta+\delta+1)\ell_{e^*} \cdot x - m(\zeta+\delta+1)\right|^{2d}\\ &\leq C_0^{2d} \fint_{B_1(0)} |\nabla u(y) - (\zeta+\delta+1)\ell_{e^*}|^{2d}\, dy\end{aligned} \tag{6.65}$$

for every $x \in B_1(0)$. We estimate the integral in (6.65) by splitting it into two sets.

Let $\varepsilon_0 > 0$ be a constant that we choose later. Since by assumption $|\nabla u(x)|_E \leq \zeta+\delta+1$ for every $x \in B_1(0)$, and in addition

$$\begin{aligned}\big\{x \in B_1 : \partial_{e^*} u(x) \geq (1-\eta)\zeta+\delta+1\big\} \subseteq \big\{x \in B_1 : e^* \cdot \nabla u(x)\\ \geq (1-\eta)\big(\zeta+\delta+1\big)\big\},\end{aligned}$$

we apply (6.54) with $e = \nabla u(x)/(\zeta+\delta+1)$ to deduce that

$$\begin{aligned}&\frac{1}{|B_1(0)|}\int_{\{(\partial_{e^*}u-(1+\delta))_+ \geq (1-\eta)d\}} |\nabla u(y) - (\zeta+\delta+1)\ell_{e^*}|^{2d}\, dy\\ &\leq (\zeta+\delta+1)^{2d}\varepsilon_0^{2d},\end{aligned}$$

provided $\eta \leq \eta(\varepsilon_0)$.

On the other hand, since the complement has measure less than $\eta|B_1(0)|$, we simply estimate the integrand there with $C_E(\zeta+\delta+1)^{2d}$, where C_E is a constant depending only on E.

Hence, by choosing first ε_0 so that $C_0^{2d}\varepsilon_0^{2d} \leq \varepsilon^{2d}/2$, and then $\eta \leq \eta(\varepsilon_0)$ sufficiently small so that so that $C_0^{2d}C_E^{2d}\eta \leq \varepsilon^{2d}/2$, from (6.65) we easily obtain (6.63). □

Lemma 6.12. *Let $\delta > 0$, and let $a_{ij} \in C^0(\mathbb{R}^d)$ be bounded coefficients uniformly elliptic in $B_\delta(0)$, namely there exist $\lambda, \Lambda > 0$ such that*

$$\lambda I \leq a_{ij}(v) \leq \Lambda I \qquad \forall\, v \in B_\delta(0).$$

Then, for every $\tau > 0$ there exist $\sigma(\tau) > 0$, $\mu(\tau) > 0$, which depend only on τ and on the modulus of continuity of a_{ij}, such that the following holds: For every $\theta \leq \sigma(\tau)$, $f \in C^0(B_1(0))$ such that $\|f\|_{L^d(B_1(0))} \leq \mu(\tau)$, and $w \in C^2(B_1(0))$ such that $\|w\|_{L^\infty(B_1(0))} \leq 1$ and

$$a_{ij}(\theta\nabla w)\partial_{ij} w = f \qquad \text{in } B_1(0),$$

there exists $v : B_1(0) \to \mathbb{R}$ such that

$$a_{ij}(0)\partial_{ij} v = 0 \qquad \text{in } B_1(0) \tag{6.66}$$

and

$$\|v - w\|_{L^\infty(B_{1/2}(0))} \le \tau.$$

Proof. By contradiction, there exists $\tau > 0$ and sequences $\theta_m \to 0$, $\mu_m \to 0$ and functions $w_m, f_m : B_1(0) \to \mathbb{R}$ such that $\|w_m\|_{L^\infty(B_1(0))} \le 1$, $\|f_m\|_{L^d(B_1(0))} \le \mu_m$,

$$a_{ij}(\theta_m \nabla w_m)\partial_{ij} w_m = f_m \qquad \text{in } B_1(0), \tag{6.67}$$

but for every function $v : B_1(0) \to \mathbb{R}$ satisfying (6.66) we have that

$$\|v - w_m\|_{L^\infty(B_{1/2}(0))} \ge \tau \qquad \forall m \in \mathbb{N}. \tag{6.68}$$

We prove that up to subsequence (not relabeled)

$$w_m \to w_\infty \qquad \text{locally uniformly in } B_1(0) \tag{6.69}$$

and that w_∞ satisfies (6.66), which contradicts (6.68).

Consider $\Omega \Subset B_1(0)$, let $d_\Omega = \operatorname{dist}(\Omega, \mathbb{R}^d \setminus B_1(0))$, and for every $m \in \mathbb{N}$ and $x_0 \in \Omega$ we consider the function

$$u_m(x) := \frac{\theta_m}{d_\Omega}\big(w_m(x_0 + d_\Omega x) - w_m(x_0)\big) \qquad \forall x \in B_1(0),$$

which solves

$$a_{ij}(\nabla u_m(x))\partial_{ij} u_m(x) = \theta_m d_\Omega f_m(d_\Omega x) \qquad \forall x \in B_1(0).$$

We apply Theorem 6.5 to $F(x, z, p, M) = a_{ij}(p) M_{ij}$ (which satisfies all the assumptions) and let $\nu, \varepsilon, \kappa, \rho > 0$ be the constants introduced in that theorem. Thus, if $\delta' > 0$ and $k \in \mathbb{N}$ satisfy

$$\operatorname*{osc}_{B_1(0)} u_m \le \delta' \le \rho^{-k}\kappa\delta, \qquad \|\theta_m d_\Omega f_m(d_\Omega x)\|_{L^d(B_1(0))} \le \varepsilon\delta' \tag{6.70}$$

then

$$\operatorname*{osc}_{B_{\rho^s}(0)} u_m \le (1-\nu)^s \delta' \qquad \forall s = 0, ..., k+1.$$

We want to apply it with $\delta' = \theta_m$. Hence, define k_m to be the biggest positive integer such that $\theta_m \le 2^{-k_m}\kappa\delta$. Since

$$\|f_m\|_{L^d(B_1(0))} \le \varepsilon$$

for m sufficiently large, we get

$$\begin{aligned}\|\theta_m d_\Omega f_m(d_\Omega x)\|_{L^d(B_1(0))} &= \|\theta_m f_m(x)\|_{L^d(B_{d_\Omega}(0))} \\ &\le \theta_m \|f_m(x)\|_{L^d(B_1(0))} \le \varepsilon\theta_m.\end{aligned}$$

Hence (6.70) is satisfied, and we get

$$\operatorname*{osc}_{B_{\rho^s}(0)} u_m \le (1-\nu)^s \theta_m \qquad \forall s = 0, ..., k_m + 1$$

which can be rewritten in terms of w_m as

$$\operatorname*{osc}_{B_{\rho^s}(0)} (w_m(x_0 + d_\Omega x)) \le (1-\nu)^s d_\Omega \qquad \forall s = 0, ..., k_m + 1. \qquad (6.71)$$

Let $\alpha := -\log_\rho(1-\nu)$. From (6.71) we obtain that, for every m large enough, w_m is α-Hölder on points at distance at least $\rho^{-k_m} d_\Omega$, namely there exists C independent on m such that for every m large enough

$$|w_m(x) - w_m(y)| \le C|x-y|^\alpha \quad \forall x, y \in \Omega : |x-y| \ge 2^{-k_m} d_\Omega. \quad (6.72)$$

Since $k_m \to \infty$ as $m \to \infty$, it can be easily seen, with the same proof as the one of Ascoli-Arzela theorem, that the family $\{w_m\}_{m\in\mathbb{N}}$ of functions satisfying $\|w_m\|_{L^\infty(B_1(0))} \le 1$ and (6.72) is relatively compact with respect to the uniform convergence in Ω. Letting Ω vary in a countable family of open sets compactly supported in $B_1(0)$ which cover $B_1(0)$, with a diagonal argument we obtain (6.69).

We claim that w_∞ solves (6.66) in the viscosity sense. Indeed, assume by contradiction that w_∞ is not a supersolution of (6.66) in the viscosity sense. Then there exists a function $\varphi \in C^2(B_1(0))$ and a point $x_0 \in B_1(0)$ such that $\varphi(x_0) = w_\infty(x_0)$, $\varphi(x) < w_\infty(x)$ for every $x \in B_1(0) \setminus \{x_0\}$, and $a_{ij}(0)\partial_{ij}\varphi(x_0) > 0$. Since φ is C^2, there exists $r > 0$ such that

$$a_{ij}(0)\partial_{ij}\varphi(x) > 0 \qquad \forall x \in B_r(x_0). \qquad (6.73)$$

Since φ touches w_∞ strictly at x_0 and $w_m \to w_\infty$ uniformly, for every $m \in \mathbb{N}$ large enough there exist $c_m \in \mathbb{R}$ and $x_m \in B_r(x_0)$ such that $c_m + \varphi(x_m) = w_m(x_m)$, and $c_m + \varphi(x) \le w_m(x)$ for every $x \in B_r(x_0)$. In addition, $c_m \to 0$ and $x_m \to x_0$ as $m \to \infty$.

Let $h := \inf_{\partial B_{r/2}(x_0)}(w_\infty - \varphi)/2 > 0$. Since c_m converge to 0 and w_m converge to w_∞, for every m large enough $h \le \inf_{\partial B_{r/2}(x_0)}(w_m + c_m - \varphi)$. Let $(w_m + c_m - \varphi - h)^-$ be the negative part of the function $w_m + c_m - \varphi - h$, and let Γ_m be the convex envelope of $(w_m + c_m - \varphi - h)^-$ in $B_r(x_0)$.

Since the function $w_m + c_m - \varphi - h$ is of class C^2, it is a classical fact that Γ_m is of class $C^{1,1}$ inside $B_r(x_0)$ (see for instance [73]).

For every m let E_m be the contact set between $w_m + c_m - \varphi - h$ and Γ_m in $B_{r/2}(x_0)$, namely

$$E_m := \{x \in B_{r/2}(x_0) : w_m(x) + c_m - \varphi(x) - h = \Gamma_m(x)\}.$$

Recalling (6.73), we see that the function $w_m + c_m - \varphi - h$ solves

$$a_{ij}(\theta_m \nabla w_m)\partial_{ij}(w_m + c_m - \varphi - h) = f_m - a_{ij}(\theta_m \nabla w_m)\partial_{ij}\varphi$$
$$< f_m - [a_{ij}(\theta_m \nabla w_m) - a_{ij}(0)]\partial_{ij}\varphi \tag{6.74}$$

in $B_r(x_0)$. In addition, since Γ_m is convex, has oscillation h and vanishes on $\partial B_r(x_0)$, it is easy to see that

$$|\nabla \Gamma_m(x)| \leq \frac{2h}{r} \qquad \forall x \in B_{r/2}(0). \tag{6.75}$$

Since at the contact points the gradient of $w_m - \varphi$ coincides with the gradient of Γ_m, it follows that, for every $x \in E_m$,

$$a_{ij}(\theta_m \nabla w_m) - a_{ij}(0) = a_{ij}(\theta_m(\nabla\varphi + \nabla\Gamma_m)) - a_{ij}(0).$$

Hence the equation (6.74) is uniformly elliptic at the contact points for m large enough and in addition the term $a_{ij}(\theta_m \nabla w_m) - a_{ij}(0)$ converges uniformly to 0 on E_m as $m \to \infty$.

Hence, applying the Alexandroff-Bakelman-Pucci estimate (see Theorem 1.23) we obtain

$$\begin{aligned} h - c_m \leq & \sup_{B_{r/2}(x_0)} (w_m + c_m - \varphi - h)^- \\ \leq & Cr \left\| (f_m + (a_{ij}(\theta_m \nabla w_m) - a_{ij}(0))\partial_{ij}\varphi)^+ \right\|_{L^d(E_m)} \\ \leq & Cr \left(\|f_m\|_{L^d(B_1(0))} + \|a_{ij}(\theta_m \nabla w_m) - a_{ij}(0)\|_{L^d(E_m)} \|\varphi\|_{C^2(B_1(0))} \right), \end{aligned} \tag{6.76}$$

where $C > 0$ depends only on d, λ and Λ, and letting $m \to \infty$ we get

$$\begin{aligned} h \leq & Cr \liminf_{m \to +\infty} \left[\|f_m\|_{L^d(B_1(0))} + \|a_{ij}(\theta_m \nabla w_m) - a_{ij}(0)\|_{L^d(E_m)} \|\varphi\|_{C^2(B_1(0))} \right] \\ = & 0, \end{aligned}$$

a contradiction. A symmetric argument proves also that w_∞ is a subsolution of (6.66).

Therefore w_∞ solves (6.66) in the viscosity sense, and being (6.66) a uniformly elliptic equation with constant coefficients, w_∞ is actually a classical solution. This fact and (6.69) contradict (6.68). □

We prove an improvement of flatness result when the gradient is non-degenerate. In the following proposition the assumption $f \in L^q(B_1(0))$ for some $q > d$ plays a crucial role, and this is the optimal assumption one can make. Indeed, even for the Laplace equation $\Delta u = f$, the $C^{1,\alpha}$ regularity of the solution u is false for $f \in L^d$ (since $W^{2,d}$ does not embed into $C^{1,\alpha}$).

Proposition 6.13. *Let δ, $\mathcal{F}$, E, λ, Λ, f, u, and M be as in Theorem 6.4. There exist $\delta_0, \mu_0 > 0$, depending only on the modulus of continuity of $\nabla^2\mathcal{F}$, and on δ, λ, and Λ, such that the following holds:*

If $\|f\|_{L^q(B_2(0))} \leq \delta_0\mu_0$ and for any $x \in B_{1/2}(0)$ there exist $A_x \in \mathbb{R}^d$ and $b_x \in \mathbb{R}$ such that $1+\delta \leq |A_x|_E \leq M$ and $|u(y) - A_x \cdot y - b_x| \leq \delta_0$ for every $y \in B_1(0)$, then

$$|u(y) - u(x) - A \cdot (y - x)| \leq C|y - x|^{1+\alpha} \qquad \forall\, y \in B_1(0) \quad (6.77)$$

with $\alpha := 1 - d/q$, C depends only on δ, d, λ, and Λ, and $A \in \mathbb{R}^d$ satisfies

$$|A - A_0| \leq \frac{\widetilde{d}_E}{4}\delta. \qquad (6.78)$$

In particular $u \in C^{1,\alpha}(B_{1/4}(0))$ (with bounds depending only on the modulus of continuity of $\nabla^2\mathcal{F}$, on δ, d, λ, and Λ), and $|\nabla u|_E \geq 1+\delta/2$ inside $B_{1/4}(0)$.

Proof. We prove (6.77) for $x = 0$. Up to a vertical translation, we can assume without loss of generality that $u(0) = 0$. It suffices to show that there exists $r \in (0, 1)$ such that, for every $k \in \mathbb{N} \cup \{0\}$, there is a linear function $L_k(y) = A_k \cdot y + b_k$ satisfying

$$|u(y) - L_k(y)| \leq \delta_0 r^{k(\alpha+1)} \qquad \forall\, y \in B_{r^k}(0),$$

$$|A_k - A_{k+1}| \leq C'\delta_0 r^{k\alpha} \qquad |b_k - b_{k+1}| \leq C'\delta_0 r^{k(\alpha+1)}. \qquad (6.79)$$

For $k = 0$ the result is true by assumption.

Now we prove the result for $k + 1$ assuming it for $0, \ldots, k$. Let us consider the rescaled function

$$w(y) := \frac{u(r^k y) - L_k(r^k y)}{\delta_0 r^{k(\alpha+1)}} \qquad \forall\, y \in B_1(0). \qquad (6.80)$$

Observe that, by the inductive hypothesis, $|w| \leq 1$ inside $B_1(0)$ and w solves the equation

$$\partial_{ij}\mathcal{F}(A_k + \delta_0 r^{k\alpha}\nabla w(y))\partial_{ij}w(y) = \frac{r^{k(1-\alpha)}}{\delta_0} f(r^k y) \qquad \text{in } B_1(0).$$

Recalling that $\alpha = 1-d/q$, by a change of variable and Hölder inequality we get

$$\begin{aligned}\|r^k f(r^k y)\|_{L^d(B_1(0))} &= \|f\|_{L^d(B_{r^k}(0))}\\ &\leq |B_1(0)|^{1/q} r^{k\alpha}\|f\|_{L^q(B_{r^k}(0))} \qquad (6.81)\\ &\leq |B_1(0)|^{1/q} r^{k\alpha}\|f\|_{L^q(B_2(0))}.\end{aligned}$$

Since $\|f\|_{L^q(B_2(0))} \le \delta_0\mu_0$, we get

$$\frac{r^{k(1-\alpha)}}{\delta_0}\|f(r^k y)\|_{L^d(B_1(0))} \le |B_1(0)|^{1/q}\mu_0. \tag{6.82}$$

Recalling (6.53) and (6.79), by the inductive assumption we get

$$\begin{aligned}\widetilde{d}_E\sum_{i=0}^{k-1}|A_i - A_{i+1}|_E &\le \sum_{i=0}^{k-1}|A_i - A_{i+1}| \le C'\delta_0\sum_{i=0}^{k-1}r^{i\alpha}\\ &\le C'\delta_0\sum_{i=0}^{\infty}r^{i\alpha} \le \frac{\widetilde{d}_E}{4}\delta,\end{aligned}$$

provided we choose δ_0 small enough. Hence $A_k \notin E$, and more precisely

$$\begin{aligned}1+\frac{3}{4}\delta &\le |A_0|_E - \sum_{i=0}^{k-1}|A_i - A_{i+1}|_E \le |A_k|_E\\ &\le |A_0|_E + \sum_{i=0}^{k-1}|A_i - A_{i+1}|_E \le M + \frac{\delta}{4}.\end{aligned} \tag{6.83}$$

Define $a_{ij} : \mathbb{R}^d \to \mathbb{R}$ as $a_{ij}(v) := \partial_{ij}\mathcal{F}(A_k + v)$. Then by (6.53) and (6.83) we have

$$B_{\widetilde{d}_E\delta/4}(A_k) \subseteq \left\{|y - A_k|_E \le \frac{\delta}{4}\right\} \subseteq \left\{|y|_E \ge 1 + \frac{\delta}{2}\right\},$$

so by assumption (6.11) on $\mathcal{F}$ we get

$$\lambda I \le \nabla^2\mathcal{F}(v) \le \Lambda I \qquad \text{for any } v \in B_{\widetilde{d}_E\delta/4}(A_k),$$

which implies that the coefficients a_{ij} are uniformly elliptic inside $B_{\widetilde{d}_E\delta/4}(0)$ with constants λ, Λ.

Let σ and μ be the functions provided by Lemma 6.12. If δ_0 is small enough so that $\delta_0 r^{k\alpha} \le \sigma(r^{1+\alpha}/2)$, and μ_0 is small enough so that $|B_1(0)|^{1/q}\mu_0 \le \mu(r^{1+\alpha}/2)$, Lemma 6.12 applied to w implies the existence of a function $v : B_1(0) \to \mathbb{R}$ such that

$$\partial_{ij}\mathcal{F}(A_k)\partial_{ij}v = 0 \qquad \text{in } B_1(0)$$

and

$$|v(y) - w(y)| \le \frac{r^{1+\alpha}}{2} \qquad \forall\, y \in B_{1/2}(0). \tag{6.84}$$

In particular, since $|v(y)| \le |v(y) - w(y)| + |w(y)| \le 3/2$ in $B_{1/2}(0)$, and v solves a uniformly elliptic equation with constant coefficients, there exist $C' > 0$ (depending only on d, λ, Λ) and a linear function $L(y) = A \cdot y + b$, such that

$$|v(y) - A \cdot y - b| \leq C'|y|^2 \qquad \forall\, y \in B_{1/4}(0).$$

In particular, if $C'r^{1-\alpha} \leq 1/2$ and $r \leq 1/4$, we get

$$|v(y) - A \cdot y - b| \leq C'r^2 \leq \frac{r^{1+\alpha}}{2} \qquad \forall\, y \in B_r(0). \tag{6.85}$$

Hence, first we choose $0 < r < 1/4$ such that

$$C'r^{1-\alpha} \leq \frac{1}{2},$$

then fix δ_0 such that

$$\delta_0 r^\alpha \leq \sigma(r^{1+\alpha}/2) \qquad \text{and} \qquad C'\delta_0 \sum_{i=0}^{\infty} r^{i\alpha} \leq \frac{\widetilde{d}_E}{4}\delta,$$

and finally take μ_0 such that

$$|B_1(0)|^{1/q}\mu_0 \leq \mu(r^{1+\alpha}/2).$$

Then from (6.84) and (6.85) we get

$$|w(y) - A \cdot y - B| \leq |w(y) - v(y)| + |v(y) - A \cdot y - B| \leq r^{1+\alpha} \quad \forall\, y \in B_r(0),$$

which can be rewritten in terms of u as (see (6.80))

$$|u(y) - L_{k+1}(y)| \leq \delta_0 r^{(k+1)(\alpha+1)} \qquad \forall\, y \in B_{r^{k+1}}(0),$$

where

$$L_{k+1}(y) := L_k(y) - \delta_0 r^{k(\alpha+1)} L\left(\frac{y}{r^k}\right).$$

It is easy to check that (6.79) holds for some C' large enough independent of δ_0 and r, and this concludes the proof of the inductive step.

Also, it follows from (6.79) and the definition of δ_0 that

$$|A_k - A_0| \leq \sum_{i=0}^{k-1} |A_i - A_{i+1}| \leq \frac{\widetilde{d}_E}{4}\delta, \tag{6.86}$$

which proves (6.78) in the limit.

Finally, the fact that (6.77) implies that $u \in C^{1,\alpha}(B_{1/4}(0))$ is standard (see for instance [73, Lemma 3.1]). □

6.5. Proof of Theorems 6.4 and 6.1

Proof of Theorem 6.4. For any $x_0 \in B_1(0)$ and $r \in (0,1)$, we have

$$\int_{B_1(0)} |rf(x_0+rx)|^p\,dx = \int_{B_r(x_0)} r^{p-d}|f(x)|^p\,dx \le r^{p-d}\|f\|^p_{L^p(B_2(0))}.$$

Let μ_0 and δ_0 be as in Proposition 6.13. Fix $r < 1/2$ small enough such that

$$r^{1-d/p}\|f\|_{L^p(B_1(0))} \le \delta_0\mu_0,$$

so that

$$\|rf(x_0+rx)\|_{L^p(B_1(0))} \le \delta_0\mu_0. \tag{6.87}$$

Consider now the function $w : B_1(0) \to \mathbb{R}$ given by

$$w(x) := \frac{1}{r}u(x_0+rx) \qquad \forall x \in B_1(0),$$

which by (6.12) solves

$$\partial_i[\partial_{ij}\mathcal{F}(\nabla w(x))\partial_j w(x))] = rf(x_0+rx). \tag{6.88}$$

Our goal is to show that the quantity

$$\sup_{x\in B_{2^{-i}}(0)} \left\{|(|\nabla w(x)|_E - 1 - \delta)_+ - (|\nabla w(0)|_E - 1 - \delta)_+|\right\} \qquad \forall i \in \mathbb{N}, \tag{6.89}$$

decays geometrically.

For every $i \in \mathbb{N}$ set

$$d_i := \sup_{x\in B_{2^{-i}}(0)} (|\nabla w(x)|_E - (1+\delta))_+,$$

and let k be the smallest value of $i \in \mathbb{N}$ such that

$$\begin{aligned} \sup_{e^*\in\partial E^*} \Big|\Big\{x \in B_{2^{-2i-1}(0)} : (\partial_{e^*} w(x) - (1+\delta))_+ \\ \ge (1-\eta)d_{2i}\Big\}\Big| \ge (1-\eta)|B_{2^{-2i-1}}(0)| \end{aligned} \tag{6.90}$$

($k = \infty$ if there is no such i). By Lemma 6.10 there exists a constant $C_0 > 0$ and $\alpha_0 \in (0,1)$ such that

$$d_{2i} \le C_0 2^{-2i\alpha_0} \qquad \forall i = 0, \ldots, k. \tag{6.91}$$

If $k = \infty$, then there is nothing to prove. Assume then that k is finite.

For every $k+1 \le i \le 2k$ we estimate d_{2i} with d_{2k}, and from (6.91) applied to d_{2k} we obtain

$$d_{2i} \le d_{2k} \le C_0 2^{-2k\alpha_0} \le C_0 2^{-i\alpha_0}. \tag{6.92}$$

We now scale the function w in order to preserve its gradient:

$$v(x) := 2^{2k+1}(w(2^{-2k-1}x) - w(0)) \qquad \forall x \in B_1(0).$$

Since $\nabla v(x) = \nabla w(2^{-2k-1}x)$, from (6.90) we obtain that there exists $e^* \in \partial E^*$ such that

$$\left|\left\{x \in B_1(0) : (\partial_{e^*} v(x) - (1+\delta))_+ \ge (1-\eta) d_{2k}\right\}\right| \ge (1-\eta)|B_1(0)|. \tag{6.93}$$

Moreover, we have that $|\nabla v(x)|_E \le d_{2k} + \delta + 1 \le M/\widetilde{d}_E$ for every $x \in B_1(0)$ (recall (6.53)). Hence, from Lemma 6.11 applied to v with $\varepsilon = \delta_0 \widetilde{d}_E / M$ (with δ_0 as in Proposition 6.13) and $\zeta = d_{2k}$, there exist $A \in \mathbb{R}^d$ with $|A|_E = d_{2k+1} + \delta + 1$ and $b \in \mathbb{R}$ such that

$$|v(x) - A \cdot x - b| \le \varepsilon(d_{2k+1} + \delta + 1) \le \varepsilon M/\widetilde{d}_E = \delta_0 \quad \forall x \in B_1(0). \tag{6.94}$$

From (6.88), (6.87), and (6.94), the hypothesis of Proposition 6.13 are satisfied, so there exists a constant C_1, depending only on δ, d, λ, and Λ, such that

$$|\nabla v(x) - \nabla v(0)| \le C_1 |x|^{\alpha_1} \qquad \forall x \in B_{1/4}(0),$$

where $\alpha_1 := 1 - d/q$. Since the function $x \to (|x| - 1 - \delta)_+$ is 1-Lipschitz, we get

$$\begin{aligned} |(|\nabla w(x)| - 1 - \delta)_+ - (|\nabla w(0)| - 1 - \delta)_+| &\le |\nabla w(x) - \nabla w(0)| \\ &= |\nabla v(2^{2k+1}x) - \nabla v(0)|, \end{aligned}$$

for every $x \in B_{2^{-2k-2}}(0)$. In particular, for any $i \ge 2k+1$ and $x \in B_{2^{-2i}}(0)$ we have

$$\begin{aligned} |(|\nabla w(x)| - 1 - \delta)_+ - (|\nabla w(0)| - 1 - \delta)_+| &\le C_1 2^{(2k+1)\alpha_1} |x|^{\alpha_1} \\ &\le C_1 2^{(2k+1-2i)\alpha_1} \\ &\le C_1 2^{-i\alpha_1}. \end{aligned} \tag{6.95}$$

Setting $\bar{C} := 2\max\{C_0, C_1\}$ and $\bar{\alpha} := \min\{\alpha_0, \alpha_1\}/2$, from (6.91), (6.92), and (6.95), we obtain that for every $i \in \mathbb{N}$

$$\sup_{x \in B_{2^{-2i}}(0)} \left\{|(|\nabla w(x)|_E - 1 - \delta)_+ - (|\nabla w(0)|_E - 1 - \delta)_+|\right\} \le \bar{C} 2^{-2i\bar{\alpha}},$$

namely

$$\sup_{x\in B_{2^{-2i}r}(x_0)} \{|(|\nabla u(x)|_E - 1 - \delta)_+ - (|\nabla u(x_0)|_E - 1 - \delta)_+|\} \le \bar{C}2^{-2i\bar{\alpha}},$$

from which (6.13) follows easily. □

Proof of Theorem 6.1. Let $\Omega' \Subset \Omega'' \Subset \Omega''' \Subset \Omega$ and set $M := \|\nabla u\|_{L^\infty(\Omega''')}$ (M is finite because u is locally Lipschitz inside Ω). Recall that $\mathcal{F}$ is C^2 outside $\overline{E}$, so in particular it is C^2 for $|v| > d_E$ (recall (6.53)).

We now want to find a functional $\mathcal{G} \in C^2(\mathbb{R}^d \setminus \overline{E})$ which coincides with $\mathcal{F}$ inside $B_M(0)$ (so that $\mathcal{F}(\nabla u) = \mathcal{G}(\nabla u)$ inside Ω''') but $\mathcal{G}$ is quadratic at infinity. We follow a construction used in [1].

Let $M' = \sup\{\mathcal{F}(v) : v \in B_{M+2d_E}(0)\}$. Let $\psi : [0,\infty) \to \mathbb{R}$ be a C^∞ function such that $\psi(t) = t$ in $[0, M'+1]$, and $\psi(t) = M'+2$ in $[M'+2, \infty]$. Since $\mathcal{F}$ is coercive, the function $\psi(\mathcal{F}(v))$ is constant outside some ball. Hence

$$N := \sup_{|v|>M+d_E} |\nabla^2[\psi \circ \mathcal{F}](v)|$$

is finite. Let $\phi \in C^2(\mathbb{R}^d)$ be a convex function such that $\phi(x) = 0$ for every $x \in B_{M+d_E}(0)$, $\nabla^2\phi(x) \le (2N+1)\,\mathrm{Id}$ for every $x \in \mathbb{R}^d$ and $\nabla^2\phi(x) \ge (N+1)\,\mathrm{Id}$ for every $x \in \mathbb{R}^d \setminus B_{M+2d_E}(0)$. Define

$$\mathcal{G}(v) := \psi(\mathcal{F}(v)) + \phi(v) \qquad \forall v \in \mathbb{R}^d. \tag{6.96}$$

Computing the Hessian of $\mathcal{G}$, we obtain that $\mathcal{G}$ is convex, that $\nabla^2\mathcal{G}(v) \le (3N+1)\,\mathrm{Id}$ for every $|v| > M + d_E$ and that $\mathrm{Id} \le \nabla^2\mathcal{G}(v)$ for every $|v| > M + 2d_E$. Since $\mathcal{G} = \mathcal{F}$ inside $B_{2d_E+M}(0)$ and u solves the Euler-Lagrange equation (6.12) in the sense of distributions, u solves also the Euler-Lagrange equation for $\mathcal{G}$, and so by convexity of $\mathcal{G}$ it is a minimizer for the functional

$$\int_{\Omega'''} \mathcal{G}(\nabla u) + fu.$$

By (6.7) we have that for every $\delta > 0$ small there exist $\lambda'_\delta, \Lambda'_\delta > 0$, depending only on $\lambda_{\delta/4}, \Lambda_{\delta/4}, N$, such that

$$\lambda'_\delta\,\mathrm{Id} \le \nabla^2\mathcal{G}(v) \le \Lambda'_\delta\,\mathrm{Id} \qquad \text{for a.e. } v \text{ such that } 1 + \frac{\delta}{4} \le |v|_E. \tag{6.97}$$

Let ρ_ε be a standard mollification kernel whose support is contained in $B_\varepsilon(0)$ and let

$$\mathcal{G}_\varepsilon(x) := \rho_\varepsilon * \mathcal{G}(x) + \varepsilon|x|^2, \qquad f_\varepsilon(x) := \rho_\varepsilon * f(x),$$

$$u_\varepsilon := \operatorname{argmin}\left\{\int_{\Omega'''} \mathcal{G}_\varepsilon(\nabla u) + f_\varepsilon u : u \in W^{1,2}(\Omega''')\right\}.$$

Note that $u_\varepsilon \in C^\infty(\Omega''')$ thanks to the regularity of $\mathcal{G}_\varepsilon$ and f_ε, and thanks to the uniform convexity of $\mathcal{G}_\varepsilon$. From (6.97), for every δ small there exist $\lambda''_\delta, \Lambda''_\delta > 0$, depending only on $\lambda'_\delta, \Lambda'_\delta, N$, such that, for $\varepsilon \leq \delta/(4\widetilde{d}_E)$,

$$\lambda''_\delta \operatorname{Id} \leq \nabla^2 \mathcal{G}_\varepsilon(v) \leq \Lambda''_\delta \operatorname{Id} \qquad \text{for a.e. } v \text{ such that } 1 + \frac{\delta}{2} \leq |v|_E. \tag{6.98}$$

Differentiating the Euler equation solved by u_ε with respect to ∂_e for any $e \in \mathbb{S}^{d-1}$ we obtain that

$$\partial_i[\partial_{ij}\mathcal{G}_\varepsilon(\nabla u_\varepsilon(x))\partial_j(\partial_e u_\varepsilon(x))] = \partial_e f_\varepsilon(x). \tag{6.99}$$

Hence the function $v_\varepsilon(x) := (|\nabla u_\varepsilon(x)| - (1 + d_E))_+$ is a subsolution of the equation

$$\partial_i[\partial_{ij}\mathcal{G}_\varepsilon(\nabla u_\varepsilon(x))\partial_j v_\varepsilon] \geq \partial_e f 1_{\{1/2 - v_e(x) > 0\}}.$$

As we already observed in the proof of Lemma 6.10, this equation is uniformly elliptic because the values of $\partial_{ij}\mathcal{G}_\varepsilon(\nabla u_\varepsilon(x))$ are not important when $|\nabla u_\varepsilon(x)| \leq 1 + d_E$. Hence, we can apply [91, Theorem 8.17] to obtain

$$\begin{aligned} &\|(|\nabla u_\varepsilon(x)| - (1 + d_E))_+\|_{L^\infty(\Omega'')} \\ &\leq C'(1 + \|(|\nabla u_\varepsilon(x)| - (1 + d_E))_+\|_{L^2(\Omega''')}) \\ &\leq C'(1 + \|\nabla u_\varepsilon(x)\|_{L^2(\Omega''')}) \end{aligned} \tag{6.100}$$

for some constant C' depending only on $d, \lambda_{\delta_0}, \Lambda_{\delta_0}, \Omega'', \Omega'''$ (for some δ_0 small).

Since the function G_ε has quadratic growth at infinity, we get

$$\|\nabla u_\varepsilon(x)\|_{L^2(\Omega'')} \leq C\left(1 + \int_{\Omega'''} G_\varepsilon(\nabla u_\varepsilon(x))\, dx\right). \tag{6.101}$$

From the boundedness of energies of u_ε, (6.100), and (6.101), it follows that the functions u_ε are M'-Lipschitz for ε small.

Let E_δ be a strictly convex set such that $E \subseteq E_\delta \subseteq (1 + \delta/2)E$. Since

$$\left\{|v|_{E_\delta} > 1 + \frac{\delta}{2}\right\} = \left\{v \notin \left(1 + \frac{\delta}{2}\right) E_\delta\right\} \subseteq \left\{v \notin \left(1 + \frac{\delta}{2}\right) E\right\},$$

from (6.98) it follows that $\lambda''_\delta I \leq \nabla^2\mathcal{G}_\varepsilon(x) \leq \Lambda''_\delta$ for a.e. x such that $1 + \frac{\delta}{2} \leq |x|_{E_\delta}$. Applying Theorem 6.4 to u_ε and E_δ, by a covering argument we deduce that there exists a constant D_δ (independent of ε) such that

$$\left|(|\nabla u_\varepsilon(x)|_{E_\delta} - 1 - \delta)_+ - (|\nabla u_\varepsilon(y)|_{E_\delta} - 1 - \delta)_+\right| \leq D_\delta |x - y|^\alpha \quad \forall x, y \in \Omega'. \tag{6.102}$$

Without loss of generality, up to adding a constant to u_ε we can assume that $u_\varepsilon(0) = 0$. Hence, since $|\nabla u_\varepsilon| \le M$, we obtain that, up to adding a constant a subsequence,

$$u_\varepsilon \to u_0 \qquad \text{uniformly in } \Omega'$$

and

$$\nabla u_\varepsilon \rightharpoonup \nabla u_0 \qquad \text{weakly* in } L^\infty(\Omega') \tag{6.103}$$

for some Lipschitz function u_0. We claim that $\nabla u_0 = \nabla u$ outside E and that

$$(|\nabla u_\varepsilon(x)|_{E_\delta} - 1 - \delta)_+ \to (|\nabla u(x)|_{E_\delta} - 1 - \delta)_+ \quad \text{strongly in } L^p(\Omega') \tag{6.104}$$

for every $p < \infty$.

Indeed, from the convergence of the energies on a sequence of local minimizers, and thanks to the uniform convergence of $\mathcal{G}_\varepsilon$ to $\mathcal{G}$ on $B_{M'}(0)$, we have that

$$\begin{aligned}
\int_{\Omega'} \mathcal{G}(\nabla u(x))\,dx &= \lim_{\varepsilon\to 0} \int_{\Omega'} \mathcal{G}_\varepsilon(\nabla u_\varepsilon(x))\,dx \\
&= \lim_{\varepsilon\to 0} \int_{\Omega'} \mathcal{G}(\nabla u_\varepsilon(x))\,dx \\
&= \int_{\Omega'} \mathcal{G}(\nabla u_0(x))\,dx,
\end{aligned} \tag{6.105}$$

Since $\mathcal{G}$ is strictly convex outside E, it follows by standard results in the calculus of variations that $\nabla u_0 = \nabla u$ outside E and (6.104) holds (a possible way to show these facts, is to consider the Young measure ν_x generated by ∇u_ε, and show that $\nu_x = \delta_{\nabla u(x)}$ for a.e. x such that $\nabla u(x) \notin E$).

Hence, thanks to (6.104), we can take the limit as $\varepsilon \to 0$ in (6.102) to obtain $(|\nabla u|_{E_\delta} - 1 - \delta)_+ \in C^{0,\alpha}(\Omega')$. In particular, the set

$$A_\delta := \left\{x \in \Omega' : |\nabla u(x)|_{E_\delta} > 1 + \delta\right\}$$

is open. Moreover, from the choice of E_δ, it follows easily that

$$F_\delta := \left\{x \in \Omega' : |\nabla u(x)|_E > 1 + 2\delta\right\} \subset A_\delta \tag{6.106}$$

Since every partial derivative of u solves (6.99) (with $\varepsilon = 0$) which is uniformly elliptic inside A_δ, from De Giorgi regularity theorem it follows that $\nabla u \in C^{0,\alpha'}(F_\delta)$, with $C^{0,\alpha'}$ norm bounded by a constant which depends only on α, M, δ, λ_δ, Λ_δ, A_δ, and f. By the arbitrariness of δ, we deduce that ∇u is continuous inside the open set $\{|\nabla u|_E > 1\}$ with a universal modulus of continuity.

We also note that, since the functions $(|v|_{E_\delta} - 1 - \delta)_+$ converge uniformly to $(|v|_E - 1)_+$ on $B_{M'}(0)$, we get that $(|\nabla u|_{E_\delta} - 1 - \delta)_+$ converge uniformly to $(|\nabla u|_E - 1)_+$, so also $(|\nabla u|_E - 1)_+$ is continuous with a universal modulus of continuity.

Combining this fact with the continuity of ∇u inside $\{|\nabla u|_E > 1\}$ and the fact that $\mathcal{H}$ is continuous and vanishes on E, it is easy to check that $\mathcal{H}(\nabla u)$ is continuous (again with a universal modulus of continuity) everywhere inside Ω'. □

Chapter 7
An excess-decay result for a class of degenerate elliptic equations

As in the previous chapter, we study the local regularity of minimizers of the functional

$$\int_{\Omega} \mathcal{F}(\nabla u) + fu \tag{7.1}$$

where $\Omega \subseteq \mathbb{R}^d$ is an open set, $\mathcal{F} : \mathbb{R}^d \to \mathbb{R}$, $f : \Omega \to \mathbb{R}$, and $u : \Omega \to \mathbb{R}$. When a uniform ellipticity condition on $\mathcal{F}$ holds true, the regularity results are classical, as presented in the introductory Section 1.6. Even in the vectorial case, the picture is well understood: for instance, partial regularity of minimizers was proved under the uniform strict quasiconvexity assumption in [90, 2] (see also the references quoted therein).

To understand regularity for more degenerate elliptic problems, a natural idea is to prove Hölder regularity at points which do not see too much the degeneracy of the equation,namely points where the gradient is close to a value where the function $\mathcal{F}$ is C^2 and uniformly convex. This scheme has been carried out by Anzellotti and Giaquinta in [24] under the uniform convexity assumption for elliptic systems and in [3] if uniform strict quasiconvexity is assumed. In the latter paper it is proved that, if $u : \mathbb{R}^d \to \mathbb{R}^N$ (with $N \geq 1$) and

$$\lim_{r \to 0} \fint_{B_r(x_0)} |\nabla u(y) - \xi_0|^2 \, dy = 0 \tag{7.2}$$

for some $\xi_0 \in \mathbb{R}^{dN}$ and $x_0 \in \mathbb{R}^d$, $\mathcal{F}$ is C^2 in a neighborhood of ξ_0, and a uniform strict quasiconvexity holds true around ξ_0, then u is of class $C^{1,\alpha}$ in a neighborhood of x_0 for every $\alpha < 1$. Their proof is based on a linearization argument. They differentiate the Euler equation

$$\partial_i(\partial_i \mathcal{F}(\nabla u)) = f \qquad \text{in } \Omega$$

(here and in the following we use the Einstein's summation convention) with respect to a direction $e \in \mathbb{S}^{d-1}$ to obtain

$$\partial_i[\partial_{ij}\mathcal{F}(\nabla u(x))\partial_j(\partial_e u(x))] = \partial_e f(x) \qquad \text{in } \Omega.$$

Then, using (7.2), they prove that the solution of the differentiated operator is close, on smaller scales, to the solution v of a differential operator with constant coefficients

$$\partial_i[\partial_{ij}\mathcal{F}(\xi_0)\partial_j v(x)] = 0 \qquad \text{in } \Omega.$$

Since $\mathcal{F}$ is strictly quasiconvex in ξ_0, this equation is in turn nondegenerate. In this way, they obtain regularity of u from the regularity of the linearized operator.

In this chapter we study the regularity of minimizers of the function (7.1) in the scalar case assuming that $\mathcal{F}$ is $C^{1,1}$ and uniformly elliptic outside a ball, and ellipticity may degenerate inside. Basic examples which fall under these assumptions are $\mathcal{F}(x) = \boldsymbol{n}(x)^p$ for some $p > 1$ with $\boldsymbol{n}$ an elliptic norm (see Definition 7.4), and $\mathcal{F}(x) = (|x| - 1)_+^p$ for some $p > 1$ (notice that, since we consider Lipschitz minimizers, the behavior of $\mathcal{F}$ at infinity is not relevant). The first example arises as an anisotropic generalization of the p-laplacian, whereas the second example has been already presented in Chapter 6 and it is related to some recent problems of traffic dynamic. In the following we assume that $\mathcal{F} \in C^{1,1}$ outside the degeneracy region to prove that every locally Lipschitz minimizer is $C^{1,\alpha}$ at nondegenerate points, weakening the assumptions of Theorem 6.1 (where $\mathcal{F}$ was assumed to be of class C^2). When $\mathcal{F}$ is assumed to be $C^{1,1}$ new techniques are needed. In this respect we mention a De Giorgi type approach in a work of De Silva and Savin [74]; it looks possible to us that also their technique may lead to prove our result, but we believe that our approach in this setting has its own interest. On the contrary, the results in [3, 58] described above assumed $\mathcal{F} \in C^2$ and this assumption cannot be easily removed with their technique, since their proof is based on a linearization argument which cannot work if the second derivatives of $\mathcal{F}$ are not continuous, because the linearized operator has no reason to stay close to the nonlinear one. Our approach is still based on a blow-up argument; however, we prove that the operator can be linearized, up to subsequence, around a limit operator which is uniformly elliptic thanks to the fact that the gradient is assumed to be mainly outside the degeneracy. To obtain strong compactness of a rescaled sequence, we use an idea of De Silva and Savin [74] presented in Lemma 7.10.

The chapter is organized as follows. In Section 7.1 we present the basic estimate of decay of the excess function around nondegenerate points. Then we see that this estimate can be iterated at every scale to obtain the $C^{1,\alpha}$ regularity. Finally, we see that the smallness assumption is satisfied if u is close to a linear nondegenerate function in a certain sense, which in turn can be verified in the applications. In Section 7.2 we see how the estimate allows to prove $C^{1,\alpha}$ regularity for the solutions of the anisotropic

p-laplacian and regularity outside the degeneracy for the equations arising in the context of traffic congestion. In Section 7.3 we collect all the proofs.

7.1. Excess-decay result at nondegenerate points and consequences

First we introduce the excess function, which measures the distance of the gradient of a solution ∇u from its average. In terms of this quantity we express the smallness condition which guarantees regularity. The $C^{0,\alpha}$ regularity for ∇u is expressed in terms of the decay of the excess itself, through Campanato's Theorem.

We denote by $B_r(x)$ the open ball of center $x \in \mathbb{R}^d$ and radius $r > 0$, often shortened as B_r if $x = 0$. Given $g : \Omega \to \mathbb{R}^d$, with the notation $\fint_{B_r(x)} g$ or $(g)_{B_r(x)}$ we mean the average of g on the ball $B_r(x)$.

Let Ω be an open set and let $f \in L^q(\Omega)$ for some $q > d$. For every $u \in W^{1,2}(\Omega), x \in \Omega, r < d(x, \Omega)$ we consider the excess

$$U(u, x, r) := \left(\fint_{B_r(x)} |\nabla u(y) - (\nabla u)_{B_r(x)}|^2 \, dy\right)^{1/2} + r^{(q-d)/(2q)} \|f\|_{L^q(B_1)}.$$

The following Theorem provides an excess-decay estimate for local minimizers of the functional (7.1) at points where ∇u is nondegenerate. In order for the equation to be considered nondegenerate on a certain ball $B_r(x)$, we require that the average $(\nabla u)_{B_r(x)}$ is not in the degeneracy region, that ∇u does not oscillate too much, and that the scale r is chosen sufficiently small to make the right-hand side irrelevant. These last two informations are encoded in the smallness of the excess. As we shall show in the corollaries below, the result can be iterated on smaller scales to provide Hölder regularity for the gradient around nondegenerate points.

Theorem 7.1. *Let* $0 < \lambda \leq \Lambda$ *and let* $f \in L^q(B_1)$ *for some* $q > d \geq 2$. *Let* $\mathcal{F} : \mathbb{R}^d \to \mathbb{R}$ *be a convex function such that* $\mathcal{F} \in C^{1,1}(\mathbb{R}^d \setminus B_{1/4}(0))$ *and*

$$\lambda \,\mathrm{Id} \leq \nabla^2 \mathcal{F}(x) \leq \Lambda \,\mathrm{Id} \qquad \text{for } \mathscr{L}^d\text{-a.e. } x \in \mathbb{R}^d \setminus B_{1/4}(0). \tag{7.3}$$

Let $u \in W^{1,\infty}(B_1)$ *be a minimizer of the functional* (7.1) *and let us assume that* $|\nabla u| \leq 1$ *in* B_1.

Then there exist $\tau_0, \alpha > 0$, *depending only on* $d, q, \lambda, \Lambda, \|\nabla \mathcal{F}\|_{L^\infty(B_1)}$, *such that for every* $\tau \leq \tau_0$ *there exists* $\varepsilon = \varepsilon(\tau)$ *for which the following property holds true: If for some* $x \in B_{1/2}$ *and* $r < 1/4$ *we have*

$$\frac{3}{4} \leq |(\nabla u)_{B_r(x)}| \leq 1, \qquad U(u, x, r) \leq \varepsilon,$$

then

$$U(u,x,\tau r) \le \tau^\alpha U(u,x,r).$$

Theorem 7.1 can be iterated to obtain the decay of the excess at every scale.

Corollary 7.2. *Let $\lambda, \Lambda, q, f, \mathcal{F}$, and u be as in Theorem 7.1. Then there exist $\tau_0, \alpha > 0$, depending only on $d, q, \lambda, \Lambda, \|\nabla\mathcal{F}\|_{L^\infty(B_1)}$, such that for every $\tau \le \tau_0$ there exists $\varepsilon = \varepsilon(\tau)$ for which the following property holds true: If for some $x \in B_{1/2}$ and $r < 1/4$ we have*

$$\frac{7}{8} \le |(\nabla u)_{B_r(x)}| \le 1, \qquad U(u,x,r) \le \varepsilon, \tag{7.4}$$

then

$$U(u,x,\tau^k r) \le \tau^{\alpha k} U(u,x,r) \qquad \forall k \in \mathbb{N}. \tag{7.5}$$

The assumption in Corollary 7.3 is satisfied in a ball if the gradient of u is aligned in a fixed direction, as the following corollary states. This will be in turn useful to obtain $C^{1,\alpha}$ regularity at nondegenerate points in the applications of Section 7.2.

Corollary 7.3. *Let λ, Λ, q, f, $\mathcal{F}$, and u be as in Theorem 7.1. Then there exist $\eta, \alpha, C, \tau, r_0 > 0$, depending only on $d, q, \lambda, \Lambda, \|f\|_{L^q(B_1)}$, $\|\nabla\mathcal{F}\|_{L^\infty(B_1)}$, such that if $|\nabla u(x)| \le 1$ for every $x \in B_1$ and*

$$|\{x \in B_1 : \partial_{\mathbf{v}} u(x) \ge 1-\eta\}| \ge (1-\eta)|B_1| \tag{7.6}$$

for some $\mathbf{v} \in \mathbb{S}^{d-1}$, then

$$U(u,x,\tau^k r_0) \le \tau^{\alpha k} U(u,x,r_0) \qquad \forall k \in \mathbb{N} \qquad \forall x \in B_{1/2}. \tag{7.7}$$

In particular, we have

$$\|u\|_{C^{1,\alpha}(B_{1/2})} \le C. \tag{7.8}$$

7.2. Applications: the anisotropic p-Laplace equation and traffic models

The anisotropic p-Laplace equation. The simplest example of degenerate elliptic equation is given by the p-Laplace equation

$$\partial_i(|\nabla u|^{p-2}\partial_i u) = f,$$

corresponding to the choice $\mathcal{F}(x) = |x|^p/p$ in the minimization of the function (7.1); in this case the degeneracy consists in a single point, the origin, and it is possible to obtain $C^{1,\alpha}$ regularity of the solution (see

Section 1.6.1). In the following, we introduce a generalization of the p-laplacian which involves an anisotropic norm. We consider an open set $\Omega \subseteq \mathbb{R}^d$ and a local minimizer for the functional

$$\int_\Omega \frac{\boldsymbol{n}(\nabla u)^p}{p} + fu, \tag{7.9}$$

where $\boldsymbol{n} : \mathbb{R}^d \to \mathbb{R}^+$ is a positively 1-homogeneous convex function and $f \in L^q(\Omega)$ for some $q > d$.

To ensure the equation to be elliptic outside the origin, we need to consider only norms which satisfy an ellipticity condition in the direction orthogonal to $\nabla \boldsymbol{n}$. For example, the p-norms (namely $\boldsymbol{n}(x) = (|x_1|^p + ... + |x_d|^p)^{1/p}$ for $x = (x_1, ..., x_d) \in \mathbb{R}^d$) are not included in the following definition and indeed the problem of regularity of minimizers is, to our knowledge, open.

Definition 7.4. An "elliptic norm" $\boldsymbol{n} \in C^{1,1}_{\mathrm{loc}}(\mathbb{R}^d \setminus \{0\})$ is a convex positively 1-homogenous function with $\boldsymbol{n}(0) = 0$, positive outside the origin, for which there exist $\lambda, \Lambda > 0$ such that

$$\lambda \left| \tau - (\tau \cdot \nabla \boldsymbol{n}(v)) \frac{\nabla \boldsymbol{n}(v)}{|\nabla \boldsymbol{n}(v)|^2} \right|^2 \leq \boldsymbol{n}(v) \partial_{ij} \boldsymbol{n}(v) \tau_i \tau_j \leq \Lambda |\tau|^2 \tag{7.10}$$

for $\mathscr{L}^d$-a.e. $v \in \mathbb{R}^d$, $\tau \in \mathbb{R}^d$.[1]

In the following, we prove that every Lipschitz solution of the anisotropic p-Laplace equation is $C^{1,\alpha}$.

Theorem 7.5. *Let* $0 < \lambda \leq \Lambda$, $p > 1$, Ω *a bounded open subset of* $\mathbb{R}^d$, $d \geq 2$, *and* $f \in L^q(B_1)$ *for some* $q > d$. *Let* $\boldsymbol{n} : \mathbb{R}^d \to \mathbb{R}$ *be an elliptic norm and let* $u \in W^{1,\infty}_{\mathrm{loc}}(\Omega)$ *be a local minimizer of the functional* (7.9).

Then there exists $\alpha \in (0, 1)$, *which depends only on* $d, p, q, \lambda, \Lambda$, $\|\nabla \boldsymbol{n}\|_\infty$ *such that* $\nabla u \in C^{0,\alpha}_{\mathrm{loc}}(\Omega)$, *namely for every* $\Omega' \Subset \Omega$ *there exists a constant* $C > 0$ *such that*

$$|\nabla u(x) - \nabla u(y)| \leq C|x - y|^\alpha \qquad \forall x, y \in \Omega'.$$

This constant C *depends only on* $d, p, q, \lambda, \Lambda$, $\|\nabla \boldsymbol{n}\|_{L^\infty(\mathbb{R}^d)}$, $\mathrm{dist}(\Omega', \partial\Omega)$, $\|f\|_q$, *and* $\|\nabla u\|_\infty$ *in a neighborhood of* Ω'.

---

[1] In this definition the term "norm" is used with a slight abuse of notation: indeed we are not requiring the symmetry of $\boldsymbol{n}$, namely $\boldsymbol{n}(v) = \boldsymbol{n}(-v)$. We also observe that an equivalent formulation for (7.10) is to ask that

$$\lambda' |\tau|^2 \leq \partial_{ij} \mathcal{H}(v) \tau_i \tau_j \leq \Lambda' |\tau|^2 \qquad \forall v, \tau \in \mathbb{R}^n$$

for some $0 < \lambda' \leq \Lambda'$, where $\mathcal{H}(v) := (\boldsymbol{n}(v))^2$.

In the theorem above we assume Lipschitz regularity of the solution to prove $C^{1,\alpha}$ regularity; notice that the Lipschitz regularity follows from [84, 37, 87]. To avoid annoying details about a regularization argument, we prove the result in terms of an a-priori estimate; hence we assume that u is smooth, and so is $\boldsymbol{n}$ outside the origin (For more details about the regularization, see for instance the proof of Theorem 6.1).

The key idea to prove Theorem 7.5 is a lemma which provides a separation between degeneracy and nondegeneracy; here, there is a clear analogy with the main idea behind the proof of Theorem 6.1. The basic lemma says that the gradient of the solution ∇u is either close to a nonzero constant, or it decays on a smaller ball. When the first case happens at some scale, we obtain $C^{1,\alpha}$ regularity of u through Corollary 7.3. Otherwise, the decay of ∇u at every scale provides $C^{1,\alpha}$ regularity of u.

As we show now the dichotomy, stated at scale one in Lemma 7.11 (compare with Lemma 6.10), is based on the construction of suitable subsolutions to a uniformly elliptic equation, namely $(\partial_e u(x) - 1/2)_+$ for every $e \in \mathbb{S}^{d-1}$. Indeed, let $u : B_1 \to \mathbb{R}$ be a Lipschitz local minimizer of (7.1) with Lipschitz constant 1; then it solves the Euler equation

$$\partial_i\Big[\boldsymbol{n}\big(\nabla u(x)\big)^{p-1}\partial_i \boldsymbol{n}\big(\nabla u(x)\big)\Big] = f(x) \qquad x \in B_1. \tag{7.11}$$

Let us introduce the coefficients

$$A_{ij}(x) := \boldsymbol{n}(x)^{p-2}\Big((p-1)\partial_i \boldsymbol{n}(x)\partial_j \boldsymbol{n}(x) + \boldsymbol{n}(x)\partial_{ij}\boldsymbol{n}(x)\Big) \qquad \forall x \in \mathbb{R}^d. \tag{7.12}$$

Given $e \in \mathbb{S}^{d-1}$, we differentiate (7.11) in the direction $e \in \mathbb{S}^{d-1}$ to obtain

$$\partial_i\Big[A_{ij}\big(\nabla u(x)\big)\partial_j\big(\partial_e u(x)\big)\Big] = \partial_e f(x).$$

We notice that, setting

$$a_{ij}(x) := (p-1)\partial_i \boldsymbol{n}(x)\partial_j \boldsymbol{n}(x) + \boldsymbol{n}(x)\partial_{ij}\boldsymbol{n}(x) \qquad \forall x \in \mathbb{R}^d, \tag{7.13}$$

the coefficients a_{ij} are uniformly elliptic. Indeed, $\nabla \boldsymbol{n}$ is 0-homogeneous and since $\boldsymbol{n} \in C^{1,1}_{\mathrm{loc}}(\mathbb{R}^d \setminus \{0\})$ we have that $0 < c \le |\nabla \boldsymbol{n}| \le C < \infty$; therefore for every $\tau \in \mathbb{R}^d$ we obtain that

$$\begin{aligned} a_{ij}\tau_i\tau_j &\ge (p-1)|\nabla \boldsymbol{n}(v)|^2 \left|\tau \cdot \frac{\nabla \boldsymbol{n}(v)}{|\nabla \boldsymbol{n}(v)|}\right|^2 + \lambda\left|\tau - (\tau\cdot\nabla \boldsymbol{n}(v))\frac{\nabla \boldsymbol{n}(v)}{|\nabla \boldsymbol{n}(v)|^2}\right|^2 \\ &\ge \min\{c^2(p-1), \lambda\}|\tau|^2, \end{aligned}$$

and analogously from above. Hence the coefficients A_{ij} are uniformly elliptic in every compact region which does not contain the origin.

Since the function $t \mapsto (t-1/2)_+$ is convex and Lipschitz with derivative $\mathbf{1}_{\{t>1/2\}}$, it follows that the function

$$v_e(x) := (\partial_e u(x) - 1/2)_+ \qquad e \in \mathbb{S}^{d-1} \tag{7.14}$$

is a subsolution of the equation

$$\partial_i \Big[A_{ij}\big(\nabla u(x)\big) \partial_j v_e(x) \Big] = \partial_e f(x) 1_{\{\partial_e u > 1/2\}}(x).$$

Notice that the values of the coefficients $A_{ij}(\nabla u(x))$ are only relevant when $1/2 \le |\nabla u(x)| \le 1$. Indeed the solution satisfies $|\nabla u(x)| \le 1$ (by assumption), and when $|\nabla u(x)| \le 1/2$ we have that $v_e(x) = 0$. Therefore, thanks to the ellipticity assumption on $\boldsymbol{n}$, the equation might be assumed to be uniformly elliptic.

The idea of the proof now follows a paper by Wang [122], where Theorem 7.5 is presented for the classical p-laplacian. In this case, however, the author considers a different subsolution, namely $\boldsymbol{n}(\nabla u)^p$, which solves an elliptic equation with nondegenerate coefficients. Indeed, given a locally Lipschitz minimizer of (7.1) with $f = 0$, the coefficients a_{ij} (introduced in (7.13)) are uniformly elliptic and the function $\boldsymbol{n}(\nabla u)^p$ formally solves

$$\partial_i \Big[a_{ij}\big(\nabla u(x)\big) \partial_j \Big(\boldsymbol{n}\big(\nabla u(x)\big)^p \Big) \Big] \ge 0.$$

The choice of the subsolution in [122] leads to additional difficulties to pass from a nondegenerate slope of u in modulus to closeness to a linear function. Moreover, the regularity at nondegenerate points is carried out in [122] through the analysis of the equation in nondivergence form, proving as a key lemma that any solution of the p-laplace equation is close to the solution of the linearized problem at nondegeneracy points. Wang's scheme can be carried out for a general elliptic norm $\boldsymbol{n}$ only assuming better regularity on $\boldsymbol{n}$, namely $\boldsymbol{n} \in C^2(\mathbb{R}^d \setminus \{0\})$. Hence, as we shall see in Section 7.3, the proof of Theorem 7.5 requires the use of our Theorem 7.1.

Degenerate elliptic equations and traffic models. Corollary 7.3 can be used to prove local $C^{0,\alpha}$ regularity of the gradient of the solution of a degenerate elliptic equation outside the degeneracy region.

The following result is a generalization of Theorem 6.1 to more general functions $\mathcal{F}$ (we do not require C^2 regularity of $\mathcal{F}$). The degeneracy region is a convex set containing the origin, described, in coherence with the present chapter, as the unit ball of a convex positively 1-homogenous function which does not need to be elliptic. The variational proof is based on Corollary 7.3, which in turn uses a different technique with respect to the proof presented in Chapter 6, that is based on some ideas of Savin [115] and Wang [122].

Theorem 7.6. *Let* $0 < \lambda \leq \Lambda$, Ω *a bounded open subset of* $\mathbb{R}^d$, $d \geq 2$, $f \in L^q(\Omega)$ *for some* $q > d$. *Let* $\boldsymbol{m} : \mathbb{R}^d \to \mathbb{R}$ *be a convex positively* 1*-homogenous function with* $\boldsymbol{m}(0) = 0$ *which is positive outside the origin. Let* $\mathcal{F} : \mathbb{R}^d \to \mathbb{R}$ *be a convex nonnegative function such that* $\mathcal{F} \in C^{1,1}_{loc}(\mathbb{R}^d \setminus \overline{\{\boldsymbol{m} \leq 1\}})$, *and assume that for every* $\delta > 0$ *there exist* $\lambda_\delta, \Lambda_\delta > 0$ *such that*

$$\lambda_\delta I \leq \nabla^2 \mathcal{F}(x) \leq \Lambda_\delta I \qquad \text{for } \mathscr{L}^d\text{-a.e. } x \text{ such that } 1+\delta \leq \boldsymbol{m}(x) \leq 1/\delta.$$

Let $u \in W^{1,\infty}_{\mathrm{loc}}(\Omega)$ *be a local minimizer of the functional* (7.1). *Then, for any continuous function* $H : \mathbb{R}^d \to \mathbb{R}$ *such that* $\{\boldsymbol{m} \leq 1\} \subseteq \{H = 0\}$, *we have*

$$H(\nabla u) \in C^0(\Omega).$$

More precisely, for every open set $\Omega' \Subset \Omega$ *there exists a modulus of continuity* $\omega : [0, \infty) \to [0, \infty)$ *for* $H(\nabla u)$ *on* Ω', *which depends only on* d, *the modulus of continuity of* H, *the functions* $\delta \to \lambda_\delta, \delta \to \Lambda_\delta$, $\|\nabla u\|_\infty$ *in a neighborhood* $\Omega'' \subset \Omega$ *of* Ω', *and* $\|\nabla \mathcal{F}\|_\infty$ *in a neighborhood of* $\nabla u(\Omega'')$, *such that*

$$\big|H(\nabla u(x)) - H(\nabla u(y))\big| \leq \omega(|x - y|) \qquad \forall x, y \in \Omega'.$$

In particular, if $\mathcal{F} \in C^1(\mathbb{R}^d)$ *then* $\nabla \mathcal{F}(\nabla u) \in C^0(\Omega)$.

7.3. Proofs

Proof of Theorem 7.1. Before proving the result, we state some simple lemmas. The proof of the first lemma is an easy computation which is left to the reader.

Lemma 7.7. *Let* $p > 1$, $X \in \mathbb{R}^d$, *and let* $\mathbf{v}_1, ..., \mathbf{v}_d \in \mathbb{R}^d$ *be a family of vectors satisfying* $|\mathbf{v}_i| = 1$ *for any* $i = 1, ..., d$ *and* $\big|\det\big(\mathbf{v}_1|\dots|\mathbf{v}_d\big)\big| > c_0 > 0$ *(here* $\big(\mathbf{v}_1|\dots|\mathbf{v}_d\big)$ *denotes the matrix whose columns are given by the vectors* $\mathbf{v}_1, ..., \mathbf{v}_d \in \mathbb{R}^d$*). Then there exists a constant* $c > 0$, *which depends only on* d *and* c_0, *such that*

$$|X \cdot \mathbf{v}_j| \leq |X| \leq \frac{1}{c} \sum_{i=1}^{d} |X \cdot \mathbf{v}_i| \qquad \forall j = 1, ..., d. \tag{7.15}$$

Proof. The first inequality follows by $|X \cdot v_j| \leq |v_j||X|$. To prove the second inequality we estimate $|X|$ with $|X \cdot e_1| + ... + |X \cdot e_d|$; we write each element of the canonical basis of $\mathbb{R}^d$, namely e_j, as a linear combination of $v_1, ..., v_d$; we estimate each $|X \cdot e_j|$ with the same linear combination of $|X \cdot v_i|$. Hence we proved (7.15) with a constant c that may depend on the particular choice of $v_1, ..., v_d$. A simple contradiction argument shows that the constant depends only on c_0. □

From Lemma 7.7 we deduce that, given independent unit vectors $\mathbf{v}_1, ..., \mathbf{v}_d \in \mathbb{R}^d$ and $X \in L^2(\Omega; \mathbb{R}^d)$, we have

$$\|X \cdot \mathbf{v}_j\|_{L^2(\Omega)} \leq \|X\|_{L^2(\Omega;\mathbb{R}^d)} \leq \frac{1}{c} \sum_{i=1}^{d} \|X \cdot \mathbf{v}_i\|_{L^2(\Omega)} \qquad \forall j = 1, ..., d.$$

This implies the following lemma:

Lemma 7.8. *Let $\Omega \subseteq \mathbb{R}^d$ be an open set. Let $\{X_h\}_{h\in\mathbb{N}} \subseteq L^2(\Omega; \mathbb{R}^d)$, $X_\infty \in L^2(\Omega; \mathbb{R}^d)$, and let $\{\mathbf{v}_1, ..., \mathbf{v}_d\}$ be a basis of $\mathbb{R}^d$. Then $\{X_h\}_{h\in\mathbb{N}}$ is precompact in $L^2(\Omega; \mathbb{R}^d)$ if and only if $\{X_h \cdot \mathbf{v}_i\}_{h\in\mathbb{N}}$ is precompact in $L^2(\Omega)$ for every $i = 1, .., d$. If this happens then we have that*

$$\lim_{h\to\infty} X_h = X_\infty \quad \text{in } L^2(\Omega; \mathbb{R}^d) \qquad \text{if and only if} \qquad \lim_{h\to\infty} X_h \cdot \mathbf{v}_i = X_\infty \cdot \mathbf{v}_i \quad \text{in } L^2(\Omega)\ \forall i = 1, .., d. \tag{7.16}$$

Another useful lemma is the following:

Lemma 7.9. *Let $\Lambda > \lambda > 0$ and $r > 0$. For every $h \in \mathbb{N}$ let $A^h : B_r \to \mathbb{R}^{d\times d}$ be a sequence of measurable functions such that $A^h(x)$ is a nonnegative symmetric matrix for $\mathscr{L}^d$-a.e. $x \in B_r$, $A^h \leq \Lambda\,\mathrm{Id}$ and*

$$\lim_{h\to\infty} \left|\{A^h \leq \lambda\,\mathrm{Id}\}\right| = 0. \tag{7.17}$$

Then there exists a measurable function $A : B_r \to \mathbb{R}^{d\times d}$ such that $A(x)$ is a nonnegative symmetric matrix for $\mathscr{L}^d$-a.e. $x \in B_r$,

$$\lambda\,\mathrm{Id} \leq A(x) \leq \Lambda\,\mathrm{Id} \qquad \text{for } \mathscr{L}^n\text{-a.e. } x \in B_r, \tag{7.18}$$

and, up to subsequences,

$$A^h \to A \qquad \text{weakly in } L^2(B_r; \mathbb{R}^{d\times d}). \tag{7.19}$$

Proof. Since $0 \leq A^h \leq \Lambda\,\mathrm{Id}$ for every $h \in \mathbb{N}$ we have that there exists a function $A : B_r \to \mathbb{R}^{d\times d}$ with $0 \leq A \leq \Lambda\,\mathrm{Id}$ and such that, up to a subsequence, (7.19) holds. By (7.17), up to a further subsequence we may assume that

$$\sum_{h=1}^{\infty} \big|\{A^h \leq \lambda\,\mathrm{Id}\}\big| < \infty. \tag{7.20}$$

Setting

$$I_k = \bigcup_{k \leq h} \{A^h \leq \lambda\,\mathrm{Id}\} \qquad \forall k \in \mathbb{N}$$

we have that $|I_k| \to 0$ by (7.20) and that, by (7.19), $A^h \to A$ weakly in $L^2(B_r \setminus I_k; \mathbb{R}^{d\times d})$ for every $k \in \mathbb{N}$. The set $\{A \in \mathbb{R}^{d\times d} : \lambda\,\mathrm{Id} \leq A \leq \Lambda\,\mathrm{Id}\}$ is convex and closed in $\mathbb{R}^{d\times d}$. Since $\lambda\,\mathrm{Id} \leq A^h(x) \leq \Lambda\,\mathrm{Id}$ for every $x \in B_r \setminus I_k$ and for every $h > k$, we take the limit in the weak convergence as $h \to \infty$ and we obtain that $\lambda\,\mathrm{Id} \leq A(x) \leq \Lambda\,\mathrm{Id}$ for $\mathscr{L}^d$-a.e. $x \in B_r \setminus I_k$. Since k is arbitrary, we obtain (7.18). □

The following lemma is a Caccioppoli inequality for a subsolution of an elliptic differential operator in terms of an a priori estimate. The proof follows an idea in [74, Proposition 2.3] and it is based on the variational structure of the equation (7.22).

Lemma 7.10. *Let* $\mathbf{v} \in \mathbb{S}^{d-1}$, $\lambda > 0$, $c > 0$, *and* $f \in C^1(B_1)$. *Let* $\mathcal{F} \in C^2(\mathbb{R}^d)$ *be a convex function such that*

$$\lambda\,\mathrm{Id} \leq \nabla^2 \mathcal{F}(x) \qquad \text{for all } x \in \mathbb{R}^d \text{ such that } x \cdot \mathbf{v} \geq c. \tag{7.21}$$

Let $u \in C^2(B_1)$ *be a solution of*

$$\partial_i(\partial_i \mathcal{F}(\nabla u)) = f \qquad \text{in } B_1 \tag{7.22}$$

which is Lipschitz with constant 1 *in* B_1. *Let* $G : \mathbb{R} \to \mathbb{R}$ *be a nondecreasing* 1*-Lipschitz function which is constant on the set* $\{t \leq c\}$. *Then there exists* $C > 0$, *depending only on* d *and* λ, *such that for every* $\eta \in \mathbb{R}^d$

$$\begin{aligned} &\big\|\nabla[G(\partial_{\mathbf{v}} u)]\big\|_{L^2(B_{3/4})} \\ &\leq C\big(\|G(\partial_{\mathbf{v}} u)\|_{L^2(B_1)} + \|f\|_{L^2(B_1)} + \|\nabla \mathcal{F}(\nabla u) - \eta\|_{L^2(B_1)}\big). \end{aligned} \tag{7.23}$$

Proof. By approximation, it suffices to prove the result when $G \in C^1$. We differentiate the equation (7.22) in the direction $\mathbf{v}$ to get

$$\partial_i(\partial_{ij} \mathcal{F}(\nabla u) \partial_{j\mathbf{v}} u) = \partial_{\mathbf{v}} f \qquad \text{in } B_1.$$

Let $\zeta \in C_c^\infty(B_1)$ be a nonnegative and smooth cutoff function which is 1 in $B_{3/4}$. We test the above equation with the test function $G(\partial_{\mathbf{v}} u)\,\zeta^2$, which is Lipschitz and compactly supported, and we integrate by parts:

$$\begin{aligned}
&\int_{B_1} \partial_{ij}\mathcal{F}(\nabla u)\, \partial_{j\mathbf{v}} u\, \partial_i[G(\partial_{\mathbf{v}} u)]\, \zeta^2 \\
&= -2\int_{B_1} \partial_{ij}\mathcal{F}(\nabla u)\, \partial_{j\mathbf{v}} u\, G(\partial_{\mathbf{v}} u)\, \zeta\, \partial_i \zeta + \int_{B_1} f\, \partial_{\mathbf{v}}[G(\partial_{\mathbf{v}} u)]\, \zeta^2 \\
&\quad + 2\int_{B_1} f\, G(\partial_{\mathbf{v}} u)\, \zeta\, \partial_{\mathbf{v}} \zeta.
\end{aligned} \tag{7.24}$$

We estimate each term of (7.24). As regards the left-hand side we notice that $G'(\partial_{\mathbf{v}} u) = 0$ on the set $\{\partial_{\mathbf{v}} u \le c\}$. Hence we apply (7.21) and the fact that $0 \le G' \le 1$ to get

$$\begin{aligned}
\int_{B_1} \partial_{ij}\mathcal{F}(\nabla u)\, \partial_{j\mathbf{v}} u\, G'(\partial_{\mathbf{v}} u)\, \partial_{i\mathbf{v}} u\, \zeta^2 &\ge \lambda \int_{B_1} G'(\partial_{\mathbf{v}} u)\, |\nabla \partial_{\mathbf{v}} u|^2 \zeta^2 \\
&\ge \lambda \int_{B_1} |\nabla[G(\partial_{\mathbf{v}} u)]|^2 \zeta^2
\end{aligned} \tag{7.25}$$

To estimate the first term in the right-hand side of (7.24) we integrate by parts and, for some ε to be chosen later, we have

$$\begin{aligned}
&-2\int_{B_1} \partial_{ij}\mathcal{F}(\nabla u)\, \partial_{j\mathbf{v}} u\, G(\partial_{\mathbf{v}} u)\, \zeta\, \partial_i \zeta \\
&= -2\int_{B_1} \partial_{\mathbf{v}}[\partial_i \mathcal{F}(\nabla u) - \eta_i]\, G(\partial_{\mathbf{v}} u)\, \zeta\, \partial_i \zeta \\
&= 2\int_{B_1} [\partial_i \mathcal{F}(\nabla u) - \eta_i]\, \partial_{\mathbf{v}}[G(\partial_{\mathbf{v}} u)]\, \zeta\, \partial_i \zeta \\
&\quad + 2\int_{B_1} [\partial_i \mathcal{F}(\nabla u) - \eta_i]\, G(\partial_{\mathbf{v}} u)\, \partial_{\mathbf{v}}[\zeta\, \partial_i \zeta] \\
&\le \varepsilon \int_{B_1} |\nabla[G(\partial_{\mathbf{v}} u)]|^2 \zeta^2 + \frac{\|\nabla \zeta\|_\infty^2}{\varepsilon} \int_{B_1} |\nabla \mathcal{F}(\nabla u) - \eta|^2 \\
&\quad + \|\nabla[\zeta \nabla \zeta]\|_\infty^2 \int_{B_1} |G(\partial_{\mathbf{v}} u)|^2 + \int_{B_1} |\nabla \mathcal{F}(\nabla u) - \eta|^2
\end{aligned} \tag{7.26}$$

As regards the last two terms in (7.24) we have

$$\begin{aligned}
&\int_{B_1} f\, \partial_{\mathbf{v}}[G(\partial_{\mathbf{v}} u)] \zeta^2 + 2\int_{B_1} f\, G(\partial_{\mathbf{v}} u)\, \zeta\, \partial_{\mathbf{v}} \zeta \\
&\le \frac{\varepsilon}{2} \int_{B_1} |\nabla[G(\partial_{\mathbf{v}} u)]|^2 \zeta^2 + \frac{1}{2\varepsilon} \int_{B_1} f^2 + \|\nabla \zeta\|_\infty^2 \int_{B_1} |G(\partial_{\mathbf{v}} u)|^2 \\
&\quad + \int_{B_1} f^2.
\end{aligned} \tag{7.27}$$

We choose $\varepsilon \le \lambda/3$ and we obtain from (7.24), (7.25), (7.26), (7.27) that there exists a constant C, depending only on d and λ, such that

$$\begin{aligned}\int_{B_{3/4}} |\nabla[G(\partial_\nu u)]|^2 &\le \int_{B_1} |\nabla[G(\partial_\nu u)]|^2 \zeta^2 \\ &\le C\Big(\int_{B_1} |G(\partial_\nu u)|^2 + \int_{B_1} f^2 + \int_{B_1} |\nabla\mathcal{F}(\nabla u) - \eta|^2\Big),\end{aligned}$$

proving (7.23). □

Proof of Theorem 7.1. With a standard regularization, presented in detail in an analogous situation in the proof of Theorem 6.1, we may assume without loss of generality that $\mathcal{F} \in C^2(B_1)$, $f \in C^1(B_1)$, and that $u \in C^2(B_1)$ is a solution of

$$\partial_i(\partial_i \mathcal{F}(\nabla u)) = f \qquad \text{in } B_1. \tag{7.28}$$

By contradiction, let $\tau, \alpha > 0$ to be chosen later and let us consider sequences $\{x_h\}_{h\in\mathbb{N}} \subseteq B_{1/2}$, $\{r_h\}_{h\in\mathbb{N}} \subseteq (0, 1/4)$, and $\{u_h\}_{h\in\mathbb{N}} \subseteq C^2(B_1)$ such that u_h are solutions to (7.28) and

$$|\nabla u_h| \le 1 \qquad \text{in } B_1 \quad \forall h \in \mathbb{N}, \tag{7.29}$$

$$U(u_h, x_h, r_h) = \lambda_h \to 0 \qquad \text{as } h \to \infty, \tag{7.30}$$

$$U(u_h, x_h, \tau r_h) > \tau^\alpha U(u_h, x_h, r_h) \qquad \forall h \in \mathbb{N}, \tag{7.31}$$

$$(\nabla u_h)_{B_{r_h}(x_h)} \to \gamma_\infty \quad \text{as } h \to \infty, \quad \gamma_\infty \in \mathbb{R}^d, \ \frac{3}{4} \le |\gamma_\infty| \le 1. \tag{7.32}$$

Let us define the rescaled functions

$$\tilde{u}_h(x) := \frac{u_h(x_h + r_h x)}{r_h} \qquad x \in B_1;$$

since u_h are solutions to (7.28) we have

$$\partial_i(\partial_i \mathcal{F}(\nabla \tilde{u}_h)) = \tilde{f}_h \qquad \text{in } B_1, \tag{7.33}$$

where $\tilde{f}_h(x) := r_h f(x_h + r_h x)$ for $x \in B_1$. Moreover, setting $\gamma_h := (\nabla u_h)_{B_{r_h}(x_h)}$, we have that $\gamma_h = (\nabla \tilde{u}_h)_{B_1}$. We remark that, by a change of variables,

$$\begin{aligned}\|\tilde{f}_h\|_{L^q(B_1)} &= r_h^{(q-d)/q}\Big(\int_{B_{r_h}(x_h)} |f(y)|^q \, dy\Big)^{1/q} \\ &= r_h^{(q-d)/q} \|f\|_{L^q(B_{r_h}(x_h))}\end{aligned} \tag{7.34}$$

By the change of variable formula we rewrite (7.29), (7.30), (7.31), and (7.32) in terms of $\tilde{u}_h$:

$$|\nabla \tilde{u}_h| \leq 1 \qquad \forall h \in \mathbb{N}, \tag{7.35}$$

$$\left(\fint_{B_1} |\nabla \tilde{u}_h(y) - \gamma_h|^2 \, dy \right)^{1/2} + r_h^{(q-d)/(2q)} \|f\|_{L^q(B_1)} = \lambda_h \to 0 \quad \text{as } h \to \infty, \tag{7.36}$$

(which implies that $r_h \to 0$ as $h \to \infty$ unless $f \equiv 0$),

$$\left(\fint_{B_\tau} \frac{|\nabla \tilde{u}_h - (\nabla \tilde{u}_h)_{B_\tau}|^2}{\lambda_h^2} \right)^{1/2} + \frac{(\tau r_h)^{(q-d)/(2q)}}{\lambda_h} \|f\|_{L^q(B_1)} > \tau^\alpha \quad \forall h \in \mathbb{N}, \tag{7.37}$$

$$(\nabla \tilde{u}_h)_{B_1} = \gamma_h \to \gamma_\infty \qquad \text{as } h \to \infty, \qquad \gamma_\infty \in \mathbb{R}^d, \ \frac{3}{4} \leq |\gamma_\infty| \leq 1. \tag{7.38}$$

By Poincaré inequality and (7.36) we have that

$$\|\tilde{u}_h(x) - \tilde{u}_h(0) - \gamma_h \cdot x\|_{L^2(B_1)} \lesssim \lambda_h; \tag{7.39}$$

therefore the functions

$$\frac{\tilde{u}_h(x) - \tilde{u}_h(0) - \gamma_h \cdot x}{\lambda_h}$$

are bounded in $W^{1,2}(B_1)$. Hence there exists $u_\infty \in W^{1,2}(B_1)$ such that, up to a subsequence,

$$\frac{\tilde{u}_h(x) - \tilde{u}_h(0) - \gamma_h \cdot x}{\lambda_h} \to u_\infty(x) \qquad \text{in } L^2(B_1), \tag{7.40}$$

$$\frac{\nabla \tilde{u}_h(x) - \gamma_h}{\lambda_h} \to \nabla u_\infty(x) \qquad \text{weakly in } L^2(B_1). \tag{7.41}$$

The scheme of the proof is the following. In Step 1 we employ the Caccioppoli-type inequality of Lemma 7.10 to obtain that the (suitably rescaled) partial derivatives $\partial_{\mathbf{v}} \tilde{u}_h(x)$ in certain directions $\mathbf{v}$ are strongly precompact in L^2. In Step 2 we deduce that $(\nabla \tilde{u}_h - \gamma_h)/\lambda_h$ converges to ∇u_∞ strongly in L^2, by taking d linearly independent directions in Step 1. Next, we would like to show that, for any $\mathbf{v} \in \mathbb{S}^{d-1}$, the function $\partial_{\mathbf{v}} u_\infty$ solves a uniformly elliptic PDE. Indeed, $\partial_{\mathbf{v}} u_\infty$ is a limit of solutions $(\partial_{\mathbf{v}} \tilde{u}_h - \gamma_h \cdot \mathbf{v})/\lambda_h$ of degenerate PDEs, whose degeneracy becomes less relevant as $h \to \infty$ due to the fact that $\nabla \tilde{u}_h$ is nondegenerate on average and the excess vanishes.

However, the equation for $(\partial_{\mathbf{v}}\tilde{u}_h - \gamma_h \cdot \mathbf{v})/\lambda_h$ involves the second derivatives of $\tilde{u}_h$, and unfortunately we don't have any strong convergence at this level. We overcome this difficulty by finding the equation solved by the incremental quotients $\partial^\varepsilon_{\mathbf{v}}\tilde{u}_h = [\tilde{u}_h(\cdot + \varepsilon\mathbf{v}) - \tilde{u}_h]/\varepsilon$ and taking the limit as $h \to \infty$ with ε fixed (see Steps 3-5). Finally, in Step 6 we apply De Giorgi-Nash-Moser Theorem to obtain Hölder estimates for the incremental quotients $\partial^\varepsilon_{\mathbf{v}} u_\infty$; this provides an excess decay for u_∞ which, in turn, gives a contradiction.

Step 1: precompactness of certain rescaled partial derivatives of $\tilde{u}_h$ in L^2 via a Caccioppoli-type inequality. Let $\mathbf{v} \in \mathbb{S}^{d-1}$ be such that $5/8 < \gamma_\infty \cdot \mathbf{v}$ (so that $1/2 < \gamma_h \cdot \mathbf{v} \le 1$ for h large enough), and set

$$v_h(x) := \left(\partial_{\mathbf{v}}\tilde{u}_h(x) - \frac{\gamma_h \cdot \mathbf{v}}{2}\right)_+ - \frac{\gamma_h \cdot \mathbf{v}}{2},$$

$$w_h(x) := \partial_{\mathbf{v}}\tilde{u}_h(x) - \gamma_h \cdot \mathbf{v}. \tag{7.42}$$

From the fact that

$$v_h = w_h \qquad \text{on } \left\{x \in B_1 : \partial_{\mathbf{v}}\tilde{u}_h(x) \ge \frac{\gamma_h \cdot \mathbf{v}}{2}\right\}$$

and

$$0 > v_h = -\frac{\gamma_h \cdot \mathbf{v}}{2} > w_h \qquad \text{on } \left\{x \in B_1 : \partial_{\mathbf{v}}\tilde{u}_h(x) < \frac{\gamma_h \cdot \mathbf{v}}{2}\right\}$$

we obtain

$$\|v_h\|_{L^2(B_1)} \le \|w_h\|_{L^2(B_1)} \le C_0\lambda_h, \tag{7.43}$$

which implies

$$\lim_{h\to\infty} \|v_h\|_{L^2(B_1)} = 0, \qquad \lim_{h\to\infty} \|w_h\|_{L^2(B_1)} = 0. \tag{7.44}$$

Let $\sigma := 2d/(d-1)$. We claim that there exist constants $C_1, C_2, C_3 > 0$ such that

$$\begin{aligned}\|\nabla v_h\|_{L^2(B_{3/4})} &\le C_1\big(\|v_h\|_{L^2(B_1)} + \|\tilde{f}_h\|_{L^2(B_1)} \\ &\qquad + \|\nabla\mathcal{F}(\nabla\tilde{u}_h) - \nabla\mathcal{F}(\gamma_h)\|_{L^2(B_1)}\big) \\ &\le C_2\lambda_h,\end{aligned} \tag{7.45}$$

$$\|v_h\|_{L^\sigma(B_{3/4})} + \|w_h\|_{L^\sigma(B_{3/4})} \le C_3\lambda_h, \tag{7.46}$$

$$\lim_{h\to\infty} \left\|\frac{v_h - w_h}{\lambda_h}\right\|_{L^2(B_{3/4})} = 0. \tag{7.47}$$

Notice that from (7.45) and (7.43) we obtain that the sequence $\{v_h/\lambda_h\}_{h\in\mathbb{N}}$ is bounded in $W^{1,2}(B_{3/4})$ and therefore it is precompact in $L^2(B_{3/4})$; from (7.47) we also obtain that

$$\text{the sequence } \{w_h/\lambda_h\}_{h\in\mathbb{N}} \text{ is precompact in } L^2(B_{3/4}). \tag{7.48}$$

We now prove (7.45), (7.46), and (7.47). By Lemma 7.10 applied with $u = \tilde{u}_h$, $f = \tilde{f}_h$, $c = \gamma_h \cdot \mathbf{v}/2 > 1/4$, $\eta = \nabla\mathcal{F}(\gamma_h)$, and $G(t) = (t - \gamma_h \cdot \mathbf{v}/2)_+ - \gamma_h \cdot \mathbf{v}/2$, we obtain that

$$\|\nabla v_h\|_{L^2(B_{3/4})} \le C_1\Big(\|v_h\|_{L^2(B_1)} + \|\tilde{f}_h\|_{L^2(B_1)} + \|\nabla\mathcal{F}(\nabla\tilde{u}_h) - \nabla\mathcal{F}(\gamma_h)\|_{L^2(B_1)}\Big)$$

We claim that the three terms in the right-hand side can be estimated by the excess λ_h up to a constant. Indeed by (7.43) we estimate the first term; from (7.34) we deduce that

$$\|\tilde{f}_h\|_{L^2(B_1)} \lesssim \|\tilde{f}_h\|_{L^q(B_1)} \le r_h^{-(q-d)/(2q)} \|\tilde{f}_h\|_{L^q(B_1)} \lesssim \lambda_h.$$

Finally, for the last term we remember that $|\gamma_h| \ge 3/4$, $\mathcal{F}$ is Lipschitz in B_1 (by convexity) and $\mathcal{F} \in C^{1,1}(\mathbb{R}^d \setminus B_{1/4})$. Hence, $|\nabla\mathcal{F}(\nabla\tilde{u}_h) - \nabla\mathcal{F}(\gamma_h)|$ can be estimated thanks to the Lipschitz regularity of $\nabla\mathcal{F}$ on the set $\{|\nabla\tilde{u}_h| \ge 1/4\}$; on the complement $\{|\nabla\tilde{u}_h| < 1/4\}$ we estimate the quantity $|\nabla\mathcal{F}(\nabla\tilde{u}_h) - \nabla\mathcal{F}(\gamma_h)|$ by $2\|\nabla\mathcal{F}\|^2_{L^\infty(B_1)}$ and we notice that on that set $|\nabla\tilde{u}_h - \gamma_h| \ge 1/8$. We therefore obtain

$$\begin{aligned}
&\int_{B_1} |\nabla\mathcal{F}(\nabla\tilde{u}_h) - \nabla\mathcal{F}(\gamma_h)|^2 \\
&\le C\Big(\|\nabla^2\mathcal{F}\|^2_{L^\infty(\mathbb{R}^d\setminus B_{1/4})} + \|\nabla\mathcal{F}\|^2_{L^\infty(B_1)}\Big)\int_{B_1} |\nabla\tilde{u}_h - \gamma_h|^2
\end{aligned}$$

and we conclude the proof of the second inequality in (7.45). Since $W^{1,2}(B_{3/4})$ embeds into $L^\sigma(B_{3/4})$, by (7.45) we have that

$$\|v_h\|_{L^\sigma(B_{3/4})} \le C_4\lambda_h;$$

from the higher integrability of v_h and the fact that $\gamma_h \cdot \mathbf{v}/2 \ge 1/4$ we obtain

$$\begin{aligned}
&\left|\left\{x \in B_{3/4} : \partial_{\mathbf{v}}\tilde{u}_h(x) < \frac{\gamma_h \cdot \mathbf{v}}{2}\right\}\right| \\
&\le \left|\left\{x \in B_{3/4} : \partial_{\mathbf{v}}\tilde{u}_h(x) < \frac{\gamma_h \cdot \mathbf{v}}{2}\right\}\right| 4^\sigma \left(\frac{\gamma_h \cdot \mathbf{v}}{2}\right)^\sigma \\
&\le 4^\sigma \|v_h\|^\sigma_{L^\sigma(B_{3/4})} \le C_5\lambda_h^\sigma.
\end{aligned} \tag{7.49}$$

Then, from (7.49) and since $\tilde{u}_h$ is Lipschitz with constant 1 (see (7.35)) we get

$$\left\| \frac{v_h - w_h}{\lambda_h} \right\|^2_{L^2(B_{3/4})} \leq \frac{4}{\lambda_h^2} \left| \left\{ x \in B_{3/4} : \partial_{\mathbf{v}} \tilde{u}_h(x) < \frac{\gamma_h \cdot \mathbf{v}}{2} \right\} \right| \leq 4C_5 \lambda_h^{\sigma-2},$$

which converges to 0 by (7.36) and proves (7.47).

Finally, by (7.35), (7.49), and (7.43) we have

$$\begin{aligned} &\|w_h\|^\sigma_{L^\sigma(B_{3/4})} \\ &\leq \int_{B_{3/4} \cap \{\partial_{\mathbf{v}} \tilde{u}_h \geq \frac{\gamma_h \cdot \mathbf{v}}{2}\}} |\partial_{\mathbf{v}} \tilde{u}_h(x) - \gamma_h \cdot \mathbf{v}|^\sigma + \left| \left\{ x \in B_{3/4} : \partial_{\mathbf{v}} \tilde{u}_h(x) < \frac{\gamma_h \cdot \mathbf{v}}{2} \right\} \right| 2^\sigma \\ &\leq \|v_h\|^\sigma_{L^\sigma(B_{3/4})} + C_5 2^\sigma \lambda_h^\sigma \leq (1 + 2^\sigma C_5) \lambda_h^\sigma, \end{aligned}$$

which proves (7.46).

Step 2: strong convergence of the rescaled gradients of $\tilde{u}_h$. We claim that

$$\lim_{h \to \infty} \frac{\nabla \tilde{u}_h - \gamma_h}{\lambda_h} = \nabla u_\infty \qquad \text{in } L^2(B_{3/4}) \tag{7.50}$$

and

$$\|\nabla \tilde{u}_h - \gamma_h\|_{L^\sigma(B_{3/4})} \leq C_6 \lambda_h. \tag{7.51}$$

Indeed, let $\mathbf{v}_1, ..., \mathbf{v}_d \in \mathbb{S}^{d-1}$ be d linearly independent vectors such that $\gamma_\infty \cdot \mathbf{v}_i > 5/8$ and $\left| \det \left(\mathbf{v}_1 | ... | \mathbf{v}_d \right) \right| \geq C(d) > 0$. First, we prove that the sequence $(\nabla \tilde{u}_h - \gamma_h)/\lambda_h$ is precompact in $L^2(B_{3/4}; \mathbb{R}^d)$. Thanks to Lemma 7.8 it is enough to show that $\mathbf{v}_i \cdot (\nabla \tilde{u}_h - \gamma_h)/\lambda_h$ is precompact for every $i = 1, ..., d$, which in turn follows from (7.48), applied with $w_h = \partial_{\mathbf{v}_i} \tilde{u}_h(x) - \gamma_h \cdot \mathbf{v}_i$. The characterization of the limit of a subsequence of $(\nabla \tilde{u}_h - \gamma_h)/\lambda_h$ follows from (7.41). As a consequence, it is not necessary to consider a subsequence. Finally, from Lemma 7.7 and (7.46) we obtain that

$$\|\nabla \tilde{u}_h - \gamma_h\|_{L^\sigma(B_{3/4})} \lesssim \sum_{i=1}^d \|\mathbf{v}_i \cdot (\nabla \tilde{u}_h - \gamma_h)\|_{L^\sigma(B_{3/4})} \leq d C_3 \lambda_h,$$

which proves (7.51).

Step 3: incremental quotients for u_∞. Given a function $f : B_1 \to \mathbb{R}$, $\mathbf{v} \in \mathbb{S}^{d-1}$, and $\varepsilon > 0$, we define the discrete derivative of f as

$$[\partial^\varepsilon_{\mathbf{v}} f](x) := \frac{f(x + \varepsilon \mathbf{v}) - f(x)}{\varepsilon} \qquad x \in B_{1-\varepsilon}$$

and the discrete gradient as

$$\nabla^\varepsilon f(x) := \Big([\partial^\varepsilon_{e_1} f](x), ..., [\partial^\varepsilon_{e_d} f](x)\Big) \qquad x \in B_{1-\varepsilon}.$$

We claim that, for ε sufficiently small,

$$\|\nabla^\varepsilon u_\infty\|_{L^2(B_{3/4})} \le \|\nabla u_\infty\|_{L^2(B_1)} \lesssim 1, \tag{7.52}$$

$$\fint_{B_\tau} |\nabla u_\infty - (\nabla u_\infty)_{B_\tau}|^2\, dx \ge \frac{\tau^{2\alpha}}{4}, \tag{7.53}$$

$$\fint_{B_\tau} |\nabla^\varepsilon u_\infty - (\nabla^\varepsilon u_\infty)_{B_\tau}|^2\, dx > \frac{\tau^{2\alpha}}{8}. \tag{7.54}$$

We notice that the second inequality in (7.52) follows from (7.41) and the lower semicontinuity of the norm. To prove (7.53), we see that from the definition of λ_h

$$\frac{(\tau r_h)^{(q-d)/(2q)}}{\lambda_h}\|f\|_{L^q(B_1)} \le \tau^{(q-d)/(2q)} \le \frac{\tau^\alpha}{2}, \tag{7.55}$$

where in the last inequality we have assumed that $\alpha < (q-d)/(2q)$ and τ is sufficiently small (depending on q, d, α).

We notice now, as a general remark, that if $r \in (0, 1]$, $f_h, f \in L^2(B_r)$, and $\lim_{h\to\infty} f_h = f$ in $L^2(B_r)$, then

$$\lim_{h\to\infty} \fint_{B_r} |f_h - (f_h)_{B_r}|^2 = \fint_{B_r} |f - (f)_{B_r}|^2. \tag{7.56}$$

Applying (7.56) to $f_h := (\nabla u_h - \gamma_h)/\lambda_h$ and $r := \tau < 3/4$ (so that by (7.50) we have that $(\nabla u_h - \gamma_h)/\lambda_h \to \nabla u_\infty$ in $L^2(B_\tau)$), letting $h \to \infty$ in (7.37) and taking (7.55) into account we obtain

$$\begin{aligned}\Big(\fint_{B_\tau} |\nabla u_\infty - (\nabla u_\infty)_{B_\tau}|^2\, dx\Big)^{1/2} &= \lim_{h\to\infty}\Big(\fint_{B_\tau} \frac{|\nabla \tilde u_h - (\nabla \tilde u_h)_{B_\tau}|^2}{\lambda_h^2}\Big)^{1/2} \\ &\ge \liminf_{h\to\infty}\Big(\tau^\alpha - \frac{(\tau r_h)^{(q-d)/(2q)}}{\lambda_h}\|f\|_{L^q(B_1)}\Big) \\ &\ge \tau^\alpha - \frac{\tau^\alpha}{2} = \frac{\tau^\alpha}{2},\end{aligned}$$

which proves (7.53).

Finally, since $\lim_{\varepsilon\to 0}\nabla^\varepsilon u_\infty = \nabla u_\infty$ in $L^2(B_\tau)$, we apply (7.56) to $\nabla^\varepsilon u_\infty$ and $r = \tau$ to deduce from (7.53) that (7.54) holds true for ε sufficiently small.

Step 4: a degenerate equation solved by $\partial^\varepsilon_{\mathbf{v}}\tilde u_h$. Let $\mathbf{v} \in \mathbb{S}^{d-1}$ and for every $h \in \mathbb{N}$ let $\mathbf{w}_h = \gamma_h/|\gamma_h|$. We claim that the function $\partial^\varepsilon_{\mathbf{v}}\tilde u_h$ solves

$$\int_{B_{3/4}} A^{h,\varepsilon}_{ij}(x)\,\partial_i\partial^\varepsilon_{\mathbf{v}}\tilde u_h(x)\,\partial_j\phi(x)\,dx + \int_{B_{3/4}} \partial^\varepsilon_{\mathbf{v}}\tilde f_h(x)\,\phi(x)\,dx = 0 \quad (7.57)$$

for every $\phi \in W^{1,2}_0(B_{3/4})$, $h \in \mathbb{N}$, and $\varepsilon \in (0, 1/4)$, for some measurable coefficients $A^{h,\varepsilon}_{ij} : B_{3/4} \to \mathbb{R}$ with the property that $A^{h,\varepsilon}(x)$ is a nonnegative symmetric matrix for every $x \in B_{3/4}$ and that

$$\begin{aligned}&\lambda\,\mathrm{Id} \le (A^{h,\varepsilon}_{ij}(y)) \le \Lambda\,\mathrm{Id}\\ &\forall\, y \in \left\{z \in B_{3/4} : \partial_{\mathbf{w}_h}\tilde u_h(z) \ge \frac{1}{4}\right\} \cap \left\{z \in B_{3/4} : \partial_{\mathbf{w}_h}\tilde u_h(z+\varepsilon\mathbf{v}) \ge \frac{1}{4}\right\}.\end{aligned} \quad (7.58)$$

Indeed, since $\tilde u_h$ are solutions of (7.33), for every $\phi \in W^{1,2}_0(B_{3/4})$ and $\varepsilon < 1/4$ we have

$$\int_{B_{3/4}} \partial_i\mathcal{F}(\nabla\tilde u_h(x))\,\partial_i\phi(x) = -\int_{B_{3/4}} \tilde f_h(x)\,\phi(x),$$

$$\int_{B_{3/4}} \partial_i\mathcal{F}(\nabla\tilde u_h(x+\varepsilon\mathbf{v}))\,\partial_i\phi(x) = -\int_{B_{3/4}} \tilde f_h(x+\varepsilon\mathbf{v})\,\phi(x).$$

Subtracting the two equations and dividing by ε we obtain

$$\begin{aligned}&-\int_{B_{3/4}} \frac{\tilde f_h(x+\varepsilon\mathbf{v}) - \tilde f_h(x)}{\varepsilon}\,\phi(x)\\ &= \int_{B_{3/4}} \frac{\partial_i\mathcal{F}(\nabla\tilde u_h(x+\varepsilon\mathbf{v})) - \partial_i\mathcal{F}(\nabla\tilde u_h(x))}{\varepsilon}\,\partial_i\phi(x)\\ &= \int_{B_{3/4}} A^{h,\varepsilon}_{ij}(x)\,\partial_j\partial^\varepsilon_{\mathbf{v}}\tilde u_h(x)\,\partial_i\phi(x),\end{aligned}$$

where

$$A^{h,\varepsilon}_{ij}(x) := \int_0^1 \partial_{ij}\mathcal{F}\big((1-t)\nabla\tilde u_h(x+\varepsilon\mathbf{v}) + t\nabla\tilde u_h(x)\big)\,dt \quad \forall\, x \in B_{3/4}. \quad (7.59)$$

Notice that, if $x \in B_{3/4}$ is a point such that $\partial_{\mathbf{w}_h}\tilde{u}_h(x) \geq 1/4$ and $\partial_{\mathbf{w}_h}\tilde{u}_h(x+\varepsilon\mathbf{v}) \geq 1/4$ then for every $t \in [0,1]$

$$|(1-t)\nabla\tilde{u}_h(x+\varepsilon\mathbf{v})+t\nabla\tilde{u}_h(x)| \geq (1-t)\partial_{\mathbf{w}_h}\tilde{u}_h(x+\varepsilon\mathbf{v})+t\partial_{\mathbf{w}_h}\tilde{u}_h(x) \geq \frac{1}{4},$$

therefore (7.58) holds true thanks to (7.3).

Step 5: a uniformly elliptic equation solved by $\partial^\varepsilon_{\mathbf{v}} u_\infty$. Let $\mathbf{v} \in \mathbb{S}^{d-1}$. We claim that, for every $\varepsilon > 0$ sufficiently small, the function $\partial^\varepsilon_{\mathbf{v}} u_\infty$ solves

$$\int_{B_{3/4}} A^\varepsilon_{ij}(x)\,\partial_i\partial^\varepsilon_{\mathbf{v}} u_\infty(x)\,\partial_j\varphi(x)\,dx = 0 \tag{7.60}$$

for every $\varphi \in W^{1,2}_0(B_{3/4-\varepsilon})$, for some measurable coefficients $A^\varepsilon_{ij} : B_{3/4} \to \mathbb{R}$ with the property that

$$\lambda\,\mathrm{Id} \leq (A^\varepsilon_{ij}) \leq \Lambda\,\mathrm{Id} \qquad \forall x \in B_{3/4}. \tag{7.61}$$

Indeed, let us consider the function $\phi(x) := \varphi(x)\chi(\partial_{\mathbf{w}_h}\tilde{u}_h(x))\chi(\partial_{\mathbf{w}_h}\tilde{u}_h(x+\varepsilon\mathbf{v}))$ where $\varphi \in W^{1,2}_0(B_{3/4-\varepsilon})$ and $\chi \in C^\infty(\mathbb{R})$ is a function such that $\chi((-\infty,1/2]) = 0$ and $\chi([5/8,\infty)) = 1$. By the identity

$$\chi(\partial_{\mathbf{w}_h}\tilde{u}_h) = \chi\left(\left(\partial_{\mathbf{w}_h}\tilde{u}_h(x) - \frac{|\gamma_h|}{2}\right)_+ - \frac{|\gamma_h|}{2}\right)$$

and (7.45) applied to $v_h = \left(\partial_{\mathbf{w}_h}\tilde{u}_h(x) - |\gamma_h|/2\right)_+ + |\gamma_h|/2$ we have that $\chi(\partial_{\mathbf{w}_h}\tilde{u}_h(x)) \in W^{1,2} \cap L^\infty(B_{3/4})$ with derivative

$$\partial_j[\chi(\partial_{\mathbf{w}_h}\tilde{u}_h)] = \chi'(\partial_{\mathbf{w}_h}\tilde{u}_h)\partial_j\left[\left(\partial_{\mathbf{w}_h}\tilde{u}_h(x) - \frac{|\gamma_h|}{2}\right)_+\right].$$

Similarly $\chi(\partial_{\mathbf{w}_h}\tilde{u}_h(x+\varepsilon\mathbf{v})) \in W^{1,2} \cap L^\infty(B_{3/4})$ with derivative

$$\partial_j[\chi(\partial_{\mathbf{w}_h}\tilde{u}_h(x+\varepsilon\mathbf{v})))] = \chi'(\partial_{\mathbf{w}_h}\tilde{u}_h(x+\varepsilon\mathbf{v}))\partial_j\left[\left(\partial_{\mathbf{w}_h}\tilde{u}_h(x+\varepsilon\mathbf{v}) - \frac{|\gamma_h|}{2}\right)_+\right].$$

Hence $\phi(x) \in W^{1,2}_0 \cap L^\infty(B_{3/4})$. Notice also that from (7.45) it follows that

$$\left\|\nabla\left(\partial_{\mathbf{w}_h}\tilde{u}_h(x) - \frac{|\gamma_h|}{2}\right)_+\right\|_{L^2(B_{3/4})} \lesssim \lambda_h. \tag{7.62}$$

Moreover we have that, since $|\gamma_h| \geq 3/4$,

$$\begin{aligned}\frac{\left|\left\{x \in B_{3/4} : \partial_{\mathbf{w}_h}\tilde{u}_h(x) < \frac{5}{8}\right\}\right|^{1/2}}{|B_{3/4}|^{1/2}} \cdot \frac{1}{8} &\leq \left(\fint_{B_{3/4}} \left|\partial_{\mathbf{w}_h}\tilde{u}_h(y) - |\gamma_h|\right|^2 dy\right)^{1/2} \\ &\lesssim \left(\fint_{B_1} |\nabla\tilde{u}_h(y) - \gamma_h|^2\,dy\right)^{1/2} \lesssim \lambda_h\end{aligned}$$

and therefore

$$\lim_{h\to\infty}\left|\left\{x\in B_{3/4}:\partial_{\mathbf{w}_h}\tilde u_h(x)\ge\frac{5}{8}\right\}\right| = |B_{3/4}|. \tag{7.63}$$

Similarly

$$\lim_{h\to\infty}\left|\left\{x\in B_{3/4}:\partial_{\mathbf{w}_h}\tilde u_h(x+\varepsilon\mathbf{v})\ge\frac{5}{8}\right\}\right| = |B_{3/4}|. \tag{7.64}$$

Using ϕ as a test function in (7.57) and dividing by λ_h we obtain

$$\begin{aligned}
0 &= \int_{B_{3/4}} \frac{\partial^\varepsilon_{\mathbf{v}}\tilde f_h(x)}{\lambda_h}\phi(x)\\
&\quad+\int_{B_{3/4}} A^{h,\varepsilon}_{ij}(x)\,\partial_i\partial^\varepsilon_{\mathbf{v}}\tilde u_h(x)\,\partial_j\Big[\varphi(x)\chi(\partial_{\mathbf{w}_h}\tilde u_h(x))\chi(\partial_{\mathbf{w}_h}\tilde u_h(x+\varepsilon\mathbf{v}))\Big]\\
&= \int_{B_{3/4}} \frac{\partial^\varepsilon_{\mathbf{v}}\tilde f_h(x)}{\lambda_h}\phi(x)\\
&\quad+\int_{B_{3/4}} A^{h,\varepsilon}_{ij}(x)\,\frac{\partial_i\partial^\varepsilon_{\mathbf{v}}\tilde u_h(x)}{\lambda_h}\,\partial_j\varphi(x)\,\chi(\partial_{\mathbf{w}_h}\tilde u_h(x))\,\chi(\partial_{\mathbf{w}_h}\tilde u_h(x+\varepsilon\mathbf{v}))\\
&\quad+\int_{B_{3/4}} A^{h,\varepsilon}_{ij}(x)\,\frac{\partial_i\partial^\varepsilon_{\mathbf{v}}\tilde u_h(x)}{\lambda_h}\,\varphi(x)\,\partial_j\Big(\chi(\partial_{\mathbf{w}_h}\tilde u_h(x))\chi(\partial_{\mathbf{w}_h}\tilde u_h(x+\varepsilon\mathbf{v}))\Big).
\end{aligned} \tag{7.65}$$

We want to take the limit as $h\to\infty$ in (7.65). As regards the first term in the right-hand side, by Hölder inequality and (7.34) we have (here we can assume that $f\not\equiv 0$, so in particular $r_h\to 0$)

$$\begin{aligned}
\left|\int_{B_{3/4}}\frac{\partial^\varepsilon_{\mathbf{v}}\tilde f_h(x)}{\lambda_h}\phi(x)\right| &\le \frac{\|\partial^\varepsilon_{\mathbf{v}}\tilde f_h(x)\|_{L^1(B_{3/4})}}{\lambda_h}\|\phi(x)\|_{L^\infty(B_{3/4})}\\
&\lesssim \frac{\|\tilde f_h(x)\|_{L^q(B_1)}}{\varepsilon r_h^{(q-d)/(2q)}\|f\|_{L^q(B_1)}}\|\phi(x)\|_{L^\infty(B_{3/4})}\\
&\le \frac{r_h^{(q-d)/(2q)}}{\varepsilon}\|\phi(x)\|_{L^\infty(B_{3/4})},
\end{aligned}$$

therefore

$$\lim_{h\to\infty}\int_{B_{3/4}}\frac{\partial^\varepsilon_{\mathbf{v}}\tilde f_h(x)}{\lambda_h}\phi(x)=0. \tag{7.66}$$

Then, we apply Lemma 7.9 to $A^h(x):=A^{h,\varepsilon}(x)\chi(\partial_{\mathbf{w}_h}\tilde u_h(x))\chi(\partial_{\mathbf{w}_h}\tilde u_h(x+\varepsilon\mathbf{v}))$. For this, notice that the assumption (7.17) of the lemma is satisfied thanks to (7.58), (7.63), (7.64), and the fact that

$$\chi(\partial_{\mathbf{w}_h}\tilde u_h(x))=\chi(\partial_{\mathbf{w}_h}\tilde u_h(x+\varepsilon\mathbf{v}))=1$$

on the set

$$\left\{x \in B_{3/4} : \partial_{\mathbf{w}_h}\tilde{u}_h(x) \geq \frac{5}{8}\right\} \cap \left\{x \in B_{3/4} : \partial_{\mathbf{w}_h}\tilde{u}_h(x+\varepsilon\mathbf{v}) \geq \frac{5}{8}\right\}.$$

Moreover, for every $x \in B_{3/4}$ such that $\chi(\partial_{\mathbf{w}_h}\tilde{u}_h(x))\chi(\partial_{\mathbf{w}_h}\tilde{u}_h(x+\varepsilon\mathbf{v})) > 0$ we have that $\partial_{\mathbf{w}_h}\tilde{u}_h(x) > 1/2$ and $\partial_{\mathbf{w}_h}\tilde{u}_h(x+\varepsilon\mathbf{v}) > 1/2$ and therefore

$$\lambda\,\mathrm{Id} \leq A^{h,\varepsilon}(x) \leq \Lambda\,\mathrm{Id}.$$

This implies that

$$0 \leq A^{h,\varepsilon}(x)\,\chi(\partial_{\mathbf{w}_h}\tilde{u}_h(x))\,\chi(\partial_{\mathbf{w}_h}\tilde{u}_h(x+\varepsilon\mathbf{v})) \leq \Lambda\,\mathrm{Id} \qquad \forall x \in B_{3/4}.$$

Hence, applying Lemma 7.9 we obtain that there exist $A^\varepsilon : B_{3/4} \to \mathbb{R}^{d\times d}$ such that $\lambda\,\mathrm{Id} \leq A^\varepsilon \leq \Lambda\,\mathrm{Id}$ and, up to subsequences,

$$\begin{aligned}&A^{h,\varepsilon}(x)\,\chi(\partial_{\mathbf{w}_h}\tilde{u}_h(x))\,\chi(\partial_{\mathbf{w}_h}\tilde{u}_h(x+\varepsilon\mathbf{v})) \to A^\varepsilon(x)\\ &\text{weakly in } L^2(B_{3/4};\mathbb{R}^{d\times d}).\end{aligned} \tag{7.67}$$

From the equality

$$\frac{\partial_i\partial_{\mathbf{v}}^\varepsilon\tilde{u}_h(x)}{\lambda_h} = \frac{1}{\varepsilon}\left(e_i\cdot\left(\frac{\nabla\tilde{u}_h(x+\varepsilon\mathbf{v})-\gamma_h}{\lambda_h}\right) - e_i\cdot\left(\frac{\nabla\tilde{u}_h(x)-\gamma_h}{\lambda_h}\right)\right)$$

and by (7.50) we have

$$\begin{aligned}\lim_{h\to\infty}\frac{\partial_i\partial_{\mathbf{v}}^\varepsilon\tilde{u}_h(x)}{\lambda_h} &= \frac{\partial_i u_\infty(x+\varepsilon\mathbf{v}) - \partial_i u_\infty(x)}{\varepsilon}\\ &= \partial_i\partial_{\mathbf{v}}^\varepsilon u_\infty(x) \qquad \text{in } L^2(B_{3/4-\varepsilon}),\end{aligned} \tag{7.68}$$

so by (7.67), (7.68), and the fact that $\partial_j\varphi \in L^\infty(B_{3/4})$, we obtain

$$\begin{aligned}&\lim_{h\to\infty}\int_{B_{3/4}} A_{ij}^{h,\varepsilon}(x)\,\frac{\partial_i\partial_{\mathbf{v}}^\varepsilon\tilde{u}_h(x)}{\lambda_h}\partial_j\varphi(x)\,\chi(\partial_{\mathbf{w}_h}\tilde{u}_h(x))\chi(\partial_{\mathbf{w}_h}\tilde{u}_h(x+\varepsilon\mathbf{v}))\\ &\quad= \int_{B_{3/4}} A_{ij}^\varepsilon(x)\,\partial_i\partial_{\mathbf{v}}^\varepsilon u_\infty(x)\,\partial_j\varphi(x).\end{aligned} \tag{7.69}$$

To estimate the last term we notice that, since

$$\frac{1}{2}+\frac{1}{\sigma}+\frac{1}{2d} = 1,$$

by Hölder inequality we have that

$$\left|\int_{B_{3/4}} A_{ij}^{h,\varepsilon}(x)\frac{\partial_i\partial_{\mathbf{v}}^{\varepsilon}\tilde{u}_h(x)}{\lambda_h}\varphi(x)\chi'(\partial_{\mathbf{w}_h}\tilde{u}_h(x))\partial_j\Big(\partial_{\mathbf{w}_h}\tilde{u}_h(x)-\frac{|\gamma_h|}{2}\Big)_+\chi(\partial_{\mathbf{w}_h}\tilde{u}_h(x+\varepsilon\mathbf{v}))\right|$$

$$\le \|\varphi\|_{L^\infty(B_{3/4})}\cdot\left\|\frac{\nabla\partial_{\mathbf{v}}^{\varepsilon}\tilde{u}_h(x)}{\lambda_h}\right\|_{L^\sigma(B_{3/4})}\cdot\left\|\nabla\Big(\partial_{\mathbf{w}_h}\tilde{u}_h(x)-\frac{|\gamma_h|}{2}\Big)_+\right\|_{L^2(B_{3/4})}$$

$$\cdot\left\|A_{ij}^{h,\varepsilon}(x)\,\chi'(\partial_{\mathbf{w}_h}\tilde{u}_h(x))\,\chi(\partial_{\mathbf{w}_h}\tilde{u}_h(x+\varepsilon\mathbf{v}))\right\|_{L^{2d}(B_{3/4})}.$$

Since $0\le A_{ij}^{h,\varepsilon}(x)\le\Lambda\,\mathrm{Id}$ for every x such that $\chi'(\partial_{\mathbf{w}_h}\tilde{u}_h(x))\chi(\partial_{\mathbf{w}_h}\tilde{u}_h(x+\varepsilon\mathbf{v}))>0$, it follows that

$$\left\|A_{ij}^{h,\varepsilon}(x)\,\chi'(\partial_{\mathbf{w}_h}\tilde{u}_h(x))\,\chi(\partial_{\mathbf{w}_h}\tilde{u}_h(x+\varepsilon\mathbf{v}))\right\|_{L^{2d}(B_{3/4})}\le C(\Lambda)\|\chi'\|_\infty. \tag{7.70}$$

Thus, from (7.70), (7.51), (7.62), (7.70) we have that

$$\begin{aligned}\lim_{h\to\infty}\int_{B_{3/4}} & A_{ij}^{h,\varepsilon}(x)\,\frac{\partial_i\partial_{\mathbf{v}}^{\varepsilon}\tilde{u}_h(x)}{\lambda_h}\,\varphi(x)\,\chi'(\partial_{\mathbf{w}_h}\tilde{u}_h(x))\\ &\partial_j\Big(\partial_{\mathbf{w}_h}\tilde{u}_h(x)-\frac{|\gamma_h|}{2}\Big)_+\chi(\partial_{\mathbf{w}_h}\tilde{u}_h(x+\varepsilon\mathbf{v}))=0.\end{aligned} \tag{7.71}$$

Similarly,

$$\begin{aligned}\lim_{h\to\infty}\int_{B_{3/4}} & A_{ij}^{h,\varepsilon}(x)\,\frac{\partial_i\partial_{\mathbf{v}}^{\varepsilon}\tilde{u}_h(x)}{\lambda_h}\,\varphi(x)\,\chi(\partial_{\mathbf{w}_h}\tilde{u}_h(x))\\ &\chi'(\partial_{\mathbf{w}_h}\tilde{u}_h(x+\varepsilon\mathbf{v}))\,\partial_j\Big(\partial_{\mathbf{w}_h}\tilde{u}_h(x+\varepsilon\mathbf{v})-\frac{|\gamma_h|}{2}\Big)_+=0.\end{aligned} \tag{7.72}$$

Hence, letting $h\to\infty$ in (7.57) and taking (7.66), (7.69), (7.71), and (7.72) into account, we obtain

$$0=\int_{B_{3/4}} A_{ij}^{\varepsilon}(x)\,\partial_i\partial_{\mathbf{v}}^{\varepsilon}u_\infty(x)\,\partial_j\varphi(x).$$

Step 6: a contradiction based on the excess decay for u_∞. We find a contradiction.

Since by (7.60) the functions $\partial_{\mathbf{v}}^{\varepsilon}u_\infty\in W^{1,2}(B_{3/4})$ solve a uniformly elliptic equation for $\varepsilon>0$ small enough, by De Giorgi-Nash-Moser Theorem (see Theorem 1.19) there exists $\alpha > 0$ such that for every $\mathbf{v} \in \mathbb{S}^{d-1}$

$$\|\partial^\varepsilon_{\mathbf{v}} u_\infty\|_{C^{0,2\alpha}(B_{1/2})} \lesssim \|\partial^\varepsilon_{\mathbf{v}} u_\infty\|_{L^2(B_{3/4})} \leq \|\nabla^\varepsilon u_\infty\|_{L^2(B_{3/4})} \lesssim 1, \qquad (7.73)$$

where the last inequality follows from (7.52); in particular, applying the previous inequality to $\mathbf{v} = e_1, ..., e_d$, we obtain that

$$\|\nabla^\varepsilon u_\infty\|_{C^{0,2\alpha}(B_{1/2})} \leq C_7. \qquad (7.74)$$

Hence, by Jensen inequality and (7.74), for τ sufficiently small we have that

$$\begin{aligned} \fint_{B_\tau} |\nabla^\varepsilon u_\infty(x) - (\nabla^\varepsilon u_\infty)_{B_\tau}|^2 dx &\leq \fint_{B_\tau} \fint_{B_\tau} |\nabla^\varepsilon u_\infty(x) - \nabla^\varepsilon u_\infty(y)|^2 \, dx \, dy \\ &\leq C_7^2 (2\tau)^{4\alpha} < \frac{\tau^{2\alpha}}{8}, \end{aligned}$$

which contradicts (7.54) and concludes the proof. □

Proof of Corollaries 7.2 *and* 7.3. The proof of Corollary 7.2 relies on an iterated application of Theorem 7.1; at every scale, the assumptions are satisfied thanks to the geometric decay of the excess on larger scales. In particular, we estimate at every step the difference between the average of u on $B_{\tau^{i-1}}$ and on B_{τ^i} by means of the excess in $B_{\tau^{i-1}}$.

Proof of Corollary 7.2. Let $x \in B_{1/2}$ and $r < 1/4$; let $\tau_0, \alpha, \tau, \varepsilon(\tau)$ be as in Theorem 7.1. Let $\varepsilon \leq \varepsilon(\tau)$ be a constant to be chosen later. We prove (7.5) by induction. For $k = 1$ we apply Theorem 7.1 and we obtain (7.5). Assuming as inductive assumption that

$$U(u, x, \tau^i r) \leq \tau^{\alpha i} U(u, x, r) \qquad \forall i \leq k-1, \qquad (7.75)$$

we prove

$$U(u, x, \tau^k r) \leq \tau^{\alpha k} U(u, x, r). \qquad (7.76)$$

By (7.4) and (7.75) applied with $i = k-1$ we have that $U(u, x, \tau^{k-1} r) \leq \varepsilon \leq \varepsilon(\tau)$. In order to satisfy the assumptions of Theorem 7.1 at x with radius $\tau^{k-1} r$ we have to show that

$$\frac{3}{4} \leq |(\nabla u)_{B_{\tau^{k-1} r}(x)}| \leq 1. \qquad (7.77)$$

For every $i \in \mathbb{N}$ let us set $\gamma_i = (\nabla u)_{B_{\tau^i r}(x)}$. For every $i = 1, \ldots, k-1$ by (7.75) we have that

$$
\begin{aligned}
|\gamma_i - \gamma_{i-1}| &= \Big(\fint_{B_{\tau^i r}(x)} |\gamma_i - \gamma_{i-1}|\, dy \Big)^{1/2} \\
&\leq \Big(\fint_{B_{\tau^i r}(x)} |\nabla u(y) - \gamma_i|^2 dy \Big)^{1/2} + \Big(\fint_{B_{\tau^i r}(x)} |\nabla u(y) - \gamma_{i-1}|^2\, dy \Big)^{1/2} \\
&\leq \Big(\fint_{B_{\tau^i r}(x)} |\nabla u(y) - \gamma_i|^2 dy \Big)^{1/2} + \frac{1}{\tau^{d/2}} \Big(\fint_{B_{\tau^{i-1} r}(x)} |\nabla u(y) - \gamma_{i-1}|^2 dy \Big)^{1/2} \\
&\leq U(u, x, \tau^i r) + \frac{1}{\tau^{d/2}} U(u, x, \tau^{i-1} r) \leq \Big(\tau^\alpha + \frac{1}{\tau^{d/2}} \Big) \tau^{\alpha(i-1)} U(u, x, r).
\end{aligned}
$$

Hence, by the triangular inequality we obtain

$$
\begin{aligned}
|(\nabla u)_{B_{\tau^{k-1} r}(x)} - (\nabla u)_{B_r(x)}| &= |\gamma_{k-1} - \gamma_0| \\
&\leq \sum_{i=1}^{k-1} |\gamma_i - \gamma_{i-1}| \leq \Big(\tau^\alpha + \frac{1}{\tau^{d/2}} \Big) \Big(\sum_{i=1}^{\infty} \tau^{\alpha(i-1)} \Big) U(u, x, r) \\
&\leq C(\tau, d, \alpha)\, \varepsilon \leq \frac{1}{8}
\end{aligned}
\tag{7.78}
$$

where in the last inequality we have chosen ε small (depending on d, τ, α). From (7.78) and (7.4) we obtain (7.77). So, we can apply Theorem 7.1 with radius $\tau^{k-1} r$ to obtain

$$
U(u, x, \tau^k r) \leq \tau^\alpha U(u, x, \tau^{k-1} r), \tag{7.79}
$$

which, together with (7.75), implies (7.76). □

In order to prove Corollary 7.3, we apply Corollary 7.2; its assumptions are satisfied because the gradient of u lies close to a fixed vector $\mathbf{v}$ for a large fraction of points in B_1 (see Figure 7.1).

Proof of Corollary 7.3. Let $x \in B_{1/2}$; let $\tau = \tau_0, \alpha, \varepsilon = \varepsilon(\tau_0)$ be as in Corollary 7.2. First we prove that, if η and r_0 are chosen sufficiently small, then

$$
\frac{7}{8} \leq |(\nabla u)_{B_{r_0}(x)}| \leq 1 \qquad U(u, x, r_0) \leq \varepsilon. \tag{7.80}
$$

We choose $r_0 < 1/4$ sufficiently small so that

$$
r_0^{(q-d)/(2q)} \|f\|_{L^q(B_1)} \leq \frac{\varepsilon}{2}. \tag{7.81}
$$

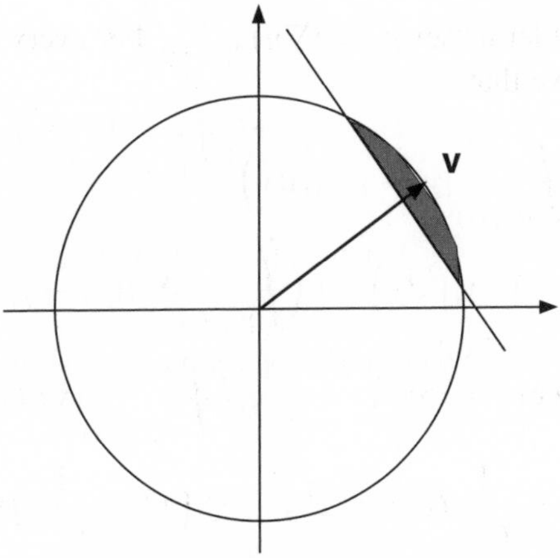

Figure 7.1. Under the assumptions of Corollary 7.3, ∇u lies in B_1 and, in particular, in the grey area for a fraction of large measure (namely, $1 - \eta$) of points in B_1.

We estimate the first term in the excess splitting the integral over $B_{r_0}(x) \cap \{\partial_{\mathbf{v}} u \geq 1-\eta\}$ and its complement. For every $y \in B_{r_0}(x) \cap \{\partial_{\mathbf{v}} u \geq 1-\eta\}$ we have that

$$|\nabla u(y) - \mathbf{v}|^2 = |\nabla u(y)|^2 + |\mathbf{v}|^2 - 2\nabla u(y) \cdot \mathbf{v} \leq 2(1 - \nabla u(y) \cdot \mathbf{v}) \leq 2\eta.$$

In the complement of $B_{r_0}(x) \cap \{\partial_{\mathbf{v}} u \geq 1 - \eta\}$ we have that $|\nabla u - \mathbf{v}| \leq |\nabla u| + |\mathbf{v}| \leq 2$. Therefore we have

$$\begin{aligned} &\fint_{B_{r_0}(x)} |\nabla u(y) - \mathbf{v}|^2 \, dy \\ &\leq \frac{1}{|B_{r_0}|} \Big(|\{y \in B_{r_0}(x) : \partial_{\mathbf{v}} u(y) \geq 1 - \eta\}| 4\eta^2 \\ &\qquad + 4|\{y \in B_{r_0}(x) : \partial_{\mathbf{v}} u(y) \leq 1 - \eta\}| \Big) \\ &\leq 4\eta^2 + \frac{1}{|B_{r_0}|} |\{y \in B_1 : \partial_{\mathbf{v}} u(y) \leq 1 - \eta\}|. \end{aligned} \tag{7.82}$$

Noticing that (7.6) implies that $|\{y \in B_1 : \partial_{\mathbf{v}} u(y) \leq 1 - \eta\}| \leq \eta |B_1|$ we obtain

$$\fint_{B_{r_0}(x)} |\nabla u(y) - \mathbf{v}|^2 \, dy \leq 4\eta^2 + \eta \frac{|B_1|}{|B_{r_0}|} \leq \frac{\varepsilon^2}{4}, \tag{7.83}$$

where in the last inequality we have chosen η sufficiently small, depending only on $\|f\|_{L^q(B_1)}$ and ε. From (7.83) it follows that

$$\begin{aligned}\fint_{B_{r_0}(x)} |\nabla u(y) - (\nabla u)_{B_{r_0}(x)}|^2\,dy &= \inf_{\gamma\in\mathbb{R}^d} \fint_{B_{r_0}(x)} |\nabla u(y) - \gamma|^2\,dy \\ &\le \fint_{B_{r_0}(x)} |\nabla u(y) - \mathbf{v}|^2\,dy \le \frac{\varepsilon^2}{4}\end{aligned} \tag{7.84}$$

and therefore by (7.84) and (7.81) we get the second inequality in (7.80). From (7.83) we have

$$\left|\fint_{B_{r_0}(x)} (\nabla u(y) - \mathbf{v})\,dy\right| \le \left(\fint_{B_{r_0}(x)} |\nabla u(y) - \mathbf{v}|^2\,dy\right)^{1/2} \le \frac{\varepsilon}{2};$$

it implies

$$|(\nabla u)_{B_{r_0}(x)}| \ge |\mathbf{v}| - |(\nabla u)_{B_{r_0}(x)} - \mathbf{v}| \ge 1 - \frac{\varepsilon}{2} \ge \frac{7}{8},$$

which proves the first inequality in (7.80). Hence the assumptions of Corollary 7.2 are satisfied and we obtain (7.7).

We are left to prove (7.8). From (7.7) and (7.80) it follows that for every $k \in \mathbb{N}$ and $x \in B_{1/2}$,

$$\begin{aligned}\left(\fint_{B_{\tau^k r_0}(x)} |\nabla u(y) - (\nabla u)_{B_{\tau^k r_0}(x)}|^2\,dy\right)^{1/2} &\le U(u, x, \tau^k r_0) \\ &\le \tau^{\alpha k} U(u, x, r_0) \le \varepsilon\tau^{\alpha k}.\end{aligned}$$

From Campanato theorem, stated in Lemma 1.24 we obtain (7.8). □

Proof of Theorem 7.5.

Lemma 7.11. *Let $\eta \in (0,1)$. Let $p, \mathbf{n}, \lambda, \Lambda, u, q, f$ be as in Theorem 7.5 with $\Omega = B_1$. Assume that $|\nabla u(x)| \le 1$ for every $x \in B_1$ and*

$$\sup_{e\in\mathbb{S}^{d-1}} |\{x \in B_{1/2} : \partial_e u(x) \ge (1-\eta)\}| \le (1-\eta)|B_{1/2}|. \tag{7.85}$$

Then there exist constants $c := c(d, p, q, \lambda, \Lambda)$ and $C := C(d, \eta, p, q, \lambda, \Lambda)$ such that if $\|f\|_{L^q(B_1)} \le C$ then

$$|\nabla u| \le 1 - c\eta^2 \qquad \forall x \in B_{1/4}. \tag{7.86}$$

Proof. Let us fix $e \in \mathbb{S}^{d-1}$ and let v_e be defined as in (7.14). We repeat the proof of [91, Theorem 8.18] (see also [114, Lemma 4]) applied to the function $1/2 - v_e(x)$, which is a nonnegative supersolution in B_1 of the equation

$$\partial_i\Big[A_{ij}(\nabla u(x))\partial_j\Big(\frac{1}{2} - v_e(x)\Big)\Big] = \partial_e f(x)1_{\{\partial_e u \geq 1/2\}}$$

(the coefficients A_{ij} are defined in (7.12); as we mentioned before, to properly justify this computation one needs to perform a suitable regularization argument in the spirit of the proof of Theorem 6.1 and [122]). This equation can be considered to be uniformly elliptic since the values of $A_{ij}(\nabla u(x))$ where $|\nabla u(x)| \leq 1/2$ are not relevant. We obtain that there exists a constant $c_0 := c_0(d, p, q, \lambda, \Lambda)$ such that a weak Harnack inequality holds

$$c_0\|1/2 - v_e\|_{L^1(B_{1/2})} \leq \inf_{x \in B_{1/4}(0)} \{1/2 - v_e(x)\} + \|f\|_{L^q(B_1)}.$$

On the set

$$\{x \in B_{1/2} : \partial_e u \leq (1 - \eta)\}$$

(whose measure is greater than $\eta|B_{1/2}|$ from (7.85)), the integrand is greater or equal to η and we obtain

$$\begin{aligned}
&\inf\{1/2 - v_e(x) : x \in B_{1/4}\} \\
&\quad \geq c_0 \int_{B_{1/2}} (1/2 - v_e(x))\, dx \ - \|f\|_{L^q(B_1)} \\
&\quad \geq c_0\eta|\{x \in B_{1/2} : (\partial_e u(x) - 1/2)_+ \leq 1 - \eta\}| - \|f\|_{L^q(B_1)} \\
&\quad \geq c_0\eta^2|B_{1/2}| - C.
\end{aligned}$$

Therefore, setting $c := c_0|B_{1/2}|/2$ and $C := c_0\eta^2|B_{1/2}|/2$, we have

$$\inf\{1/2 - v_e(x) : x \in B_{1/4}\} \geq c\eta^2,$$

which in turn can be rewritten as

$$\partial_e u(x) \leq 1 - c\eta^2 \qquad \forall x \in B_{1/4}.$$

Since this argument holds true for every direction $e \in \mathbb{S}^{d-1}$ we obtain (7.86). □

Iterating the previous lemma on smaller scales and using the scale invariance of the anisotropic p-laplacian we obtain the following result.

Lemma 7.12. *Let p, $\boldsymbol{n}$, λ, Λ, u, q, f be as in Theorem 7.5. Let $\eta > 0$ be sufficiently small, c and C as in Lemma 7.11, $\delta = c\eta^2$, and $k \in \mathbb{N}$. If $|\nabla u(x)| \leq 1$ for every $x \in B_1$,*

$$\sup_{e\in\mathbb{S}^{d-1}} |\{x \in B_{2^{-2i-1}}(0) : \partial_e u \geq (1-\eta)(1-\delta)^i\}| \leq (1-\eta)|B_{2^{-2i-1}}| \quad \forall i = 1, \dots, k, \tag{7.87}$$

and $\|f\|_{L^q(B_1)} \leq C$, then we have that

$$|\nabla u(x)| \leq (1-\delta)^i \quad \forall x \in B_{2^{-2i}} \quad \forall i = 1, \dots, k+1. \tag{7.88}$$

Proof. We prove the result by induction on i. Assuming (7.87) with $i = 0$ we obtain (7.88) with $i = 1$ from Lemma 7.11. Let us assume that the result holds true for i and let us prove it for $i+1$. Thanks to the homogeneity of the anisotropic p-laplacian, the function

$$v(x) := \frac{2^{2i} u(2^{-2i}x)}{(1-\delta)^i} \qquad x \in B_1$$

satisfies by inductive assumption $|\nabla v| \leq 1$ in B_1 and it is a minimizer of

$$\int_{B_1} \frac{\boldsymbol{n}(\nabla v)^p}{p} + \tilde{f} v, \tag{7.89}$$

where

$$\tilde{f}(x) := \frac{2^{-2i}}{(1-\delta)^{i(p-1)}} f(2^{-2i}x).$$

Hence the norm of $\tilde{f}$ is estimated by

$$\|\tilde{f}\|^q_{L^q(B_1)} = \frac{2^{-2i(p-d)}}{(1-\delta)^{i(p-1)q}} \int_{B_{2^{-2i}}} |f(y)|^q \, dy \leq \frac{2^{-2i(q-d)}}{(1-\delta)^{i(p-1)q}} \|f\|^q_{L^q(B_1)}.$$

Therefore, provided that δ is chosen small enough so that $2^{\frac{-2(q-d)}{(pq-q)}} \leq 1-\delta$, we obtain that $\|\tilde{f}\|_{L^q(B_1)} \leq \|f\|_{L^q(B_1)} \leq C$. The assumption (7.87) can be rewritten as (7.85) applied to v instead of u; therefore, Lemma 7.11 gives us that

$$|\nabla v(x)| \leq 1 - \delta \qquad \forall x \in B_{1/4},$$

which implies (7.88) with $i+1$ in place of i. □

Proof of Theorem 7.5. By a covering argument, it is enough to show that, if $u : B_1 \to \mathbb{R}$ is Lipschitz, then

$$\sup_{x \in B_{2^{-2i}}} |\nabla u(x) - \nabla u(0)| \le C_0 2^{-2\alpha i} \qquad \forall i \in \mathbb{N}, \tag{7.90}$$

for some $\alpha \in (0,1)$, $C_0 > 0$ which depends only on d, p, λ, Λ to be chosen later. Let $\eta > 0$ to be fixed later; let $c, C, \delta = c\eta^2$ as in Lemma 7.12. Up to changing u with

$$\frac{u(r_0 x) - u(0)}{r_0 \|\nabla u\|_{L^\infty(B_1)}} \qquad \forall x \in B_1$$

thanks to the homogeneity of the anisotropic p-laplacian we can assume that

$$u(0) = 0, \quad |\nabla u(x)| \le 1 \quad \forall x \in B_1, \quad \text{and} \quad \|f\|_{L^q(B_1)} \le C.$$

Let $k \in \mathbb{N} \cup \{\infty\}$ be the largest index for which (7.87) holds true. Let $\alpha_1 \in (0,\infty)$ be such that $2^{-2\alpha_1} = 1 - \delta$. If $k = \infty$ we have that by Lemma 7.12

$$\sup_{x \in B_{2^{-2i}}} |\nabla u(x)| \le (1-\delta)^i = 2^{-2\alpha_1 i} \qquad \forall i \in \mathbb{N};$$

hence (7.90) is satisfied. If $k < \infty$ set

$$v(x) := \frac{2^{2(k+1)} u(2^{-2(k+1)} x)}{(1-\delta)^{k+1}}.$$

By the maximality of k we have that there exists $e \in \mathbb{S}^{d-1}$ such that

$$|\{x \in B_{1/2} : \partial_e v(x) \ge 1 - \eta\}| \ge (1-\eta)|B_1|. \tag{7.91}$$

Thanks to Lemma 7.12 applied to u we obtain that

$$\sup_{x \in B_{2^{-2i}}} |\nabla u(x)| \le (1-\delta)^i = 2^{-2\alpha_1 i} \qquad \forall i = 1, ..., k+1. \tag{7.92}$$

and

$$|\nabla v(x)| \le 1 \qquad \forall x \in B_1.$$

We choose $\eta > 0$ so that Corollary 7.3 applies to v with $\mathcal{F}(x) = \boldsymbol{n}(x)^p/p$ (notice that assumption (7.3) is not a restriction since $|\nabla v| \le 1$); we obtain that there exist $\alpha_2, C_2 > 0$ such that for every $i \in \mathbb{N}$

$$\begin{aligned}\frac{1}{2^{-\alpha_1(k+1)}} \sup_{x \in B_{2^{-2i}}} |\nabla u(2^{-2(k+1)} x) - \nabla u(0)| &= \sup_{x \in B_{2^{-2i}}} |\nabla v(x) - \nabla v(0)| \\ &\le C_2 2^{-2\alpha_2 i},\end{aligned}$$

which can be rewritten, setting $\alpha = \min\{\alpha_1, \alpha_2\}$, as

$$\sup_{x \in B_{2^{-2(i+k+1)}}} |\nabla u(x) - \nabla u(0)| \leq C_2 2^{-2\alpha_2 i + \alpha_1(k+1)} \leq C_2 2^{-2\alpha(i+k+1)}. \quad (7.93)$$

From (7.92) we deduce that for every $i = 1, \ldots, k+1$

$$\sup_{x \in B_{2^{-2i}}} |\nabla u(x) - \nabla u(0)| \leq 2 \sup_{x \in B_{2^{-2i}}} |\nabla u(x)| \leq 2 \cdot 2^{-2\alpha_1 i} \leq 2 \cdot 2^{-2\alpha i},$$

which, together with (7.93), proves (7.90) when $k < \infty$ with $C_0 = \max\{2, C_2\}$. □

Proof of Theorem 7.6. Since the proof of this theorem is a modification of the proof of Theorem 6.1, we just outline the differences.

First we remark that all results in Section 7.1 hold replacing B_1 and $B_{1/4}$ with sets $\{\boldsymbol{m} < M\}$ and $\{\boldsymbol{m} < m\}$ for some $0 \leq m < M$ (indeed, the statements and the proofs can easily be adapted to this setting with easy modifications).

Then we regularize the equation by approximation, reducing ourselves to prove an a-priori estimate on a regular solution as in the proof Theorem 6.1. Finally, to prove regularity at nondegenerate points we use Corollary 7.3 instead of Lemma 6.11 and Proposition 6.13. □

Chapter 8
The Vlasov-Poisson system

The d-dimensional Vlasov-Poisson system describes the evolution of a nonnegative distribution function $f : (0, \infty) \times \mathbb{R}^d \times \mathbb{R}^d \to [0, \infty)$ according to Vlasov's equation, under the action of a self-consistent force determined by the Poisson's equation:

$$\begin{cases} \partial_t f_t + v \cdot \nabla_x f_t + E_t \cdot \nabla_v f_t = 0 & \text{in } (0, \infty) \times \mathbb{R}^d \times \mathbb{R}^d \\ \rho_t(x) = \displaystyle\int_{\mathbb{R}^d} f_t(x, v)\, dv & \text{in } (0, \infty) \times \mathbb{R}^d \\ E_t(x) = \sigma\, c_d \displaystyle\int_{\mathbb{R}^d} \rho_t(y) \frac{x - y}{|x - y|^d}\, dy & \text{in } (0, \infty) \times \mathbb{R}^d. \end{cases} \tag{8.1}$$

Here $f_t(x, v)$ stands for the density of particles having position x and velocity v at time t, $\rho_t(x)$ is the distribution of particles in the physical space, $E_t = -\sigma \nabla(\Delta^{-1} \rho_t)$ is the force field, $c_d > 0$ is a dimensional constant chosen in such a way that $c_d \mathrm{div}\left(\frac{x}{|x|^d}\right) = \delta_0$, and $\sigma \in \{\pm 1\}$. The case $\sigma = 1$ corresponds to the case of electrostatic forces between charged particles with the same sign (repulsion) while $\sigma = -1$ corresponds to the gravitational case (attraction).

This system appears in several physical models. For instance, when $\sigma = 1$ it describes in plasma physics the evolution of charged particles under their self-consistent electric field, while when $\sigma = -1$ the same system is used in astrophysics to describe the motion of galaxy clusters under the gravitational field. Many different models have been developed in connection with the Vlasov-Poisson equation: amongst others, we mention the relativistic version of (8.1) (where the velocity of particles is given by $v/\sqrt{1 + |v|^2}$) and the Vlasov-Maxwell system (which takes into account both the electric and magnetic fields of the Maxwell equations). The latter can be written as

$$\begin{cases} \partial_t f_t + v\cdot\nabla_x f_t + (E_t + v\times B_t)\cdot\nabla_v f_t = 0 & \text{in } (0,\infty)\times\mathbb{R}^3\times\mathbb{R}^3 \\ \partial_t E_t - \nabla\times B_t = -j_t, \qquad \nabla\cdot E_t = \rho_t & \text{in } (0,\infty)\times\mathbb{R}^3 \\ \partial_t B_t + \nabla\times E_t = 0, \qquad \nabla\cdot B_t = 0 & \text{in } (0,\infty)\times\mathbb{R}^3 \\ j_t = \displaystyle\int_{\mathbb{R}^3} v f_t\,dv, \qquad \rho_t = \int_{\mathbb{R}^3} f_t\,dv & \text{in } (0,\infty)\times\mathbb{R}^3. \end{cases} \tag{8.2}$$

Regarding the existence of classical solutions of the Vlasov-Poisson system, namely, solutions where all the relevant derivatives exist, the first contributions were given by Iordanskii [99] for the existence of solutions in dimension 1, by Ukai and Okabe [120] in dimension 2, and by Bardos and Degond [26] in dimension 3 for small data. For symmetric initial data, more existence results have been proven in [29, 123, 95, 116] (see also the presentation in [112] for an overview of the topic and the references quoted therein). Finally, in 1989 Pfaffelmöser [111] and Lions and Perthame [105] were able to prove global existence of classical solutions starting from general data. In [105] the authors consider an initial datum $f_0 \in L^1\cap L^\infty(\mathbb{R}^6)$ with finite moments $|v|^m f_0(x,v) \in L^1(\mathbb{R}^6)$ for some $m > 3$, and, thanks to an a priori estimate on the propagation of moments, they show the existence of a distributional solution $f \in C((0,\infty); L^p(\mathbb{R}^6)) \cap L^\infty((0,\infty); L^\infty(\mathbb{R}^6))$ for every $1 \le p < \infty$. Moreover, in [105] the problem of uniqueness is also addressed; under more restrictive assumptions on the initial datum, the authors show that there is uniqueness in the class of solutions with bounded space densities in $[0,\infty)\times\mathbb{R}^3$. Uniqueness is achieved by considering the Lagrangian flow associated to the vector field $\boldsymbol{b}_t(x,v) := (v, E_t(x))$, which is regular enough under a global bound on the space density (see also [108] for a different proof based on stability in the Wasserstein metric).

As one can see, the above results require strong assumptions on the initial data. However, it would be very desirable to get global existence of solutions under much weaker conditions. In the classical paper [25], Arsen'ev proved global existence of weak solutions under the assumption that the initial datum is bounded and has finite kinetic energy (see also [97]). This result has then been improved in [96], where the authors relaxed the boundedness assumption on an L^p bound for some suitable $p > 1$.

Notice that these higher integrability assumptions are needed even to give a meaning to the equation in the distributional sense: indeed, when f_t is merely L^1 the product $E_t f_t$ does not belong to L^1_{loc} (when $d = 3$, for the term $E_t f_t$ to belong to L^1_{loc} one needs to have $f_t \in L^p$ with $p \ge$

$(12 + 2\sqrt{5})/11$, see for instance [78]). To overcome this difficulty, in [78] the authors considered the concept of renormalized solutions and obtained global existence in the case $\sigma = 1$ under the assumption that the total energy is finite and $f_0 \log(1 + f_0) \in L^1$ (in the case $\sigma = -1$ they still need some L^p assumption on f). Also, under some suitable integrability assumptions on f_t, they can show that the concepts of weak and renormalized solutions are equivalent.

In order to conclude this general introduction about the Vlasov-Poisson system, we mention a surprising regularizing effect of the equation, that was used in a similar context, namely for the Vlasov-Maxwell system (8.2), to build distributional solutions. Indeed, given an equation of the type

$$\partial_t f_t + v \cdot \nabla_x f_t = \nabla_v \cdot g \qquad \text{in } (0, \infty) \times \mathbb{R}^3 \times \mathbb{R}^3$$

where $f \in L^2(\mathbb{R} \times \mathbb{R}^3 \times \mathbb{R}^3)$ and $g \in L^2(\mathbb{R} \times \mathbb{R}^3 \times \mathbb{R}^3; \mathbb{R}^3)$, a *velocity averaging lemma* [80] says that the velocity averages of f_t with respect to all smooth weight functions $\psi(v) \in C_c^\infty(\mathbb{R}^3)$ gain a fractional derivative, namely

$$\int_{\mathbb{R}^d} f_t(x, v)\psi(v)\, dv \in H^{1/4}(\mathbb{R} \times \mathbb{R}^3).$$

Similar results were first proved in [93, 92, 88].

A velocity average lemma was used, together with the transport arguments presented below in the context o the Vlasov-Poisson system, to show existence of weak solutions of the Vlasov-Maxwell system (8.2). More precisely, Di Perna and Lions [80] proved that, given arbitrary initial data f_0, B_0, E_0 with finite mass and total energy, there exists a distributional solution of the Vlasov-Maxwell system (8.2) with initial data f_0, B_0, and E_0. The assumptions on the initial data are the following: $f_0 \geq 0$,

$$\int_{\mathbb{R}^6} (1+|v|^2) f_0\, dx\, dv < \infty, \qquad \int_{\mathbb{R}^6} f_0^2\, dx\, dv + \int_{\mathbb{R}^3} \left(|E_0|^2 + |B_0|^2\right) dx < \infty,$$

together with the compatibility conditions

$$\operatorname{div} B_0 = 0, \qquad \operatorname{div} E_0 = \int_{\mathbb{R}^3} f_0\, dv.$$

It is important to observe that the Vlasov-Poisson system has a transport structure which allows one to prove that, when the solutions is sufficiently smooth, f_t is transported along the characteristics of the vector field $\boldsymbol{b}_t(x, v) = (v, E_t(x))$. However, when dealing with weak or renormalized solutions, it is not clear that such a vector field defines a flow on the phase-space and, at least a priori, one loses the relation between the Eulerian and Lagrangian picture.

The goal of this chapter is twofold: on the one hand we show that the Lagrangian picture is still valid even for weak/renormalized solutions, and secondly we obtain global existence of weak solutions under minimal assumptions on the initial data. Under such generality, we need to employ a suitable notion of Lagrangian solution (see the precise Definition 4.2 below) that allows for blow up in finite time of the trajectories. More precisely, the idea is that particles evolve along integral curves of the vector field $\boldsymbol{b}_t$, and they can escape to infinity and/or appear from infinity in finite time. However, under some suitable assumptions on the initial data, we can prevent such a finite-time blow-up phenomenon.

Both main results of this chapter rely on a combination of the following tools:

(i) the local version of the DiPerna-Lions theory developed in Chapters 2, 3, and 4;
(ii) the uniqueness of bounded compactly supported solutions to the continuity equation for a special class of vector fields obtained by convolving a singular kernel with a measure (see Section 1.4);
(iii) the fact that the concept of Lagrangian solution is stronger than the one of renormalized solution (see Section 4.2);
(iv) a general superposition principle stating that every nonnegative solution of the continuity equation has a Lagrangian structure without any regularity or growth assumption on the vector field (see Section 4.3).

The above machinery, developed in the first four chapters of this thesis, was needed to prove Theorem 4.9, a general result on the renormalization property for solutions of transport equations which is crucial in our proof. However, from a PDE viewpoint all we shall need is the statement of Theorem 4.9 and the renormalization property, which will be presented again in the context of the Vlasov-Poisson equation (see Definition 8.1 below). Therefore, we keep the presentation as much as possible independent of the heavy machinery of Chapters 2, 3, and 4, so that the statements of the next section and the PDE proofs can be read with the sole reference to Theorem 4.9.

8.1. Statement of the results

As already observed in the introduction, the Vlasov-Poisson system has a transport structure: indeed we can rewrite it as

$$\partial_t f_t + \boldsymbol{b}_t \cdot \nabla_{x,v} f_t = 0, \tag{8.3}$$

where the vector field $\boldsymbol{b}_t(x, v) = (v, E_t(x)) : \mathbb{R}^{2d} \to \mathbb{R}^{2d}$ is divergence-free, and is coupled to f_t via the relation $E_t = \sigma c_d \rho_t * (x/|x|^d)$. Recalling that $c_d \mathrm{div}\left(\frac{x}{|x|^d}\right) = \delta_0$, the vector field E_t can also be found as $E_t = -\nabla_x V_t$ where the potential $V_t : (0, \infty) \times \mathbb{R}^d \to \mathbb{R}$ solves[1]

$$-\Delta V_t = \sigma \rho_t \qquad \text{in } \mathbb{R}^d, \qquad \lim_{|x| \to \infty} V_t(x) = 0. \tag{8.4}$$

Notice that, because the kernel $x/|x|^d$ is locally integrable, the electric field E_t belongs to $L^1_{\mathrm{loc}}(\mathbb{R}^d; \mathbb{R}^d)$, therefore $\boldsymbol{b}_t \in L^1_{\mathrm{loc}}(\mathbb{R}^{2d}; \mathbb{R}^{2d})$.

Now, since $\boldsymbol{b}_t$ is divergence-free, the above equation can be rewritten as

$$\partial_t f_t + \mathrm{div}_{x,v}(\boldsymbol{b}_t f_t) = 0.$$

In order to apply the theory of flows of vector fields to this equation, however, one needs to face three difficulties.

- The equation can be reinterpreted in the distributional sense provided the product $\boldsymbol{b}_t f_t$ belongs to L^1_{loc}. Unfortunately, as mentioned before, this is not true if f_t is merely L^1.
- The vector field $\boldsymbol{b}_t$ is not in general Lipschitz, so one cannot use the standard Cauchy-Lipschitz theory to construct a flow for such a vector field, and not even $W^{1,1}_{\mathrm{loc}}$ or BV_{loc}, the regularity assumptions of the DiPerna-Lions and Ambrosio theory.
- The theory of flows of non-smooth vector fields requires usually the a priori assumption that the trajectories of the flow do not blow up in finite time, which is expressed in terms of the vector field by the following global hypothesis:

$$\frac{|\boldsymbol{b}_t|(x, v)}{1 + |x| + |v|} \in L^1\big((0, T); L^1(\mathbb{R}^{2d})\big) + L^1\big((0, T); L^\infty(\mathbb{R}^{2d})\big). \tag{8.5}$$

  For Vlasov-Poisson (or more in general for any Hamiltonian system where $\boldsymbol{b}_t(x, v)$ is of the form $(v, -\nabla V_t(x))$) the above assumption is satisfied if and only if

$$E_t = E_t^1 + E_t^2, \quad \text{with } |E_t^1| \in L^1\big((0, T); L^d(\mathbb{R}^d)\big)$$
$$\text{and } \frac{|E_t^2|(x)}{1 + |x|} \in L^1\big((0, T); L^\infty(\mathbb{R}^d)\big)$$

[1] This description is correct in dimension $d \geq 3$ since the fundamental solution of the Laplacian decays at infinity, while in dimension 2 the function V_t is given by the convolution of ρ_t with $-\frac{1}{2\pi} \log |x|$.

(see Lemma 8.15 in the appendix). Unfortunately this is a very restrictive assumption, as it requires both some integrability and moment (in v) conditions on f_t.

To overcome the first difficulty, one notices that if f_t is a smooth solution of (8.3) then also $\beta(f_t)$ is a solution for all C^1 functions $\beta : \mathbb{R} \to \mathbb{R}$; indeed

$$\partial_t \beta(f_t) + \boldsymbol{b}_t \cdot \nabla_{x,v} \beta(f_t) = \left[\partial_t f_t + \boldsymbol{b}_t \cdot \nabla_{x,v} f_t\right] \beta'(f_t) = 0,$$

or equivalently (since $\operatorname{div}_{x,v}(\boldsymbol{b}_t) = 0$)

$$\partial_t \beta(f_t) + \operatorname{div}_{x,v}(\boldsymbol{b}_t \beta(f_t)) = 0. \tag{8.6}$$

This motivates the introduction of the concept of renormalized solution of the Vlasov-Poisson system [78], which requires to interpret the first equation of (8.1) in a renormalized sense.

Definition 8.1. A function $f \in L^\infty([0,T]; L^1(\mathbb{R}^{2d}))$ is a *renormalized solution of the Vlasov-Poisson system* (8.1) (starting from f_0) if, setting

$$\begin{aligned} \rho_t(x) &:= \int_{\mathbb{R}^d} f_t(x,v)\,dv, \\ E_t &:= \sigma\, c_d \int_{\mathbb{R}^d} \rho_t(y) \frac{x-y}{|x-y|^d}\,dy, \\ \boldsymbol{b}_t(x,v) &:= (v, E_t(x)), \end{aligned}$$

for every $\beta \in C^1 \cap L^\infty(\mathbb{R})$ we have that (8.6) holds in the sense of distributions, namely, for every $\phi \in C_c^\infty([0,T) \times \mathbb{R}^{2d})$,

$$\begin{aligned} &\int_{\mathbb{R}^{2d}} \phi_0(x,v)\beta(f_0(x,v))\,dx\,dv \\ &+ \int_0^T \int_{\mathbb{R}^{2d}} \left[\partial_t \phi_t(x,v) + \nabla_{x,v}\phi_t(x,v)\boldsymbol{b}_t(x,v)\right]\beta(f_t(x,v))\,dx\,dv\,dt = 0. \end{aligned} \tag{8.7}$$

Notice that because β is bounded by assumption, $\beta(f_t) \in L^\infty$ so $\boldsymbol{b}_t \beta(f_t) \in L^1_{\text{loc}}$ (recall that $\boldsymbol{b}_t \in L^1_{\text{loc}}$) and (8.7) makes always sense.

In order to deal with the second difficulty listed above, by a modification of the argument in [32], we proved in Section 1.4 that for any vector field of the form $(v, \mu_t * x/|x|^d)$ with μ_t a time-dependent measure there is uniqueness of bounded compactly supported solutions of the continuity equation (see Theorem 1.14). By Remark 1.8, this property is enough to replace the regularity assumption on the vector field in Theorem 1.5.

In Chapters 2, 3, and 4, we developed a local version of the DiPerna-Lions' theory under no global assumptions on the vector field, and this will be a crucial tool for us to give a Lagrangian description of solutions and to overcome the third difficulty. More precisely, in Theorem 4.9 we proved that every bounded nonnegative solution of a continuity equation can be always represented as a superposition of mass transported along integral curves of the vector field (and these curves cannot split/intersect by the regularity of the vector field). Combining these facts we can show that all bounded/renormalized solutions of Vlasov-Poisson are Lagrangian.

As mentioned before, to express the fact that solutions are Lagrangian we need the concept of Maximal Regular Flow. Roughly speaking, the reader of this chapter who is not familiar with the first part of this thesis may think that the (uniquely defined) incompressible flow on the phase-space is composed of integral curves of $\boldsymbol{b}_t$ that "transport" the density f_t (notice that, since trajectories may blow-up in finite time, mass of f_t can disappear at infinity and/or come from infinity, but it has to follow the integral curves of $\boldsymbol{b}_t$).

Our first main result shows that *bounded or renormalized solutions of Vlasov-Poisson are Lagrangian*. As shown in Theorem 4.6, the concept of Lagrangian solutions is stronger than the one of renormalized solutions, as *all Lagrangian solutions of Vlasov-Poisson are renormalized.* We recall the notation L^1_+ to denote the space of nonnegative integrable functions and that by weakly continuous solutions we mean that the map $t \mapsto \int_{\mathbb{R}^{2d}} f_t \, \varphi \, dx \, dv$ is continuous for any $\varphi \in C_c(\mathbb{R}^{2d})$.

Theorem 8.2. *Let $T > 0$, $f_0 \in L^1_+(\mathbb{R}^{2d})$, and $f_t \in L^\infty((0,T); L^1_+(\mathbb{R}^{2d}))$ be a weakly continuous function on* $[0, T)$. *Assume that:*
(i) *either $f_t \in L^\infty((0,T); L^\infty(\mathbb{R}^{2d}))$ and f_t is a distributional solution of the Vlasov-Poisson equation* (8.1)*;*
(ii) *or f_t is a renormalized solution of the Vlasov-Poisson equation* (8.1) *(according to Definition* 8.1*).*
Then f_t is a Lagrangian solution transported by the Maximal Regular Flow associated to $\boldsymbol{b}_t(x, v) = (v, E_t(x))$. In particular f_t is renormalized.

The next corollary provides conditions in dimension $d = 2, 3, 4$ in order to avoid the finite-time blow up of the flow that transports f_t. The finiteness of kinetic energy is usually satisfied when we consider the repulsive case and the energy is assumed to be finite at time 0. As we shall explain in Remark 8.10, the case $d = 2$ is slightly different from $d \geq 3$ because of the slower decay at infinity of the kernel $x/|x|^d$. For this reason we restrict the next statements to the case $d \geq 3$, while in Remark 8.10 we mention a possible way to deal with the case $d = 2$.

Corollary 8.3. *Let $d = 3$ or $d = 4$, $T > 0$, and let $f_t \in L^\infty((0, T); (L^1_+ \cap L^q_+)(\mathbb{R}^{2d}))$ be a renormalized solution of the Vlasov-Poisson equation* (8.1) *(according to Definition* 8.1*) with*

$$q = \begin{cases} 1 & \text{if } d = 2 \text{ or } d = 3, \\ 2 & \text{if } d = 4. \end{cases} \tag{8.8}$$

Let us assume that the kinetic energy is integrable in time, that is

$$\int_0^T \int_{\mathbb{R}^{2d}} |v|^2 f_t(x, v)\, dx\, dv\, dt < \infty, \tag{8.9}$$

*Then the flow associated to $\mathbf{b}_t(x, v) = (v, E_t(x))$ is globally defined on $[0, T]$ (*i.e., *trajectories do not blow-up) for f_0-a.e. $(x, v) \in \mathbb{R}^{2d}$. In particular f_t is the image of f_0 through an incompressible flow, hence, for all $\psi : [0, \infty) \to [0, \infty)$ Borel,*

$$[0, T] \ni t \mapsto \int_{\mathbb{R}^{2d}} \psi\big(f_t(x, v)\big)\, dx\, dv$$

is constant in time.

The next statement improves the exponent q of the previous Corollary in the case $d = 4$, by further assuming the finiteness of the potential energy.

Corollary 8.4. *Let $T > 0$ and let $f_t \in L^\infty((0, T); L^1_+(\mathbb{R}^8))$ be a renormalized solution of the Vlasov-Poisson equation* (8.1) *(according to Definition* 8.1*). Let us assume that the kinetic energy and the potential energy are integrable in time, that is*

$$\int_0^T \int_{\mathbb{R}^8} |v|^2 f_t(x, v)\, dx\, dv\, dt + \int_0^T \int_{\mathbb{R}^4} |E_t(x)|^2\, dx\, dt < \infty, \tag{8.10}$$

Then, as in Corollary 8.3*, the flow associated to $\mathbf{b}_t(x, v) = (v, E_t(x))$ is globally defined on* $[0, T]$.

Remark 8.5. The energy is formally conserved along solutions of the Vlasov-Poisson system; whether this property holds also for distributional/renormalized solutions is an important open problem in the theory. However, many weak solutions built by approximation satisfy that the energy at time t is at least controlled from above by the initial energy. Hence, when $\sigma = 1$ the validity of (8.9) is often guaranteed by the assumption on the initial datum

$$\int_{\mathbb{R}^{2d}} |v|^2 f_0\, dx\, dv + \int_{\mathbb{R}^d} H * \rho_0\, \rho_0\, dx < \infty,$$

where $H(x) := \frac{c_d}{d-2}|x|^{2-d}$ (see also Corollary 8.9 below). In the case $\sigma = -1$ a bound on the total energy does not provide in general a bound on the kinetic energy, since the potential energy is negative. For instance, when $d = 3$ and $\sigma = -1$ one needs the additional hypothesis that $f_0 \in L^{9/7}(\mathbb{R}^6)$ (see [79, Equation (38)]). A similar result can also be given when $d = 4$, $\sigma = -1$ and $f_0 \in L^2(\mathbb{R}^8)$ with small L^2 norm, while in the case $d = 2$, $\sigma = -1$ and $f_0 \in L \log L(\mathbb{R}^4)$ one would need to slightly change the form of the electric field (see Remark 8.10 below). Indeed, in dimension 3 the solution $V_t = H * \rho_t$ of the equation $\Delta V_t = -\rho_t$ satisfies, by Calderón-Zygmund estimates and the Sobolev embedding, $\|V_t\|_{L^6(\mathbb{R}^3)} \le C\|D^2 V_t\|_{L^{6/5}(\mathbb{R}^3)} \le C\|\rho_t\|^2_{L^{6/5}(\mathbb{R}^3)}$. Thanks to this fact, Hölder inequality, and Lemma 8.16 below applied with $\alpha = 0$, $q = 9/7$, $p_0 = 6/5$, we estimate

$$\begin{aligned} \frac{1}{4\pi}\int_{\mathbb{R}^3} H * \rho_t\, \rho_t\, dx &\le C\|H * \rho_t\|_{L^6(\mathbb{R}^3)}\|\rho_t\|_{L^{6/5}(\mathbb{R}^3)} \\ &\le C\|\rho_t\|^2_{L^{6/5}(\mathbb{R}^3)} \\ &\le C\|f_t\|^{3/2}_{L^{9/7}(\mathbb{R}^6)}\left(\int_{\mathbb{R}^6} |v|^2 f_t\, dx\, dv\right)^{1/2}, \end{aligned} \tag{8.11}$$

where C is a universal constant. If the total energy is bounded by a constant C_0, we deduce that

$$\begin{aligned} C_0 &\ge \int_{\mathbb{R}^6} |v|^2 f_t\, dx\, dv - \frac{1}{4\pi}\int_{\mathbb{R}^3} H * \rho_t\, \rho_t\, dx \\ &\ge \int_{\mathbb{R}^6} |v|^2 f_t(x, v)\, dx\, dv - C\left(\int_{\mathbb{R}^6} |v|^2 f_t\, dx\, dv\right)^{1/2}, \end{aligned}$$

where C depends only on $\sup_{t\in[0,\infty)} \|f_t\|_{L^{9/7}(\mathbb{R}^6)}$, which in turn is often bounded (for instance, on solutions built by approximation) by $\|f_0\|_{L^{9/7}(\mathbb{R}^6)}$. Hence if the energy at time t is bounded, we deduce a control of the kinetic energy, and therefore of the full energy, thanks to (8.11), at time t.

In dimension $d = 4$, the same argument works except for the fact that the estimate of the potential energy in terms of the kinetic energy (8.11) uses different exponents and we need to require a smallness condition on the L^2-norm of the solution at time t. Indeed, the solution $V_t = H * \rho_t$ of Calderón-Zygmund estimates and the Sobolev embedding imply this time that

$$\|V_t\|_{L^4(\mathbb{R}^3)} \le C\|D^2V_t\|_{L^{4/3}(\mathbb{R}^3)} \le C\|\rho_t\|_{L^{4/3}(\mathbb{R}^3)}.$$

By Lemma 8.16 below applied with $\alpha = 0$, $q = 9/7$, $p_0 = 6/5$, we estimate

$$\begin{aligned}\frac{1}{4\pi}\int_{\mathbb{R}^4} H * \rho_t \, \rho_t \, dx &\le C\|H * \rho_t\|_{L^6(\mathbb{R}^4)}\|\rho_t\|_{L^{4/3}(\mathbb{R}^4)}\\ &\le C\|\rho_t\|^2_{L^{4/3}(\mathbb{R}^4)}\\ &\le C\|f_t\|_{L^2(\mathbb{R}^8)}\left(\int_{\mathbb{R}^8}|v|^2 f_t \, dx\, dv\right).\end{aligned}$$

If the total energy is bounded by a constant C_0, we deduce that

$$\begin{aligned}C_0 &\ge \int_{\mathbb{R}^8}|v|^2 f_t \, dx\, dv - \frac{1}{4\pi}\int_{\mathbb{R}^4} H * \rho_t \, \rho_t \, dx\\ &\ge \int_{\mathbb{R}^8}|v|^2 f_t(x,v)\, dx\, dv - C \sup_{t\in[0,\infty)}\|f_t\|_{L^2(\mathbb{R}^8)}\left(\int_{\mathbb{R}^8}|v|^2 f_t \, dx\, dv\right).\end{aligned}$$

Since $\sup_{t\in[0,\infty)}\|f_t\|_{L^2(\mathbb{R}^8)} \le \|f_0\|_{L^2(\mathbb{R}^8)}$ (at least when the solution is built by approximation), we deduce that, provided $\|f_0\|_{L^2(\mathbb{R}^8)} < 1/C$ we have a control of the kinetic (and also total) energy. Then Theorem 8.4 can be applied.

Remark 8.6. Since we proved in Theorem 8.2 that all bounded distributional solutions are renormalized, one may wonder when the converse is true, namely if renormalized solutions are also distributional solutions. This happens basically as soon as we ask enough integrability for the term $E_t f_t$ to belong to $L^1_{\mathrm{loc}}(\mathbb{R}^{2d})$: for instance, in dimension $d = 3$ it is enough to consider solutions $f \in L^\infty((0,T); L^q(\mathbb{R}^6))$ with $q = (12 + 2\sqrt{5})/11$ (see [78, Theorem 1], whose proof is based on Lemma 8.16 below applied with $\alpha = 0$).

Our second result deals with existence of global Lagrangian solutions under minimal assumptions on the initial data. In this case the sign of σ (*i.e.*, whether the potential is attractive or repulsive) plays a crucial role, since in the repulsive case the total energy controls the kinetic part, while in the attractive case the loss of an a priori bound of the kinetic energy prevents us for showing such a result. However we can state a general existence theorem that holds both in the attractive and repulsive case, and then show that in the repulsive case it gives us what we want.

The basic idea is the following: when proving existence of solutions by approximation it may happen that, in the approximating sequence, there are some particles that move at higher and higher speed while still remaining localized in a compact set in space (think of a family of particle rotating faster and faster along circles around the origin). Then, while in the limit these particles will disappear from the phase-space (having infinite velocity), the electric field generated by them will survive, since they are still in the physical space. Hence the electric field is not anymore generated by the marginal of f_t in the v-variable, instead it is generated by an "effective density" $\rho_t^{\text{eff}}(x)$ that is larger than $\rho_t(x)$.

So, our strategy will be first to prove global existence of Lagrangian (hence renormalized) solutions for a generalized Vlasov-Poisson system where the electric field is generated by ρ_t^{eff} and then show that, in the particular case $\sigma = 1$, if the initial datum has finite total energy, then $\rho_t^{\text{eff}} = \rho_t$ and our solution solves the classical Vlasov-Poisson system.

We begin by introducing the concept of generalized solutions to Vlasov-Poisson. We use the notation $\mathscr{M}_+$ to denote the space of non-negative measures with finite total mass.

Definition 8.7 (Generalized solution of the Vlasov-Poisson equation). Given $\overline{f} \in L^1(\mathbb{R}^{2d})$, let $f_t \in L^\infty((0,\infty); L^1_+(\mathbb{R}^{2d}))$ and $\rho_t^{\text{eff}} \in L^\infty((0,\infty); \mathscr{M}_+(\mathbb{R}^d))$. We say that the couple $(f_t, \rho_t^{\text{eff}})$, is a (global in time) generalized solution of the Vlasov-Poisson system starting from $\overline{f}$ if, setting

$$
\begin{aligned}
\rho_t(x) &:= \int_{\mathbb{R}^d} f_t(x, v)\, dv, \\
E_t^{\text{eff}} &:= \sigma\, c_d \int_{\mathbb{R}^d} \rho_t^{\text{eff}}(y) \frac{x-y}{|x-y|^d}\, dy, \qquad (8.12)\\
\boldsymbol{b}_t(x, v) &:= (v, E_t^{\text{eff}}(x)),
\end{aligned}
$$

f_t is a renormalized solution of the continuity equation with vector field $\boldsymbol{b}_t$ starting from $\overline{f}$,

$$\rho_t \le \rho_t^{\text{eff}} \qquad \text{as measures for } \mathscr{L}^1\text{-a.e. } t \in (0,\infty), \qquad (8.13)$$

and

$$|\rho_t^{\text{eff}}|(\mathbb{R}^d) \le \|f_0\|_{L^1(\mathbb{R}^{2d})} \qquad \text{for } \mathscr{L}^1\text{-a.e. } t \in (0,\infty). \qquad (8.14)$$

Notice that since $\|\rho_t\|_{L^1(\mathbb{R}^d)} = \|f_t\|_{L^1(\mathbb{R}^{2d})}$, it follows by (8.13) and (8.14) that whenever the mass of f_t is conserved in time, that is $\|f_t\|_{L^1(\mathbb{R}^{2d})} =$

$\|f_0\|_{L^1(\mathbb{R}^{2d})}$ for $\mathscr{L}^1$-a.e. $t \in (0,\infty)$, then $\rho_t^{\mathrm{eff}} = \rho_t$ and generalized solutions of the Vlasov-Poisson system are just standard renormalized solutions. The notion of generalized solution suggests that the only way of possible failure for the existence of renormalized solutions of the Vlasov-Poisson system is by losing mass at large velocities; this phenomenon is well-known in the analysis of kinetic equations.

We prove here that generalized solutions of the Vlasov-Poisson equation exist globally for any L^1 initial datum, both in the attractive and in the repulsive case.

Theorem 8.8. *Let us consider $f_0 \in L^1_+(\mathbb{R}^{2d})$. Then there exists a generalized solution $(f_t, \rho_t^{\mathrm{eff}})$ of the Vlasov-Poisson system starting from f_0. Moreover, f_t belongs to $C([0,\infty); L^1_{\mathrm{loc}}(\mathbb{R}^{2d}))$ and it is transported by the Maximal Regular Flow associated to $\boldsymbol{b}_t(x,v) = (v, E_t^{\mathrm{eff}}(x))$.*

As observed before, if $\rho_t^{\mathrm{eff}} = \rho_t$ then f_t is a renormalized solution of the Vlasov-Poisson system. When $\sigma = 1$ (*i.e.*, in the repulsive case) the equality $\rho^{\mathrm{eff}} = \rho_t$ is satisfied in many cases of interest, for instance whenever the initial energy is finite (namely $|v|^2 f_0 \in L^1(\mathbb{R}^{2d})$ and $E_0 \in L^2(\mathbb{R}^d)$, see Corollary 8.9 below), or in the case of infinite energy if other weaker conditions are satisfied as it happens in the context of [126] and [105] (see Remark 8.26).

The following result improves the result announced in [78], generalizing their statement to any dimension and with weaker conditions on the initial datum. In the first part of the statement, we show the global-in-time existence and some natural properties of renormalized solutions of the Vlasov-Poisson system in the repulsive case under only the finite-energy assumption on the initial datum. In the last part of the statement, we show that the trajectories of the Regular Lagrangian Flow associated to the renormalized solution starting from $t = 0$ cannot blow up if $d = 3, 4$.

Corollary 8.9. *Let $d \geq 3$, and let $f_0 \in L^1_+(\mathbb{R}^{2d})$ satisfy*

$$\int_{\mathbb{R}^{2d}} |v|^2 f_0 \, dx \, dv + \int_{\mathbb{R}^d} H * \rho_0 \, \rho_0 \, dx < \infty,$$

where $\rho_0(x) := \int_{\mathbb{R}^d} f_0(x,v)\,dv$ and $H(x) := \frac{c_d}{d-2}|x|^{2-d}$. Assume that $\sigma = 1$. Then there exists a global Lagrangian (hence renormalized) solution $f_t \in C([0,\infty); L^1_{\mathrm{loc}}(\mathbb{R}^{2d}))$ of the Vlasov-Poisson system (8.1) *with initial datum f_0.*

Moreover, the solution f_t, the associated density ρ_t, and the electric field E_t satisfy the following properties:

(i) *the density ρ_t and the electric field E_t are strongly continuous in $L^1_{\mathrm{loc}}(\mathbb{R}^d)$;*

(ii) *for every* $t \geq 0$*, we have the energy bound*

$$\begin{aligned} &\int_{\mathbb{R}^{2d}} |v|^2 f_t \, dx \, dv + \int_{\mathbb{R}^d} H * \rho_t \, \rho_t \, dx \\ &\leq \int_{\mathbb{R}^{2d}} |v|^2 f_0 \, dx \, dv + \int_{\mathbb{R}^d} H * \rho_0 \, \rho_0 \, dx; \end{aligned} \tag{8.15}$$

(iii) *if* $d = 3$ *or* $d = 4$ *the flow is globally defined on* $[0, T]$ *for* f_0*-a.e.* $(x, v) \in \mathbb{R}^{2d}$ (i.e., *trajectories do not blow-up) and* f_t *is the image of* f_0 *through an incompressible flow.*

According to the definition of generalized solution of the Vlasov-Poisson system (see Definition 8.7), the function f and the densities ρ and ρ^{eff} are defined only for $\mathscr{L}^1$-a.e. t. In Theorem 8.8 we build solutions with better properties, namely with f_t strongly continuous in $L^1_{\text{loc}}(\mathbb{R}^{2d})$. At the level of generality of this result, we cannot say the continuity of ρ^{eff}. Assuming also the finiteness of energy, instead, the function $\rho^{\text{eff}} = \rho$ turns out to be strongly $L^1_{\text{loc}}(\mathbb{R}^d)$ continuous, as well as the force field. This justifies the fact that the energy bound (8.15) holds not only for $\mathscr{L}^1$-a.e. $t \geq 0$, but for *all* $t \geq 0$.

Remark 8.10. In dimension $d = 2$, even with an initial datum $f_0 \in C^\infty_c(\mathbb{R}^d)$, the electric field E_0 cannot belong to L^2 (this is due to the fact that the kernel $x/|x|^d$ does not belong to L^2 at infinity). Moreover, the potential energy does not have a definite sign. However, one can show that an analogous statement of Corollary 8.9 holds also for solutions of a slightly modified equation, which has the form

$$\begin{cases} \partial_t f_t + v \cdot \nabla_x f_t + E_t \cdot \nabla_v f_t = 0 & \text{in } (0, \infty) \times \mathbb{R}^d \times \mathbb{R}^d \\ \rho_t(x) = \displaystyle\int_{\mathbb{R}^d} f_t(x, v) \, dv & \text{in } (0, \infty) \times \mathbb{R}^d \\ E_t(x) = \sigma c_d \displaystyle\int_{\mathbb{R}^d} (\rho_t(y) - \rho_b(y)) \frac{x - y}{|x - y|^d} \, dy & \text{in } (0, \infty) \times \mathbb{R}^d, \end{cases} \tag{8.16}$$

where f_t, ρ_t, E_t play the same role as in the standard Vlasov-Poisson equation (8.1), and $\rho_b \in L^1_+(\mathbb{R}^d)$ represents a fixed background satisfying

$$\int_{\mathbb{R}^d} \rho_b(x) \, dx = \int_{\mathbb{R}^d} \rho_0(x) \, dx.$$

This allows for cancellations in the expression for the L^2 norm of E_0, which turns out to be finite if ρ_b and ρ_0 are sufficiently nice.

Remark 8.11. When $d = 3$, the above result can be generalized to the attractive case $\sigma = -1$ under the additional assumption $f_0 \in L^{9/7}(\mathbb{R}^6)$.

Indeed, as already mentioned in Remark 8.5, this allows one to prove the the kinetic energy is uniformly bounded in time, and then by standard interpolation inequalities one obtains that also the potential energy is bounded.

In [33], Bohun, Bouchut, and Crippa gave a different proof of Corollary 8.9 in dimension $d = 2$ and $d = 3$. Their proofs are outlined in the remarks below. In both cases, the basic idea is to prove an a priori estimate on solutions with bounded energy, which shows that the flow cannot blow up in finite time. This, in turn, allows to apply the classical DiPerna-Lions theory (with the further difficulty that the vector field is not $W^{1,1}_{\text{loc}}$), instead of the theory of maximal regular flows developed in the first part of this thesis.

Remark 8.12. In dimension $d = 2$, the key observation is that any solution of the modified Vlasov-Poisson system (8.16) with $\sigma = 1$ and finite energy (more precisely, it is enough to require $E \in L^\infty((0,\infty); L^2(\mathbb{R}^4)))$ satisfies the standard growth conditions on the vector field in (8.5), which prevent the finite-time blow-up of the flow. Indeed, clearly

$$\frac{|v|}{1+|x|+|v|} \in L^1\big((0,T); L^\infty(\mathbb{R}^4)\big)$$

and, decomposing E_t as

$$\begin{aligned}\frac{E_t(x)}{1+|x|+|v|} &= \frac{E_t(x)1_{\{|v|\le E_t(x)\}}}{1+|x|+|v|} + \frac{E_t(x)1_{\{|v|> E_t(x)\}}}{1+|x|+|v|}\\ &=: E_{1t}(x,v) + E_{2t}(x,v)\end{aligned}$$

we have that $E_{2t}(x,v) \in L^\infty(\mathbb{R}^4)$ uniformly in t and

$$\begin{aligned}\int_{\mathbb{R}^4} |E_{2t}(x,v)|\,dx\,dv &\le \int_{\mathbb{R}^2} |E_t(x)| \int_{\{|v|\le|E_t(x)|\}} \frac{1}{|v|}\,dv\,dx\\ &= 2\pi \int_{\mathbb{R}^2} |E_t(x)|^2\,dx\end{aligned}$$

for every $t \ge 0$. Hence, we see that

$$\frac{|\boldsymbol{b}_t|(x,v)}{1+|x|+|v|} \le \frac{|v|+|E_t(x)|}{1+|x|+|v|} \in L^1\big((0,T); L^1(\mathbb{R}^4)\big) + L^1\big((0,T); L^\infty(\mathbb{R}^4)\big).$$

Remark 8.13. In dimension $d = 3$ any solution of the Vlasov-Poisson system with $\sigma = 1$ and finite energy satisfies the following property: any regular lagrangian flow $\boldsymbol{X} : [a,b] \times \mathbb{R}^6 \to \mathbb{R}^6$ relative to $\boldsymbol{b}_t(x,v) =$

$(v, E_t(x))$, where $[a, b] \subset [0, \infty)$, verifies the inequality (a kind of local equi-integrability)

$$\mathscr{L}^6\big(B_r \setminus \{(x, v) \in \mathbb{R}^6 : |\boldsymbol{X}(t, x, v)| \le \lambda\}\big) \le g(r, \lambda) \qquad (8.17)$$

for every $r, \lambda > 0$ and for a function $g(r, \lambda)$ which converges to 0 as $\lambda \to \infty$ at fixed r. This property of the vector field $\boldsymbol{b}$ can replace the assumption (8.5) and it is enough to guarantee a stability property of the regular lagrangian flow in the classical DiPerna-Lions setting. The proof of (8.17) is obtained by showing that for every $r > 0$

$$\int_{B_r} \sup_{s\in[a,b]} \Big(1 + \log(1 + |\boldsymbol{X}^2(s, x, v)|)\Big)^{\alpha} \, dx\, dv < \infty,$$

where $\alpha \in (0, 1/3)$ and $\boldsymbol{X} = (\boldsymbol{X}^1, \boldsymbol{X}^2) \in \mathbb{R}^3 \times \mathbb{R}^3$. This estimate is based on the finiteness of energy, which in turn implies by the Sobolev embedding that the potential V_t belongs to $L^6(\mathbb{R}^3)$.

Remark 8.14. In this chapter we restricted ourselves to the Vlasov-Poisson equation but the argument and techniques introduced here generalize to other equations. For instance, a minor modification of our proofs allows one to obtain the same results in the context of the relativistic Vlasov-Poisson system.

The proofs of Theorems 8.2 and 8.8 and Corollaries 8.3, 8.4, and 8.9 are given in the next sections.

8.2. The assumptions of DiPerna-Lions for Hamiltonian ODEs

We characterize Hamiltonian-type systems that fall under the assumptions of the classical DiPerna Lions theory.

We recall that, in the seminal paper [81], DiPerna and Lions showed that for Sobolev vector fields one can introduce a suitable notion of flow provided the trajectories of the flow do not blow up in finite time. This is is expressed in terms of the vector field by the following global hypothesis:

$$\frac{|\boldsymbol{b}_t|(x, v)}{1 + |x| + |v|} \in L^1\big((0, T); L^1(\mathbb{R}^{2d})\big) + L^1\big((0, T); L^\infty(\mathbb{R}^{2d})\big). \qquad (8.18)$$

In the case when $b_t(x, v)$ takes the form $(v, E_t(x))$ for some time dependent vector-field $E_t : (0, T) \times \mathbb{R}^d \to \mathbb{R}^d$, we see that the first term $\frac{|v|}{1+|x|+|v|}$ is bounded. Hence, one needs to understand under which assumptions on E_t the term $\frac{|E_t|(x)}{1+|x|+|v|}$ satisfies (8.18). The next result gives a complete answer to this question.

Lemma 8.15. *Let $T > 0$, $E : (0,T) \times \mathbb{R}^d \to \mathbb{R}^d$. Then the following three conditions are equivalent:*

(i) $\frac{|E_t|(x)}{1+|x|+|v|} \in L^1\big((0,T); L^1(\mathbb{R}^{2d})\big) + L^1\big((0,T); L^\infty(\mathbb{R}^{2d})\big)$,

(ii) $E_t = E_t^1 + E_t^2$ *with* $|E_t^1| \in L^1\big((0,T); L^d(\mathbb{R}^d)\big)$ *and* $\frac{|E_t^2|(x)}{1+|x|} \in L^1\big((0,T); L^\infty(\mathbb{R}^d)\big)$,

(iii) *there exists* $C(t) \in L^1((0,T))$ *such that* $\big(|E_t|(x) - C(t)(1+|x|)\big)_+ \in L^1\big((0,T); L^d(\mathbb{R}^d)\big)$.

Proof. We first prove the equivalence between (ii) and (iii). If (ii) holds, write $E_t = E_t^1 + E_t^2$ as in (ii) and define $C(t) := \big\| \frac{E_t^2(x)}{1+|x|} \big\|_{L^\infty(\mathbb{R}^d)}$. Then

$$\big(|E_t| - C(t)(1+|x|)\big)_+ = \big(|E_t^1| + |E_t^2| - C(t)(1+|x|)\big)_+ \leq |E_t^1| \in L^d\big((0,T); L^d(\mathbb{R}^d)\big).$$

On the other hand, if (iii) holds we choose

$$E_t^1(x) := \frac{E_t(x)}{|E_t(x)|}\big(|E_t|(x) - C(t)(1+|x|)\big)_+,$$

$$E_t^2(x) := \frac{E_t(x)}{|E_t(x)|}\Big[C(t)(1+|x|) - \big(|E_t|(x) - C(t)(1+|x|)\big)_-\Big],$$

and it is easily seen that they satisfy (ii).

We now show that (ii) implies (i). Writing $E_t = E_t^1 + E_t^2$ as in (ii), we notice that, for any function $D \in L^1((0,T))$ with $D(t) \geq 1$,

$$\frac{E_t(x)}{1+|x|+|v|} = \frac{E_t^1(x)}{1+|x|+|v|} 1_{\{(x,v): D(t)|v| \leq |E_t^1|(x)\}} + \frac{E_t^1(x)}{1+|x|+|v|} 1_{\{(x,v): D(t)|v| > |E_t^1|(x)\}} + \frac{E_t^2(x)}{1+|x|+|v|},$$

and the last two terms belong to $L^1\big((0,T); L^\infty(\mathbb{R}^{2d})\big)$. To show that the first term belongs to $L^1\big((0,T); L^1(\mathbb{R}^{2d})\big)$, we pass in polar coordinates in v and notice that the function $r^{d-1}/(1+|x|+r)$ is increasing in r for any $x \in \mathbb{R}^d$, to get

$$\begin{aligned}
&\int_0^T \int_{\mathbb{R}^d} \int_{\big\{v: |v| \leq \frac{|E_t^1|(x)}{D(t)}\big\}} \frac{|E_t^1|(x)}{1+|x|+|v|}\, dv\, dx\, dt \\
&\leq C_d \int_0^T \int_{\mathbb{R}^d} |E_t^1|(x) \int_0^{\frac{|E_t^1|(x)}{D(t)}} \frac{r^{d-1}}{1+|x|+r}\, dr\, dx\, dt \\
&\leq C_d \int_0^T \int_{\mathbb{R}^d} \frac{|E_t^1|(x)^2}{D(t)} \frac{\frac{|E_t^1|(x)^{d-1}}{D(t)^{d-1}}}{1+|x|+\frac{|E_t^1|(x)}{D(t)}}\, dx\, dt \\
&\leq C_d \int_0^T \frac{1}{D(t)^{d-1}} \int_{\mathbb{R}^d} |E_t^1|(x)^d\, dx\, dt.
\end{aligned}$$

In particular, choosing $D(t) := 1 + \|E^1_t\|_{L^d(\mathbb{R}^d)}$, we can bound the last term above by

$$\int_0^T \|E^1_t\|_{L^d(\mathbb{R}^d)}\, dt < \infty,$$

which concludes the proof that (iii) implies (i).

Finally, we assume that (i) holds and show (iii). Indeed, write $\frac{|E_t|(x)}{1+|x|+|v|}$ as a sum as in (i), and denote by $C(t) \in L^1((0,T))$ a bound for the $L^\infty(\mathbb{R}^{2d})$-norm of the second addend. With no loss of generality, we can assume that $C(t) \geq 1$. By assumption, with this choice of $C(t)$ we have

$$\int_0^T \int_{\mathbb{R}^{2d}} \left(\frac{|E_t|(x)}{1+|x|+|v|} - C(t) \right)_+ dv\, dx\, dt < \infty.$$

We first rewrite this integral as

$$\begin{aligned}
&\int_0^T \int_{\mathbb{R}^{2d}} \left(\frac{|E_t|(x)}{1+|x|+|v|} - C(t) \right)_+ dv\, dx\, dt \\
&= \int_0^T \int_{\left\{x : \frac{|E_t|(x)}{1+|x|} \geq C(t)\right\}} \int_{\left\{v : |v| \leq \frac{|E_t|(x)}{C(t)} - 1 - |x|\right\}} \left(\frac{|E_t|(x)}{1+|x|+|v|} - C(t) \right)_+ dv\, dx\, dt.
\end{aligned} \tag{8.19}$$

Then, we note that

$$\begin{aligned}
A &:= \left\{x : \frac{|E_t|(x)}{1+|x|} \geq 4C(t)\right\} \cap \left\{\frac{|E_t|(x)}{4C(t)} \leq |v| \leq \frac{|E_t|(x)}{2C(t)}\right\} \\
&\subset \left\{x : \frac{|E_t|(x)}{1+|x|} \geq C(t)\right\} \cap \left\{0 \leq |v| \leq \frac{|E_t|(x)}{C(t)} - 1 - |x|\right\},
\end{aligned}$$

and that, inside A,

$$\frac{|E_t|(x)}{1+|x|+|v|} \geq \frac{|E_t|(x)}{\frac{|E_t|(x)}{4C(t)} + \frac{|E_t|(x)}{2C(t)}} = \frac{4}{3}C(t),$$

therefore

$$\frac{|E_t|(x)}{1+|x|+|v|} - C(t) \geq \frac{|E_t|(x)}{4(1+|x|+|v|)} \geq \frac{|E_t|(x)}{8|v|} \qquad \text{inside } A.$$

Thus, we can bound from below the second integral in (8.19) by

$$\begin{aligned}
&\int_0^T \int_{\left\{x:\frac{|E_t|(x)}{1+|x|}\geq 4C(t)\right\}} \int_{\left\{v:\frac{|E_t|(x)}{4C(t)}\leq |v|\leq \frac{|E_t|(x)}{2C(t)}\right\}} \frac{|E_t|(x)}{8|v|}\,dv\,dx\,dt \\
&= c_d \int_0^T \int_{\left\{x:\frac{|E_t|(x)}{1+|x|}\geq 4C(t)\right\}} \int_{\frac{|E_t|(x)}{2C(t)}}^{\frac{|E_t|(x)}{2C(t)}} \frac{|E_t|(x)}{8} r^{d-2}\,dr\,dx\,dt \\
&= \hat{c}_d \int_0^T \int_{\left\{x:\frac{|E_t|(x)}{1+|x|}\geq 4C(t)\right\}} \frac{|E_t|(x)^d}{C(t)^{d-1}}\,dx\,dt \\
&\geq \hat{c}_d \int_0^T \frac{1}{C(t)^{d-1}} \int_{\mathbb{R}^d} \big(|E_t|(x) - 4C(t)(1+|x|)\big)_+^d\,dx\,dt.
\end{aligned} \tag{8.20}$$

Since, by Hölder inequality,

$$\begin{aligned}
&\int_0^T \|\big(|E_t|(x) - 4C(t)(1+|x|)\big)_+\|_{L^d(\mathbb{R}^d)}\,dt \\
&\leq \left(\int_0^T \frac{1}{C(t)^{d-1}} \int_{\mathbb{R}^d} \big(|E_t|(x) - 4C(t)(1+|x|)\big)_+^d\,dx\,dt\right)^{1/d} \left(\int_0^T C(t)\,dt\right)^{\frac{d-1}{d}},
\end{aligned}$$

it follows by (8.19) and (8.20) that $\big(|E_t|(x) - 4C(t)(1+|x|)\big)_+ \in L^1\big((0,T); L^d(\mathbb{R}^d)\big)$, which proves (iii). □

8.3. The flow associated to Vlasov-Poisson: proof of Theorem 8.2 and Corollaries 8.3 and 8.4

Before proving the result, we recall a classical interpolation lemma (see for instance [78], where the lemma is stated in the case $\alpha = 0$).

Lemma 8.16. *Let $\alpha \in [0,\infty)$, $f \in L^1_+(\mathbb{R}^{2d})$, and assume that $f \in L^q(\mathbb{R}^{2d})$ for some $q \geq 1$ and that $|v|^2 f \in L^1(\mathbb{R}^{2d})$. Set $p_\alpha := \frac{d(q-1)+(2+\alpha)q}{d(q-1)+2+\alpha}$. Then $\rho_\alpha(x) := \int_{\mathbb{R}^d} \frac{f(x,v)}{(1+|v|)^\alpha}\,dv$ belongs to $L^{p_\alpha}(\mathbb{R}^d)$ and there exists a constant $C > 0$, depending only on n, α and q, such that*

$$\|\rho_\alpha\|_{L^{p_\alpha}(\mathbb{R}^d)} \leq C \||v|^2 f\|^{\theta_\alpha}_{L^1(\mathbb{R}^{2d})} \|f\|^{1-\theta_\alpha}_{L^q(\mathbb{R}^{2d})}$$

where $\theta_\alpha \in [0,1]$ is given by $\theta_\alpha = \frac{d(q-1)}{d(q-1)+(2+\alpha)q}$.

Proof. We prove here the case $q < \infty$, the case $q = \infty$ being completely analogous.

By Hölder's inequality, for every $x \in \mathbb{R}^d$ and $R > 0$ we estimate

$$\begin{aligned}\rho_\alpha(x) &= \int_{\{|v|<R\}} \frac{f(x,v)}{(1+|v|)^\alpha}\,dv + \int_{\{|v|\geq R\}} \frac{f(x,v)}{(1+|v|)^\alpha}\,dv \\ &\leq R^{d(q-1)/q}\Big(\int_{\mathbb{R}^d} f(x,v)^q\,dv\Big)^{1/q} + \frac{1}{R^{2+\alpha}}\int_{\mathbb{R}^d} |v|^2 f(x,v)\,dv.\end{aligned}$$

Minimizing the right-hand side with respect to R, for every $x \in \mathbb{R}^d$ we deduce that

$$\rho_\alpha(x) \leq \Big(\int_{\mathbb{R}^d} f(x,v)^q\,dv\Big)^{\frac{2+\alpha}{d(q-1)+(2+\alpha)q}} \Big(\int_{\mathbb{R}^d} |v|^2 f(x,v)\,dv\Big)^{\frac{d(q-1)}{d(q-1)+(2+\alpha)q}}.$$

Taking the L^{p_α}-norm of ρ_α and using Hölder's inequality, we find the result. □

We can now proceed with the proof of Theorem 8.2. Notice that the vector field $\boldsymbol{b}$ satisfies assumption (a) of Section 3.5 and is divergence-free. Also, by Theorem 1.14 it satisfies assumption (b). Therefore by Theorem 4.9 we deduce that f_t (resp. $\beta(f_t)$ with $\beta(s) = \arctan(s)$ if f_t is not bounded but is renormalized) is a Lagrangian solution. In particular Theorem 4.6 ensures that f_t is a renormalized solution.

Proof of Corollary 8.3. We assume that (8.9) holds and that $f_t \in L^\infty((0,T); L^q(\mathbb{R}^{2d}))$ with the choice of q given by (8.8). By Theorem 8.2, the solution is transported by the maximal regular flow associated to $\boldsymbol{b}_t$. In order to prove that trajectories do not blow up, we apply the criterion stated in Proposition 4.7 to $g_t := 2\pi^{-1}\arctan f_t : (0,T) \times \mathbb{R}^d \to [0,1]$. Since f_t is a renormalized solution of the continuity equation with vector field $\boldsymbol{b}$, by definition of renormalization g_t is a solution of the continuity equation; we need to verify that

$$\int_0^T \int_{\mathbb{R}^{2d}} \frac{|\boldsymbol{b}_t(x,v)| g_t(x,v)}{(1+(|x|^2+|v|^2)^{1/2})\log(2+(|x|^2+|v|^2)^{1/2})}\,dx\,dv\,dt < \infty. \tag{8.21}$$

To this end, let p be the integrability exponent provided by Lemma 8.16

$$p = \frac{d(q-1)+2q}{d(q-1)+2} = \begin{cases} 1 & \text{if } d = 2 \text{ or } d = 3, \\ 4/3 & \text{if } d = 4. \end{cases} \tag{8.22}$$

In the rest of the proof we denote by C any constant which depends only on d, on the quantity in (8.9), and on the norm of f_t in $L^q(\mathbb{R}^{2d})$. Thanks to Lemma 8.16 applied with $\alpha = 0$, for $\mathscr{L}^1$-a.e. $t \in (0,T)$,we have that

$$\|\rho_t\|_{L^p(\mathbb{R}^d)} \leq C \||v|^2 f_t\|_{L^1(\mathbb{R}^{2d})}^{\theta_0} \|f_t\|_{L^q(\mathbb{R}^{2d})}^{1-\theta_0} \leq C. \tag{8.23}$$

By (8.23), Sobolev inequality, and Calderón-Zygmund estimates (see for instance [91, Corollary 9.10]) we deduce that, for $\mathscr{L}^1$-a.e. $t \in (0, T)$,

$$\|E_t\|_{L^{dp/(d-p)}(\mathbb{R}^d)} \le C\|\nabla E_t\|_{L^p(\mathbb{R}^d)} \le C\|\rho_t\|_{L^p(\mathbb{R}^d)} \le C. \tag{8.24}$$

Then we consider $\gamma \in (0, 2)$ to be fixed later and, by Young inequality with exponents $\gamma/2$ and $(2-\gamma)/2$, we estimate

$$\begin{aligned}\frac{|E_t|g_t}{(1+|v|)\log(2+|v|)} &= \frac{|E_t|}{(1+|v|)^{1+\gamma}\log(2+|v|)}\Big((1+|v|)^\gamma g_t\Big)\\ &\le \left(\frac{|E_t|}{(1+|v|)^{1+\gamma}\log(2+|v|)}\right)^{2/(2-\gamma)}\\ &\quad + \Big((1+|v|)^\gamma g_t\Big)^{2/\gamma}.\end{aligned}$$

For every $\gamma \in (0, 2)$, by $g_t \le \arctan f_t \le \min\{f_t, 1\}$, we have that $g_t^{2/\gamma} \le f_t$. Hence, the last term in the right-hand side of the previous display has finite integral since f_t has finite kinetic energy (by (8.9))

$$\int_0^T\int_{\mathbb{R}^{2d}} (1+|v|)^2 g_t^{2/\gamma}\,dx\,dv\,dt \le \int_0^T\int_{\mathbb{R}^{2d}} (1+|v|)^2 f_t\,dx\,dv\,dt < \infty$$

As regards the first term, we rewrite it with Fubini's theorem

$$\begin{aligned}&\int_0^T\int_{\mathbb{R}^{2d}}\left(\frac{|E_t|}{(1+|v|)^{1+\gamma}\log(2+|v|)}\right)^{2/(2-\gamma)} dx\,dv\,dt\\ &= \left(\int_{\mathbb{R}^d}\frac{1}{(1+|v|)^{2(1+\gamma)/(2-\gamma)}\log(2+|v|)^{2/(2-\gamma)}}\,dv\right)\\ &\quad\times\left(\int_0^T\int_{\mathbb{R}^d}|E_t|^{2/(2-\gamma)}\,dx\,dt\right)\end{aligned} \tag{8.25}$$

and we choose γ as

$$\frac{2}{2-\gamma} = \frac{pd}{d-p} \qquad \Longleftrightarrow \qquad \gamma = \frac{2(pd+p-d)}{pd}.$$

With this choice, thanks to (8.24) the second integral in the right-hand side of (8.25) is finite. Recalling the choice of p in (8.22), in dimension $d = 2, 3$, and 4 it is easily checked that $\gamma = 1, 2/3$, and 1 respectively. In all three cases, we see that the first integral in the right-hand side of (8.25) is finite with these choices of d and γ. Hence, since $|\boldsymbol{b}_t(x, v)| \le$

$|v| + |E_t(x)|$, we find that

$$\int_0^T \int_{\mathbb{R}^{2d}} \frac{|\boldsymbol{b}_t| g_t}{(1 + (|x|^2 + |v|^2)^{1/2}) \log(2 + (|x|^2 + |v|^2)^{1/2})} \, dx \, dv \, dt$$

$$\leq \int_0^T \int_{\mathbb{R}^{2d}} f_t \, dx \, dv \, dt + \int_0^T \int_{\mathbb{R}^{2d}} |E_t| \frac{g_t}{(1 + |v|) \log(2 + |v|)} \, dx \, dv \, dt$$

$$\leq \left(\int_{\mathbb{R}^d} \frac{1}{(1 + |v|)^{2(1+\gamma)/(2-\gamma)} \log(2 + |v|)^{2/(2-\gamma)}} \, dv \right)$$

$$\times \left(\int_0^T \int_{\mathbb{R}^d} |E_t|^{2/(2-\gamma)} \, dx \, dt \right)$$

$$+ 2 \int_0^T \int_{\mathbb{R}^{2d}} (1 + |v|)^2 f_t \, dx \, dv \, dt$$

As explained above, each term in the previous sum is bounded by our choice of γ. This proves (8.21). By the no blow-up criterion stated in Proposition 4.7, it follows that the Maximal Regular Flow $\boldsymbol{X}$ of $\boldsymbol{b}$ is globally defined on $[0, T]$, namely its trajectories $\boldsymbol{X}(\cdot, x, v)$ belong to $AC([0, T]; \mathbb{R}^{2d})$ for f_0-a.e. $(x, v) \in \mathbb{R}^{2d}$, and $f_t = \boldsymbol{X}(t, \cdot)_\# f_0 = f_0 \circ \boldsymbol{X}(t, \cdot)^{-1}$. Also, no integral curves of $\boldsymbol{b}_t$ that transport f_t can appear from and/or disappear at infinity during the time interval $[0, T]$. In particular, for all Borel functions $\psi : [0, \infty) \to [0, \infty)$ we have

$$\int_{\mathbb{R}^{2d}} \psi(f_t) \, dx \, dv = \int_{\mathbb{R}^{2d}} \psi(f_0) \circ \boldsymbol{X}(t, \cdot)^{-1} \, dx \, dv = \int_{\mathbb{R}^{2d}} \psi(f_0) \, dx \, dv,$$

where the second equality follows by the incompressibility of the flow.

□

Remark 8.17. In the previous proof, the logarithm in the denominator of (8.21) is needed only to deal with the case $d = 4, q = 2$. In all other cases (namely, $d = 2$ or 3 and $p = 1, d = 4$ and $q > 2$), it would have been enough to verify the condition

$$\int_0^T \int_{\mathbb{R}^{2d}} \frac{|\boldsymbol{b}_t|(x, v) g_t(x, v)}{1 + (|x|^2 + |v|^2)^{1/2}} \, dx \, dv \, dt < \infty.$$

Remark 8.18. Another strategy to prove Corollary 8.3 which leads to worse bounds on q with respect to (8.8). More precisely, in this Remark we sketch a proof of Corollary 8.3 when we assume that $f_t \in$

$L^\infty((0,T); L^q(\mathbb{R}^{2d}))$ with the choice of q given by

$$q = \begin{cases} \dfrac{23+\sqrt{145}}{24} \approx 1.46 & \text{if } d = 2, \\ \dfrac{10+\sqrt{37}}{7} \approx 2.30 & \text{if } d = 3, \\ 13 + 3\sqrt{17} \approx 25.37 & \text{if } d = 4. \end{cases}$$

Let

$$p = \frac{d(q-1)+2q}{d(q-1)+2} = \begin{cases} \approx 1.31 & \text{if } d = 2, \\ \approx 1.44 & \text{if } d = 3, \\ \approx 1.48 & \text{if } d = 4, \end{cases}$$

and

$$r = \frac{d(q-1)+3q}{d(q-1)+3} = \begin{cases} \approx 1.35 & \text{if } d = 2, \\ \approx 1.56 & \text{if } d = 3, \\ \approx 1.72 & \text{if } d = 4. \end{cases}$$

With this choice, the integrability exponent p_α provided by Lemma 8.16 is precisely p if $\alpha = 0$ and r if $\alpha = 1$. In addition, by the choice of q in (8.8), with some elementary computations one can check that, for $d = 2, 3, 4$, the exponents p and r satisfy the relation $1 + d^{-1} = p^{-1} + r^{-1}$.

Thanks to Lemma 8.16 applied with $\alpha = 0$ and $\alpha = 1$, we have that $\|\rho_t\|_{L^p(\mathbb{R}^d)} \le C$ (as in (8.23)) and

$$\|\eta_t\|_{L^r(\mathbb{R}^d)} \le C \||v|^2 f_t\|^{\theta_1}_{L^1(\mathbb{R}^{2d})} \|f_t\|^{1-\theta_1}_{L^q(\mathbb{R}^{2d})} \le C,$$

where $\eta_t := \int_{\mathbb{R}^d} \frac{f_t(x,v)}{1+|v|}\, dv$. As in (8.24) we deduce that $\|E_t\|_{L^{dp/(d-p)}(\mathbb{R}^d)} \le C$ for $\mathscr{L}^1$-a.e. $t \in (0,T)$. Thus, noticing that our choices of p and r imply $\frac{r}{r-1} = \frac{dp}{d-p}$, using Hölder's inequality we find that, for every $t \in [0,T]$,

$$\begin{aligned} \int_{\mathbb{R}^{2d}} |E_t(x)| \frac{f_t(x,v)}{1+|v|}\, dx\, dv &= \int_{\mathbb{R}^d} |E_t(x)| \eta_t(x)\, dx \\ &\le \|E_t\|_{L^{r/(r-1)}(\mathbb{R}^d)} \|\eta_t\|_{L^r(\mathbb{R}^d)} \\ &= \|E_t\|_{L^{dp/(d-p)}(\mathbb{R}^d)} \|\eta_t\|_{L^r(\mathbb{R}^d)} \le C. \end{aligned} \tag{8.26}$$

Integrating (8.26) with respect to time, we get

$$\begin{aligned} \int_0^T \int_{\mathbb{R}^{2d}} \frac{|\boldsymbol{b}_t| f_t}{1 + (|x|^2 + |v|^2)^{1/2}}\, dx\, dv\, dt \le& \int_0^T \int_{\mathbb{R}^{2d}} f_t\, dx\, dv\, dt \\ &+ \int_0^T \int_{\mathbb{R}^{2d}} |E_t| \frac{f_t}{1+|v|}\, dx\, dv\, dt \\ &< \infty. \end{aligned}$$

We finally apply Proposition 4.7 to deduce that the Maximal Regular Flow $\boldsymbol{X}$ of $\boldsymbol{b}$ is globally defined on $[0, T]$.

Proof of Corollary 8.4. As for the proof of Corollary 8.3, setting $g_t := 2\pi^{-1}\arctan f_t$, we need only to verify (8.21) to prove that trajectories do not blow up. The proof is a simple variant of the proof of Corollary 8.3; this time we don't employ the information $E_t \in L^{dp/(d-p)}$, coming from the higher integrability of ρ_t and from the Sobolev embedding, but we know that $E_t \in L^2$ by the finiteness of potential energy. We observe that $g_t^2 \le \arctan f_t \le f_t$; hence

$$\begin{aligned}
&\int_0^T\int_{\mathbb{R}^{2d}} \frac{|\boldsymbol{b}_t| g_t}{(1+(|x|^2+|v|^2)^{1/2})\log(2+(|x|^2+|v|^2)^{1/2})}\,dx\,dv\,dt \\
&\le \int_0^T\int_{\mathbb{R}^{2d}} f_t\,dx\,dv\,dt + \int_0^T\int_{\mathbb{R}^{2d}} |E_t|\,\frac{g_t}{(1+|v|)\log(2+|v|)}\,dx\,dv\,dt \\
&\le \int_0^T\int_{\mathbb{R}^{2d}} f_t\,dx\,dv\,dt \\
&\quad + \int_0^T\int_{\mathbb{R}^{2d}} \left(\frac{|E_t|^2}{(1+|v|)^4\log^2(2+|v|)} + (1+|v|)^2 g_t^2\right)dx\,dv\,dt \\
&\le \left(\int_{\mathbb{R}^d} \frac{1}{(1+|v|)^4\log^2(2+|v|)}\,dv\right)\left(\int_0^T\int_{\mathbb{R}^d} |E_t|^2\,dx\,dt\right) \\
&\quad + 2\int_0^T\int_{\mathbb{R}^{2d}} (1+|v|)^2 f_t\,dx\,dv\,dt.
\end{aligned}$$

We notice that, if $d \le 4$,

$$\int_{\mathbb{R}^d} \frac{1}{(1+|v|)^4\log^2(2+|v|)}\,dv < \infty.$$

By the finiteness of kinetic and potential energy, each term in the previous sum is controlled by the total mass and energy of f_t, which is bounded by (8.10). This proves (8.21) also in this case. □

8.4. Global existence for the Vlasov-Poisson system: proof of Theorem 8.8 and Corollary 8.9

In this section we shall prove Theorem 8.8 and Corollary 8.9.

Proof of Theorem 8.8. To prove existence of global generalized Lagrangian solutions of Vlasov-Poisson we shall use an approximation procedure. Since the argument is rather long and involved, we divide the proof in five steps that we now describe briefly: In Step 1 we start from approximate solutions f^n, obtained by smoothing the initial datum and the kernel, and we decompose them along their level sets. Exploiting the incompressibility of the flow, these functions are still solutions of the continuity equation with the same vector field and, when n varies, they are uniformly bounded. This allows us to take their limit as $n \to \infty$ in Step 2, and show that the limit belongs to L^1. In Step 3 we introduce ρ^{eff} as the limit as $n \to \infty$ of the approximate densities ρ^n, and we motivate its properties. In Step 4 we show that the vector fields E^n converge to the vector field obtained by convolving ρ^{eff} with the Poisson kernel. Finally, in Step 5 we employ the stability results for the continuity equation and the results of Section 4.3 to take the limit in the approximate Vlasov-Poisson equation and show that the limiting solution is transported by the limiting incompressible flow. We now enter into the details of the proof.

Step 1: approximating solutions. Let $K(x) := \sigma c_d\, x/|x|^d$ and let us consider approximating kernels $K_n := K * \psi_n$, where $\psi_n(x) = n^d \psi(nx)$ and $\psi \in C_c^\infty(\mathbb{R}^d)$ is a standard convolution kernel in $\mathbb{R}^d$. Let $f_0^n \in C_c^\infty(\mathbb{R}^{2d})$ be a sequence of functions such that

$$f_0^n \to f_0 \qquad \text{in } L^1(\mathbb{R}^{2d}). \tag{8.27}$$

Let f_t^n be distributional solutions of the Vlasov system with initial datum f_0^n and kernel K_n

$$\begin{cases} \partial_t f_t^n + v \cdot \nabla_x f_t^n + E_t^n \cdot \nabla_v f_t^n = 0 & \text{in } (0,\infty) \times \mathbb{R}^d \times \mathbb{R}^d \\ \rho_t^n(x) = \displaystyle\int_{\mathbb{R}^d} f_t^n(x,v)\,dv & \text{in } (0,\infty) \times \mathbb{R}^d \\ E_t^n(x) = \sigma\, c_d \displaystyle\int_{\mathbb{R}^d} \rho_t^n(y) K_n(x,y)\,dy & \text{in } (0,\infty) \times \mathbb{R}^d. \end{cases}$$

(see [82] for this classical construction based on a fixed point argument in the Wasserstein metric, and [112]). Notice that since K_n is smooth and decays at infinity, both E_t^n and ∇E_t^n are bounded on $[0,\infty) \times \mathbb{R}^d$ (with a bound that depends on n). Hence $\boldsymbol{b}_t^n$ is a Lipschitz divergence-free vector field, and by standard theory for the transport equation we obtain that, for every $t \in (0,\infty)$,

$$f_t^n = f_0^n \circ \boldsymbol{X}^n(t)^{-1} \tag{8.28}$$

where $\boldsymbol{X}^n(t) : \mathbb{R}^{2d} \to \mathbb{R}^{2d}$ is the flow of the vector field $\boldsymbol{b}_t^n(x,v) = (v, E_t^n(x))$, and

$$\|\rho_t^n\|_{L^1(\mathbb{R}^d)} = \|f_t^n\|_{L^1(\mathbb{R}^{2d})} = \|f_0^n\|_{L^1(\mathbb{R}^{2d})}. \tag{8.29}$$

Assuming without loss of generality that $\mathscr{L}^{2d}(\{f_0 = k\}) = 0$ for every $k \in \mathbb{N}$ (otherwise we consider as level sets the values $R + k$ in place of k for some $R \in [0, 1]$), from (8.27) we deduce that

$$f_0^{n,k} \to f_0^k := 1_{\{k \le f_0 < k+1\}} f_0 \qquad \text{in } L^1(\mathbb{R}^{2d}). \tag{8.30}$$

We then consider $f_t^{n,k} := 1_{\{k \le f_t^n < k+1\}} f_t^n$ for every $k, n \in \mathbb{N}$, and by (8.28) we notice that, for every $t \in (0, \infty)$,

$$f_t^{n,k} = 1_{\{k \le f_0^n \circ X^n(t)^{-1} < k+1\}} f_0^n \circ X^n(t)^{-1} \tag{8.31}$$

is the image of $f_0^{n,k} := 1_{\{k \le f_0^n < k+1\}} f_0^n$ through the flow $X^n(t)$, that $f_t^{n,k}$ is a distributional solution of the continuity equation with vector field $b_t^n(x, v)$, and that

$$\| f_t^{n,k} \|_{L^1(\mathbb{R}^{2d})} = \| f_0^{n,k} \|_{L^1(\mathbb{R}^{2d})} \qquad \text{for every } t \in (0, \infty). \tag{8.32}$$

Step 2: limit in the phase-space. By construction the functions $\{f^{n,k}\}_{n \in \mathbb{N}}$ are nonnegative and bounded by $k + 1$ in $L^\infty((0, \infty) \times \mathbb{R}^{2d})$, hence there exists $f^k \in L^\infty((0, \infty) \times \mathbb{R}^{2d})$ nonnegative such that, up to subsequences,

$$f^{n,k} \rightharpoonup f^k \quad \text{weakly* in } L^\infty((0, \infty) \times \mathbb{R}^{2d}) \text{ as } n \to \infty \text{ for every } k \in \mathbb{N}. \tag{8.33}$$

Moreover, for any K compact subset of $\mathbb{R}^{2d}$ and any nonnegative function $\phi \in L^\infty(0, \infty)$ with compact support, using the test function $\phi(t) 1_K(x, v) \mathrm{sign}(f_t^k)(x, v)$ in the previous weak convergence, by Fatou's Lemma, (8.32), and (8.30), we get

$$\begin{aligned} \int_0^\infty \phi(t) \| f_t^k \|_{L^1(K)} \, dt &\le \liminf_{n \to \infty} \int_0^\infty \phi(t) \| f_t^{n,k} \|_{L^1(K)} \, dt \\ &\le \liminf_{n \to \infty} \int_0^\infty \phi(t) \| f_t^{n,k} \|_{L^1(\mathbb{R}^{2d})} \, dt \\ &= \liminf_{n \to \infty} \int_0^\infty \phi(t) \| f_0^{n,k} \|_{L^1(\mathbb{R}^{2d})} \, dt \\ &= \int_0^\infty \phi(t) \| f_0^k \|_{L^1(\mathbb{R}^{2d})} \, dt. \end{aligned} \tag{8.34}$$

Hence, taking the supremum among all compact subsets $K \subset \mathbb{R}^{2d}$, this proves that

$$\| f_t^k \|_{L^1(\mathbb{R}^{2d})} \le \| f_0^k \|_{L^1(\mathbb{R}^{2d})} \qquad \text{for } \mathscr{L}^1\text{-a.e. } t \in (0, \infty), \tag{8.35}$$

so, in particular, $f^k \in L^\infty((0, \infty); L^1(\mathbb{R}^{2d}))$.

Thanks to (8.35), we can define $f \in L^\infty((0,\infty); L^1(\mathbb{R}^{2d}))$ by

$$f := \sum_{k=0}^{\infty} f^k \qquad \text{in } (0,\infty)\times\mathbb{R}^{2d}, \tag{8.36}$$

where, for $\mathscr{L}^1$-a.e. $t \in [0,\infty)$, the global bound on the L^1-norm of f_t comes from

$$\|f_t\|_{L^1(\mathbb{R}^{2d})} \le \sum_{k=0}^{\infty} \|f_t^k\|_{L^1(\mathbb{R}^{2d})} \le \sum_{k=0}^{\infty} \|f_0^k\|_{L^1(\mathbb{R}^{2d})} = \|f_0\|_{L^1(\mathbb{R}^{2d})}. \tag{8.37}$$

We now claim that, for every $T > 0$,

$$f^n \rightharpoonup f \qquad \text{weakly in } L^1((0,T)\times\mathbb{R}^{2d}), \tag{8.38}$$

that is, for every $\varphi \in L^\infty((0,T)\times\mathbb{R}^{2d})$,

$$\lim_{n\to\infty} \int_0^T \int_{\mathbb{R}^{2d}} \varphi f^n \, dx\, dv\, dt = \int_0^T \int_{\mathbb{R}^{2d}} \varphi f \, dx\, dv\, dt. \tag{8.39}$$

Indeed, noticing that $f^n = \sum_{k=0}^{\infty} f^{n,k}$ and $f = \sum_{k=0}^{\infty} f^k$, by the triangle inequality we have that, for every $k_0 \ge 1$,

$$\begin{aligned}
\left| \int_0^T \int_{\mathbb{R}^{2d}} \varphi (f^n - f)\, dx\, dv\, dt \right| &= \left| \sum_{k=0}^{\infty} \int_0^T \int_{\mathbb{R}^{2d}} \varphi (f^{n,k} - f^k)\, dx\, dv\, dt \right| \\
\le \left| \sum_{k=0}^{k_0-1} \int_0^T \int_{\mathbb{R}^{2d}} \varphi (f^{n,k} - f^k)\, dx\, dv\, dt \right| & \\
+ \sum_{k=k_0}^{\infty} \int_0^T \int_{\mathbb{R}^{2d}} |\varphi| |f^{n,k}|\, dx\, dv\, dt &+ \sum_{k=k_0}^{\infty} \int_0^T \int_{\mathbb{R}^{2d}} |\varphi| |f^k|\, dx\, dv\, dt.
\end{aligned}$$

Using (8.32) and (8.35), the last two terms can be estimated

$$\begin{aligned}
&\sum_{k=k_0}^{\infty} \int_0^T \int_{\mathbb{R}^{2d}} |\varphi| |f^{n,k}|\, dx\, dv\, dt + \sum_{k=k_0}^{\infty} \int_0^T \int_{\mathbb{R}^{2d}} |\varphi| |f^k|\, dx\, dv\, dt \\
&\le T\|\varphi\|_\infty \sum_{k=k_0}^{\infty} \int_{\mathbb{R}^{2d}} |f_0^{n,k}|\, dx\, dv + T\|\varphi\|_\infty \sum_{k=k_0}^{\infty} \int_{\mathbb{R}^{2d}} |f_0^k|\, dx\, dv \\
&\le T\|\varphi\|_\infty \int_{\{f_0^n \ge k_0\}} |f_0^n|\, dx\, dv + T\|\varphi\|_\infty \int_{\{f_0 \ge k_0\}} |f_0|\, dx\, dv \\
&= T\|\varphi\|_\infty \Big(\|f_0^n 1_{\{f_0^n \ge k_0\}}\|_{L^1(\mathbb{R}^{2d})} + \|f_0 1_{\{f_0 \ge k_0\}}\|_{L^1(\mathbb{R}^{2d})} \Big).
\end{aligned}$$

Notice that, thanks to (8.30) and (8.27), it follows that

$$f_0^n 1_{\{f_0^n \geq k_0\}} \to f_0 1_{\{f_0 \geq k_0\}} \qquad \text{in } L^1(\mathbb{R}^{2d}),$$

so by letting $n \to \infty$ and using (8.33) we deduce that

$$\begin{aligned}
&\limsup_{n\to\infty} \left| \int_0^T \int_{\mathbb{R}^{2d}} \varphi(f^n - f)\, dx\, dt \right| \\
&\leq \limsup_{n\to\infty} \left| \sum_{k=0}^{k_0-1} \int_0^T \int_{\mathbb{R}^{2d}} \varphi(f^{n,k} - f^k)\, dx\, dv\, dt \right| \\
&\quad + 2T \|\varphi\|_\infty \| f_0 1_{\{f_0 \geq k_0\}} \|_{L^1(\mathbb{R}^{2d})} \\
&= 2T \|\varphi\|_\infty \| f_0 1_{\{f_0 \geq k_0\}} \|_{L^1(\mathbb{R}^{2d})}.
\end{aligned}$$

Finally, letting $k_0 \to \infty$ we deduce (8.39), which proves the claim.

Step 3: limit of physical densities. Since by (8.29) the sequence $\{\rho^n\}_{n\in\mathrm{N}}$ is bounded in $L^\infty((0,\infty); \mathscr{M}_+(\mathbb{R}^d)) \subset \left[L^1((0,\infty), C_0(\mathbb{R}^d))\right]^*$, there exists

$$\rho^{\text{eff}} \in L^\infty((0,\infty); \mathscr{M}_+(\mathbb{R}^d))$$

such that

$$\rho^n \rightharpoonup \rho^{\text{eff}} \qquad \text{weakly* in } L^\infty((0,\infty); \mathscr{M}_+(\mathbb{R}^d)). \tag{8.40}$$

Moreover, by the lower semicontinuity of the norm under weak* convergence, using (8.29) again we deduce that

$$\begin{aligned}
\operatorname*{ess\,sup}_{t\in(0,\infty)} |\rho_t^{\text{eff}}|(\mathbb{R}^d) &\leq \lim_{n\to\infty} \left(\sup_{t\in(0,\infty)} \|\rho_t^n\|_{L^1(\mathbb{R}^d)} \right) \\
&= \lim_{n\to\infty} \|f_0^n\|_{L^1(\mathbb{R}^{2d})} = \|f_0\|_{L^1(\mathbb{R}^{2d})}.
\end{aligned} \tag{8.41}$$

Now, let us consider any nonnegative function $\varphi \in C_c((0,\infty) \times \mathbb{R}^d)$. By (8.40) and (8.38) we obtain that, for any $R > 0$,

$$\begin{aligned}
\int_0^\infty \int_{\mathbb{R}^d} \varphi_t(x)\, d\rho_t^{\text{eff}}(x)\, dt &= \lim_{n\to\infty} \int_0^\infty \int_{\mathbb{R}^d} \rho_t^n(x) \varphi_t(x)\, dx\, dt \\
&= \lim_{n\to\infty} \int_0^\infty \int_{\mathbb{R}^{2d}} f_t^n(x,v) \varphi_t(x)\, dv\, dx\, dt \\
&\geq \liminf_{n\to\infty} \int_0^\infty \int_{\mathbb{R}^d \times B_R} f_t^n(x,v) \varphi_t(x)\, dv\, dx\, dt \\
&= \int_0^\infty \int_{\mathbb{R}^d \times B_R} f_t(x,v) \varphi_t(x)\, dv\, dx\, dt,
\end{aligned}$$

so by letting $R \to \infty$ we get

$$\int_0^\infty \int_{\mathbb{R}^d} \varphi_t(x)\, d\rho_t^{\mathrm{eff}}(x)\, dt \geq \int_0^\infty \int_{\mathbb{R}^{2d}} f_t(x, v)\varphi_t(x)\, dv\, dx\, dt$$
$$= \int_0^\infty \int_{\mathbb{R}^d} \varphi_t(x)\, d\rho_t(x)\, dt.$$

By the arbitrariness of φ we deduce that

$$\rho_t \leq \rho_t^{\mathrm{eff}} \qquad \text{as measures for } \mathscr{L}^1\text{-a.e. } t \in (0, \infty), \tag{8.42}$$

as desired.

Step 4: limit of vector fields. Set $E_t^{\mathrm{eff}} := K * \rho_t^{\mathrm{eff}}$ and $\boldsymbol{b}_t(x, v) := (v, E_t^{\mathrm{eff}}(x))$. We claim that

$$\boldsymbol{b}^n \rightharpoonup \boldsymbol{b} \qquad \text{weakly in } L^1_{\mathrm{loc}}((0, \infty) \times \mathbb{R}^{2d}; \mathbb{R}^{2d}) \tag{8.43}$$

and that, for every ball $B_R \subset \mathbb{R}^d$,

$$\begin{aligned} &[\rho_t^n * K_n](x + h) \to [\rho_t^n * K_n](x) \\ &\text{as } |h| \to 0 \text{ in } L^1_{\mathrm{loc}}((0, \infty); L^1(B_R)), \text{ uniformly in } n. \end{aligned} \tag{8.44}$$

To show this we first prove that the sequence $\{\boldsymbol{b}^n\}_{n\in\mathbb{N}}$ is bounded in $L^p_{\mathrm{loc}}((0, \infty) \times \mathbb{R}^{2d}; \mathbb{R}^{2d})$ for every $p \in [1, d/(d-1))$. Indeed, using Young's inequality, for every $t \geq 0$, $n \in \mathbb{N}$, and $r > 0$,

$$\begin{aligned} &\|\rho_t^n * K_n\|_{L^p(B_r)} = \|(\rho_t^n * \psi_n) * K\|_{L^p(B_r)} \\ &\leq \|(\rho_t^n * \psi_n) * (K 1_{B_1})\|_{L^p(B_r)} + \|(\rho_t^n * \psi_n) * (K 1_{\mathbb{R}^d \setminus B_1})\|_{L^p(B_r)} \\ &\leq \|(\rho_t^n * \psi_n) * (K 1_{B_1})\|_{L^p(\mathbb{R}^d)} \\ &\quad + \mathscr{L}^d(B_r)^{1/p} \|(\rho_t^n * \psi_n) * (K 1_{\mathbb{R}^d \setminus B_1})\|_{L^\infty(\mathbb{R}^d)} \\ &\leq \|\rho_t^n\|_{L^1(\mathbb{R}^d)} \|\psi_n\|_{L^1(\mathbb{R}^d)} \|K\|_{L^p(B_1)} \\ &\quad + \mathscr{L}^d(B_r)^{1/p} \|\rho_t^n\|_{L^1(\mathbb{R}^d)} \|\psi_n\|_{L^1(\mathbb{R}^d)} \|K\|_{L^\infty(\mathbb{R}^d \setminus B_1)} \end{aligned}$$

hence, up to subsequences, the sequence $\{\boldsymbol{b}^n\}_{n\in\mathbb{N}}$ converges locally weakly in L^p. In order to identify the limit, we claim that for every $\varphi \in C_c((0, \infty) \times \mathbb{R}^d)$

$$\lim_{n\to\infty} \int_0^\infty \int_{\mathbb{R}^d} \rho_t^n * K_n\, \varphi_t\, dx\, dt = \int_0^\infty \int_{\mathbb{R}^d} \rho_t^{\mathrm{eff}} * K\, \varphi_t\, dx\, dt.$$

Indeed, by standard properties of convolution,

$$
\begin{aligned}
&\Big|\int_0^\infty \int_{\mathbb{R}^d} \rho_t^n * K_n\, \varphi_t\, dx\, dt - \int_0^\infty \int_{\mathbb{R}^d} \rho_t^{\mathrm{eff}} * K\, \varphi_t\, dx\, dt\Big| \\
&= \Big|\int_0^\infty \int_{\mathbb{R}^d} \rho_t^n\, \varphi_t * K_n\, dx\, dt - \int_0^\infty \int_{\mathbb{R}^d} \rho_t^{\mathrm{eff}}\, \varphi_t * K\, dx\, dt\Big| \\
&\le \Big|\int_0^\infty \int_{\mathbb{R}^d} (\rho_t^n - \rho_t^{\mathrm{eff}})\, \varphi_t * K\, dx\, dt\Big| \\
&\quad + \Big|\int_0^\infty \int_{\mathbb{R}^d} \rho_t^n\, (\varphi_t * K - \varphi_t * K * \psi_n)\, dx\, dt\Big| \\
&\le \Big|\int_0^\infty \int_{\mathbb{R}^d} (\rho_t^n - \rho_t^{\mathrm{eff}})\, \varphi_t * K\, dx\, dt\Big| \\
&\quad + \Big(\sup_{t\in(0,\infty)} \|\rho_t^n\|_{L^1(\mathbb{R}^d)}\Big)\, \|\varphi_t * K - \varphi_t * K * \psi_n\|_{L^\infty((0,\infty)\times\mathbb{R}^d)}.
\end{aligned}
$$

Letting $n \to \infty$, the first term converges to 0 thanks to the weak convergence (8.40) of ρ_t^n to ρ_t^{eff} and thanks to the fact that $\varphi * K = \varphi * (1_{B_1} K) + \varphi * (1_{\mathbb{R}^d \setminus B_1} K)$ is a bounded continuous function, compactly supported in time and decaying at infinity in space. The second term, in turn, converges to 0 since the first factor is bounded by (8.41) and $\varphi_t * K * \psi_n$ converges to $\varphi_t * K$ uniformly in $(0, \infty) \times \mathbb{R}^d$.

This computation identifies the weak limit of $\rho_t^n * K_n$ in $L^1_{\mathrm{loc}}([0, T] \times \mathbb{R}^{2d})$, showing that it coincides with $\rho_t^{\mathrm{eff}} * K$ and proving (8.43).

We now prove (8.44). First of all, since $K \in W^{\alpha,p}_{\mathrm{loc}}(\mathbb{R}^d; \mathbb{R}^d)$ for every $\alpha < 1$ and $p < d/(d-1+\alpha)$,[2] using Young's inequality we deduce that, for any $t \in (0, \infty)$,

$$
\|\rho_t^n * K_n\|_{W^{\alpha,p}(B_R;\mathbb{R}^d)} = \|(\rho_t^n * \psi_n) * K\|_{W^{\alpha,p}(B_R;\mathbb{R}^d)} \le C(R)\|\rho_t^n * \psi_n\|_{L^1(\mathbb{R}^d)}.
$$

Since $\|\psi_n\|_{L^1(\mathbb{R}^d)} = 1$, thanks to (8.29) we deduce that the last term is bounded independently of t and n, that is, for every $R > 0$,

$$
\sup_{t\in(0,\infty)} \sup_{n\in\mathbb{N}} \|\rho_t^n * K_n\|_{W^{\alpha,p}(B_R;\mathbb{R}^d)} < \infty.
$$

Hence, by a classical embedding between fractional Sobolev spaces and Nikolsky spaces (see for instance [100, Lemma 2.3]) we find that, for $|h| \le R$,

$$
\begin{aligned}
&\int_{B_R} |\rho_t^n * K_n(x+h) - \rho_t^n * K_n(x)|^p\, dx \\
&\qquad \le C(p, \alpha, R, \|\rho_t^n * K_n\|_{W^{\alpha,p}(B_{2R};\mathbb{R}^d)})|h|^{\alpha p},
\end{aligned}
$$

from which (8.44) follows.

[2] This can be seen by a direct computation, using the definition of fractional Sobolev spaces.

Step 5: conclusion. Thanks to (8.43) and (8.44), we can apply the stability result from [81, Theorem II.7] (which does not require any growth condition on the vector fields, see also Theorem 3.2 for the stability of the associated flows) to deduce that, for every $k \in \mathbb{N}$, f^k is a weakly continuous distributional solution of the continuity equation starting from f_0^k. Since the continuity equation is linear, we deduce that also $F^m := \sum_{k=1}^m f^k$ is a distributional solution for every $m \in \mathbb{N}$.

Since F^m is bounded, Theorem 4.9 gives that F^m is a renormalized solution for every $m \in \mathbb{N}$. Letting $m \to \infty$, since $F^m \to f$ strongly in $L^1_{\mathrm{loc}}((0,\infty) \times \mathbb{R}^{2d})$, also f is a renormalized solution of the continuity equation starting from f_0 with vector field $\boldsymbol{b}$. Together with (8.42), (8.37), and (8.41) this proves that $(f_t, \rho_t^{\mathrm{eff}})$ is a generalized solution of the Vlasov-Poisson equation starting from f_0 according to Definition 8.7.

Finally, the fact that f is transported by the Maximal Regular Flow associated to $\boldsymbol{b}_t$ simply follows by the fact that each density f^k is transported by Maximal Regular Flow associated to $\boldsymbol{b}_t$ (thanks to Theorem 8.2) and that $f = \sum_{k=0}^{\infty} f^k$ is an absolutely convergent series (see (8.37)). Finally, this implies that f_t belongs to $C([0,\infty); L^1_{\mathrm{loc}}(\mathbb{R}^{2d}))$ by Theorem 4.6. □

Remark 8.19. We remark that the existence of global solutions for the Vlasov-Poisson equation starting from regular initial data is known only in dimension up to 3 (see for instance [112, Theorem 6.1]). Therefore, the smoothing of the kernel K performed in the proof of Theorem 8.8 is essential in order to be able to build smooth solutions of the approximating problems in dimension $d \geq 4$; in dimension $d = 2$ or 3 one could avoid this part of the approximation argument. Indeed the stability of the scheme allows to prove that the solutions obtained by smoothing only the initial datum (and not the kernel) converge, when $d = 2$ or 3, to a generalized solution of the Vlasov-Poisson equation.

The proof of existence of renormalized solutions in Corollary 8.9 is an easy adaptation of the proof of Theorem 8.8, obtained by approximating the initial datum with a sequence of smooth data with bounded energy. In turn, this bound ensures that the approximating sequence of phase-space distributions is tight in the v variable uniformly in time, allowing us to show that $\rho_t^{\mathrm{eff}} = \rho_t$ for $\mathscr{L}^1$-a.e. $t \in (0,\infty)$. The approximation of the initial datum with a smooth sequence having uniformly bounded energy is a technical task that we describe in the next lemma.

Lemma 8.20. *Let $d \geq 3$, let ψ be a standard convolution kernel, and set $\psi_k(x) := k^d \psi(kx)$ for every $k \geq 1$. Let $f_0 \in L^1(\mathbb{R}^{2d})$ be an initial datum* *of finite energy, namely*

$$\int_{\mathbb{R}^{2d}} |v|^2 f_0(x,v)\,dx\,dv + \int_{\mathbb{R}^d} [H * \rho_0](x)\,\rho_0(x)\,dx < \infty,$$

where $\rho_0(x) := \int_{\mathbb{R}^d} f_0(x,v)\,dv$ *and* $H(x) := c_d(d-2)^{-1}|x|^{2-d}$ *for every* $x \in \mathbb{R}^d$. *Then there exist a sequence of functions* $\{f_0^n\}_{n\in\mathbb{N}} \subset C_c^\infty(\mathbb{R}^{2d})$ *and a sequence* $\{k_n\}_{n\in\mathbb{N}}$ *such that* $k_n \to \infty$ *and, setting* $\rho_0^n(x) = \int_{\mathbb{R}^d} f_0^n(x,v)dv$,

$$\begin{aligned}\lim_{n\to\infty}\left(\int_{\mathbb{R}^{2d}} |v|^2 f_0^n\,dx\,dv + \int_{\mathbb{R}^d} H * \psi_{k_n} * \rho_0^n\,\rho_0^n\,dx\right)\\ = \int_{\mathbb{R}^{2d}} |v|^2 f_0\,dx\,dv + \int_{\mathbb{R}^d} H * \rho_0\,\rho_0\,dx.\end{aligned} \tag{8.45}$$

Proof. We split the approximation procedure in three steps. We use the notation L_c^∞ to denote the space of bounded functions with compact support.

Step 1: approximation of the initial datum when $f_0 \in L_c^\infty(\mathbb{R}^{2d})$**.** Assuming that $f_0 \in L_c^\infty(\mathbb{R}^{2d})$, we claim that there exists $\{f_0^n\}_{n\in\mathbb{N}} \subset C_c^\infty(\mathbb{R}^{2d})$ such that

$$\begin{aligned}\lim_{n\to\infty}\left(\int_{\mathbb{R}^{2d}} |v|^2 f_0^n\,dx\,dv + \int_{\mathbb{R}^d} H * \rho_0^n\,\rho_0^n\,dx\right)\\ = \int_{\mathbb{R}^{2d}} |v|^2 f_0\,dx\,dv + \int_{\mathbb{R}^d} H * \rho_0\,\rho_0\,dx.\end{aligned} \tag{8.46}$$

To this end, consider smooth functions f_0^n which converge to f_0 pointwise, whose L^∞ norms are bounded by $\|f_0\|_{L^\infty(\mathbb{R}^{2d})}$, and whose supports are all contained in the same ball. By construction the densities ρ_0^n are bounded as well and their supports are also contained in a fixed ball; moreover, the functions $H * \rho_0^n$ are bounded and converge to $H * \rho_0$ locally in every L^p. These observations show the validity of (8.46), by dominated convergence.

Step 2: approximation of the initial datum when $f_0 \in L^1(\mathbb{R}^{2d})$**.** Assuming that $f_0 \in L^1(\mathbb{R}^{2d})$, we claim that there exists a sequence of functions $\{f_0^n\}_{n\in\mathbb{N}} \subset C_c^\infty(\mathbb{R}^{2d})$ such that (8.46) holds.

Indeed, by Step 1 it is enough to approximate f_0 with a sequence in $L_c^\infty(\mathbb{R}^{2d})$ and converging energies. To this aim, for every $n \in \mathbb{N}$ we define the truncations of f_0 given by

$$f_0^n(x,v) := \min\{n, 1_{B_n}(x,v) f_0(x,v)\} \qquad (x,v) \in \mathbb{R}^{2d}.$$

Since $H \geq 0$ the integrands in the left-hand side of (8.46) converge monotonically, hence the integrals converge by monotone convergence.

Step 3: approximation of the kernel. We conclude the proof of the lemma. In order to approximate the kernel, we notice that, given the sequence of functions $f_0^n \in C_c^\infty(\mathbb{R}^d)$ provided by Steps 1-2, for $n \in \mathbb{N}$ fixed we have

$$\lim_{k\to\infty} \int_{\mathbb{R}^d} H * \psi_k * \rho_0^n \, \rho_0^n \, dx = \int_{\mathbb{R}^d} H * \rho_0^n \, \rho_0^n \, dx.$$

Hence, choosing k_n sufficiently large so that

$$\left| \int_{\mathbb{R}^d} H * \psi_{k_n} * \rho_0^n \, \rho_0^n \, dx - \int_{\mathbb{R}^d} H * \rho_0^n \, \rho_0^n \, dx \right| \leq \frac{1}{n},$$

we conclude the proof of the approximation lemma. □

Proof of Corollary 8.9, *existence of renormalized solutions*. Given f_0 of finite energy, let $\{f_0^n\}_{n\in\mathbb{N}} \subset C_c^\infty(\mathbb{R}^{2d})$ and $\{k_n\}_{n\in\mathbb{N}}$ be as in Lemma 8.20. Also let $K := c_d \, x/|x|^d$ and $K_n := K * \psi_{k_n}$. Applying verbatim the arguments in Steps 1-3 in the proof of Theorem 8.8 we get a sequence f_n of smooth solutions with kernels K_n such that

$$f^n \rightharpoonup f \qquad \text{weakly in } L^1([0,T] \times \mathbb{R}^{2d}) \quad \text{for any } T > 0, \tag{8.47}$$

and

$$\rho^n \rightharpoonup \rho^{\text{eff}} \qquad \text{weakly* in } L^\infty((0,T); \mathscr{M}_+(\mathbb{R}^d)), \tag{8.48}$$

where $\rho_t^n(x) := \int_{\mathbb{R}^d} f_t^n(x,v)\, dv$.

In addition, the conservation of the energy along classical solutions gives that, for every $n \in \mathbb{N}$ and $t \in [0,\infty)$

$$\begin{aligned} &\int_{\mathbb{R}^{2d}} |v|^2 f_t^n \, dx\, dv + \int_{\mathbb{R}^d} H * \psi_{k_n} * \rho_t^n \, \rho_t^n \, dx \\ &= \int_{\mathbb{R}^{2d}} |v|^2 f_t^n \, dx\, dv + \int_{\mathbb{R}^d} H * \psi_{k_n} * \rho_0^n \, \rho_0^n \, dx \leq C, \end{aligned} \tag{8.49}$$

Hence, since $H \geq 0$ we deduce that

$$\sup_{n\in\mathbb{N}} \sup_{t\in[0,\infty)} \int_{\mathbb{R}^{2d}} |v|^2 f_t^n \, dx\, dv \leq C, \tag{8.50}$$

and by lower semicontinuity of the kinetic energy we deduce that, for every $T > 0$,

$$\int_0^T \int_{\mathbb{R}^{2d}} |v|^2 f_t \, dx \, dv \, dt \leq \liminf_{n \to \infty} \int_0^T \int_{\mathbb{R}^{2d}} |v|^2 f_t^n \, dx \, dv \, dt \leq CT. \tag{8.51}$$

We now exploit (8.50) and (8.51) to show that $\rho^{\text{eff}} = \rho$, where $\rho_t(x) := \int_{\mathbb{R}^d} f_t(x, v) \, dv \in L^\infty((0, T); L^1(\mathbb{R}^d))$. For this, we want to show that for any $\varphi \in C_c((0, \infty) \times \mathbb{R}^d)$

$$\lim_{n \to \infty} \int_0^\infty \int_{\mathbb{R}^d} \varphi \rho_t^n \, dx \, dt = \int_0^\infty \int_{\mathbb{R}^d} \varphi \rho_t \, dx \, dt. \tag{8.52}$$

To prove this, for every $k \in \mathbb{N}$ we consider a continuous nonnegative function $\zeta_k : \mathbb{R}^d \to [0, 1]$ which equals 1 inside B_k and 0 outside B_{k+1}, and observe that

$$\begin{aligned}
&\int_0^\infty \int_{\mathbb{R}^d} \varphi(\rho_t^n - \rho_t) \, dx \, dt \\
&= \int_0^\infty \int_{\mathbb{R}^{2d}} \varphi_t(x) f_t^n(x, v)(1 - \zeta_k(v)) \, dx \, dv \, dt \\
&\quad + \int_0^\infty \int_{\mathbb{R}^{2d}} \varphi_t(x)(f_t^n(x, v) - f(x, v))\zeta_k(v) \, dx \, dv \, dt \\
&\quad + \int_0^\infty \int_{\mathbb{R}^{2d}} \varphi_t(x) f_t(x, v)(\zeta_k(v) - 1) \, dx \, dv \, dt.
\end{aligned}$$

The second term in the right-hand side converges to 0 by the weak convergence of f^n to f in L^1, while the other two terms are estimated by the finiteness of energy (8.50) and (8.51) as

$$\begin{aligned}
&\left| \int_0^\infty \int_{\mathbb{R}^{2d}} \varphi f_t^n(x, v)(1 - \zeta_k(v)) \, dx \, dv \, dt \right| \\
&\leq \frac{\|\varphi\|_\infty}{k^2} \int_0^T \int_{\mathbb{R}^{2d}} f_t^n(x, v)|v|^2 \, dx \, dv \, dt \leq \frac{CT\|\varphi\|_\infty}{k^2},
\end{aligned}$$

and similarly

$$\left| \int_0^\infty \int_{\mathbb{R}^{2d}} \varphi f_t(x, v)(1 - \zeta_k(v)) \, dx \, dv \, dt \right| \leq \frac{CT\|\varphi\|_\infty}{k^2}.$$

Letting $k \to \infty$, this proves (8.52). Thanks to this fact, the conclusion of the proof proceeds exactly as in Steps 4 and 5 in the proof of Theorem 8.8 with $\rho_t^{\text{eff}} = \rho_t$. □

The proof of the energy inequality (8.15) is based on the conservation of energy along approximate solutions and on a lower semicontinuity argument. In the following basic lemmas, we prove some properties of the potential energy, namely the lower semicontinuity and an estimate from below. For $d \geq 3$, a formal integration by parts, rigorously justified in the case that μ has smooth, compactly supported density with respect to the Lebesgue measure, suggest that for every $\mu \in \mathscr{M}_+(\mathbb{R}^d)$

$$\int_{\mathbb{R}^d} H * \mu(x)\, d\mu(x) = \int_{\mathbb{R}^d} |\nabla H * \mu(x)|^2\, dx \tag{8.53}$$

(meaning that, if one of the two sides is finite, than so is the other and they coincide). This would immediately imply the convexity of the potential energy and its lower semicontinuity with respect to the weak* convergence of measures. However, since the justification of (8.53) seems to require some work, we prove directly the lower semicontinuity with a simpler trick.

Lemma 8.21. *Let* $H(x) := c_d(d-2)^{-1}|x|^{2-d}$ *for every* $x \in \mathbb{R}^d$. *Then the functional*

$$\mathcal{F}(\mu) = \int_{\mathbb{R}^d} H * \mu(x)\, d\mu(x) \qquad \mu \in \mathscr{M}_+(\mathbb{R}^d)$$

is lower semicontinuous with respect to the weak topology of* $\mathscr{M}(\mathbb{R}^d)$.

Proof. Given a sequence of nonnegative measures μ^n weakly converging to μ in $\mathscr{M}(\mathbb{R}^d)$, the measures $\mu^n(x)\mu^n(y) \in \mathscr{M}(\mathbb{R}^{2d})$ weakly converge to $\mu(x)\mu(y)$. Hence, since the function $H(x, y) = c_d(d-2)^{-1}|x-y|^{2-d}$ is lower semicontinuous and nonnegative in $\mathbb{R}^{2d}$, we deduce that

$$\begin{aligned}&\int_{\mathbb{R}^d}\int_{\mathbb{R}^d} \frac{1}{|x-y|^{d-2}}\, d\mu(x)\, d\mu(y)\\ &\quad\leq \liminf_{n\to\infty} \int_{\mathbb{R}^d}\int_{\mathbb{R}^d} \frac{1}{|x-y|^{d-2}}\, d\mu^n(x)\, d\mu^n(y).\end{aligned}$$

This proves the lower semicontinuity. □

The following lemma adapts the previous one to the time-dependent framework and its proof is very similar to the previous one. In particular, it takes care of a further approximation of the kernel in the right-hand side of (8.55) below and involves the time dependence of the functional. We need this kind of lemma since, at the level of generality of Theorem 8.8, the weak convergence of the approximating solutions is not pointwise in time, but it happens only as functions in space-time.

Lemma 8.22. *Let $d \geq 3$, $T > 0$, $\phi \in C_c((0,T))$ nonnegative, $\psi \geq 0$ an even convolution kernel, and $\psi_n(x) := n^d \psi(nx)$ for every $n \geq 1$. Then, for every sequence $\{\rho^n\}_{n\in\mathbb{N}} \subset C^0([0,T];\mathscr{M}_+(\mathbb{R}^d))$ such that $\sup_{n\in\mathbb{N}} \sup_{t\in[0,T]} \rho^n_t(\mathbb{R}^d) < \infty$ and*

$$\lim_{n\to\infty} \sup_{t\in[0,T]} \left| \int_{\mathbb{R}^d} \varphi \, d(\rho^n_t - \rho_t) \right| = 0 \qquad \text{for every } \varphi \in C^\infty_c(\mathbb{R}^d), \quad (8.54)$$

we have

$$\begin{aligned} &\int_0^T \phi(t) \int_{\mathbb{R}^d} H * \rho_t(x) \, d\rho_t(x) \, dt \\ &\qquad \leq \liminf_{n\to\infty} \int_0^T \phi(t) \int_{\mathbb{R}^d} H * \psi_n * \rho^n_t(x) \, d\rho^n_t(x) \, dt. \end{aligned} \quad (8.55)$$

Proof. We observe that, given any function $\phi_1 \in C^\infty_c(\mathbb{R}^d)$, we have

$$\begin{aligned} \left| \int_{\mathbb{R}^d} \phi_1 \, d(\rho^n_t * \psi^n - \rho_t) \right| &\leq \left| \int_{\mathbb{R}^d} (\phi_1 * \psi^n - \phi_1) \, d\rho^n_t \right| \\ &\quad + \left| \int_{\mathbb{R}^d} \phi_1 \, d(\rho^n_t - \rho_t) \right| \\ &\leq \|\phi_1 * \psi^n - \phi_1\|_{C^0(\mathbb{R}^d)} \rho^n_t(\mathbb{R}^d) \\ &\quad + \left| \int_{\mathbb{R}^d} \phi_1 \, d(\rho^n_t - \rho_t) \right|. \end{aligned}$$

Hence, since $\phi_1 * \psi^n$ converge uniformly to ϕ_1 as $n \to \infty$, $\rho^n_t(\mathbb{R}^d)$ is uniformly bounded in n and t, and (8.54) holds, we get

$$\lim_{n\to\infty} \sup_{t\in[0,T]} \left| \int_{\mathbb{R}^d} \phi_1 \, d(\rho^n_t * \psi^n - \rho_t) \right| = 0. \quad (8.56)$$

Now, let us consider the sequence of measures $\psi_n * \rho^n_t(x) \, \rho^n_t(y) \, dt \in \mathscr{M}((0,T)\times\mathbb{R}^{2d})$. We know that they are weakly* precompact because they have bounded masses; we claim that the weak* limit is $\rho_t(x) \, \rho_t(y) \, dt$. Indeed, testing with functions of the form $\phi_1(x)\phi_2(y)\psi(t)$, with $\phi_1, \phi_2 \in C^\infty_c(\mathbb{R}^d)$ and $\psi \in C^0([0,T])$ and using (8.54) and (8.56), we find that

$$\begin{aligned} &\lim_{n\to\infty} \int_0^T \int_{\mathbb{R}^d} \int_{\mathbb{R}^d} \psi(t)\phi_1(x)\phi_2(y)\psi_n * \rho^n_t(x) \, dx \, d\rho^n_t(y) \, dt \\ &= \lim_{n\to\infty} \int_0^T \psi(t) \Big(\int_{\mathbb{R}^d} \phi_1(x)\psi_n * \rho^n_t(x) \, dx \Big) \Big(\int_{\mathbb{R}^d} \phi_2(y) \, d\rho^n_t(y) \Big) \, dt \\ &= \int_0^T \psi(t) \Big(\int_{\mathbb{R}^d} \phi_1(x) \, d\rho_t(x) \Big) \Big(\int_{\mathbb{R}^d} \phi_2(y) \, d\rho_t(y) \Big) \, dt. \end{aligned}$$

Since the function $\phi(t)H(x-y)$ is continuous as a map from $(0,T)\times\mathbb{R}^{2d}$ to $[0,+\infty]$, this implies that (8.55) holds. □

In the following lemma we establish an inequality, under no assumptions on ρ, between the potential energy and the L^2-norm of the force field. It will be used to show the third property in Corollary 8.9.

Lemma 8.23. *Let $H(x) := c_d(d-2)^{-1}|x|^{2-d}$ for every $x \in \mathbb{R}^d$. Then for every nonnegative $\rho \in L^1(\mathbb{R}^d)$ we have*

$$\int_{\mathbb{R}^d} H * \rho(x)\, d\rho(x) \geq \int_{\mathbb{R}^d} |\nabla H * \rho(x)|^2\, dx. \tag{8.57}$$

Proof. We split the approximation procedure in three steps.

Step 1: Proof of the equality between the quantities in (8.57) **for smooth, compactly supported** ρ**.** Let ρ be a smooth, compactly supported function. For every $R > 0$, the integration by parts formula gives

$$\int_{B_R} H*\rho\,\rho\, dx = \int_{B_R} |\nabla H*\rho|^2\, dx - \int_{\partial B_R} H*\rho\,\nabla(H*\rho)\cdot\nu_{B_R}\, d\mathcal{H}^{d-1}.$$

Letting $R \to \infty$ the boundary term in the previous equality disappears for $d \geq 3$, since $H*\rho$ and $\nabla H*\rho$ decay as R^{2-d} and R^{1-d}, respectively, when evaluated on ∂B_R. This proves that

$$\int_{\mathbb{R}^d} H * \rho\rho\, dx = \int_{\mathbb{R}^d} |\nabla H * \rho|^2\, dx.$$

Step 2: Proof of (8.57) **for** $\rho \in L^\infty_c(\mathbb{R}^d)$**.** Let $\rho \in L^\infty_c(\mathbb{R}^d)$ and let us approximate ρ with a sequence $\{\rho^n\}_{n\in\mathbb{N}}$ obtained by convolution. By construction the densities ρ^n are bounded as well as their supports; moreover, the functions $H * \rho^n$ are bounded and converge to $H * \rho$ locally in every L^p. Hence, by Step 1 we have

$$\begin{aligned}\int_{\mathbb{R}^d} H * \rho(x)\rho(x)\, dx &= \lim_{n\to\infty}\int_{\mathbb{R}^d} H * \rho^n(x)\rho^n(x)\, dx\\ &= \lim_{n\to\infty}\int_{\mathbb{R}^d} |\nabla H * \rho^n(x)|^2\, dx.\end{aligned}$$

Therefore, the sequence $\{\nabla H * \rho^n\}_{n\in\mathbb{N}}$ is bounded in $L^2(\mathbb{R}^d)$ and hence it weakly converges to $\nabla H * \rho$. By the lower semicontinuity of the norm with respect to weak convergence, we find (8.57).

Step 3: Proof of (8.57) **for** $\rho \in L^1(\mathbb{R}^d)$. Let $\rho \in L^1(\mathbb{R}^d)$ and for every $n \in \mathbb{N}$ consider the truncations of ρ given by $\rho^n := \min\{n, 1_{B_n}\rho\}$ in $\mathbb{R}^d$. Since $H \geq 0$, by monotone convergence and Step 2

$$\begin{aligned}\int_{\mathbb{R}^d} H * \rho(x)\rho(x)\,dx &= \lim_{n\to\infty}\int_{\mathbb{R}^d} H * \rho^n(x)\rho^n(x)\,dx\\ &\geq \lim_{n\to\infty}\int_{\mathbb{R}^d} |\nabla H * \rho^n(x)|^2\,dx.\end{aligned}$$

Hence the sequence $\{\nabla H*\rho^n\}_{n\in\mathbb{N}}$ weakly converges in $L^2(\mathbb{R}^d)$ to $\nabla H*\rho$; the lower semicontinuity of the norm with respect to weak convergence implies (8.57). □

Proof of Corollary 8.9, *properties of renormalized solutions*. In order to prove (8.15) we perform a lower semicontinuity argument on the energy. Since the convergence of the approximate solutions f^n to f is only in space-time, in Step 1 we obtain the energy inequality integrated in time, and therefore we deduce that (8.15) holds for $\mathscr{L}^1$-a.e. $t \in [0,\infty)$. Then, in Step 2, 3, and 4, we employ the bound on the kinetic energy to prove the strong L^1_{loc} continuity of ρ_t and E_t in time. We remark that this does not imply, by itself, the conservation of mass of ρ_t in time, since we don't have any information on the compactness of f_t in the x variable, but only in v (see Remark 8.24 below). In Step 5, we use again the lower semicontinuity of the energy to deduce that the energy inequality (8.15) holds for every $t \in [0,\infty)$. Finally, in Step 6 we show the existence of a global measure-preserving flow associated to our solution f_t; this implies in particular the conservation of mass, namely that $\rho_t(\mathbb{R}^d) = \rho_0(\mathbb{R}^d)$ for every $t \in [0,\infty)$.

Step 1: bound on the total energy for $\mathscr{L}^1$-almost every time. Let us consider $T > 0$ and a nonnegative function $\phi \in C_c((0,T])$. Testing the weak convergence (8.47) of f^n with $\phi(t)|v|^2\chi_r(x,v)$, where $\chi_r \in C_c^\infty(\mathbb{R}^{2d})$ is a nonnegative cutoff function between B_r and B_{r+1}, we find that for every $r > 0$

$$\begin{aligned}&\int_0^\infty\int_{\mathbb{R}^{2d}} \phi(t)|v|^2\chi_r(x,v) f_t\,dx\,dv\,dt\\ &= \lim_{n\to\infty}\int_0^\infty\int_{\mathbb{R}^{2d}} \phi(t)|v|^2\chi_r(x,v) f_t^n\,dx\,dv\,dt\\ &\leq \liminf_{n\to\infty}\int_0^\infty \phi(t)\int_{\mathbb{R}^{2d}} |v|^2 f_t^n\,dx\,dv\,dt.\end{aligned}$$

Taking the supremum in r, we deduce that

$$\int_0^\infty \phi(t) \int_{\mathbb{R}^{2d}} |v|^2 f_t \, dx \, dv \, dt \le \liminf_{n\to\infty} \int_0^\infty \phi(t) \int_{\mathbb{R}^{2d}} |v|^2 f_t^n \, dx \, dv \, dt. \tag{8.58}$$

By the arbitrariness of ϕ it follows that (8.15) holds for $\mathscr{L}^1$-a.e. $t \in (0, \infty)$ and that $|v|^2 f_t \in L^1_{\mathrm{loc}}(\mathbb{R}^{2d})$ with a uniform bound in t for a.e. t (notice also that the same bound applies uniformly in n to $|v|^2 f_t^n$). In particular this allows us to integrate the transport equation $\mathcal{P}_t f_t + \mathrm{div}_{x,v}(\boldsymbol{b}_t f_t) = 0$ with respect to v on the whole $\mathbb{R}^d$ and obtain

$$\partial_t \rho_t + \mathrm{div}_x(J_t) = 0,$$

$$J_t(x) := \int_{\mathbb{R}^d} v \, f_t(x, v) \, dv \in L^2_{\mathrm{loc}}(\rho_t \mathscr{L}^d dt, (0, \infty) \times \mathbb{R}^d).$$

By classical results on continuity equations, this implies that ρ_t has a weakly* continuous representative (see for instance [19, Lemma 8.1.2]). In addition, since ρ_t^n satisfy a similar continuity equation with $J_t^n(x) := \int_{\mathbb{R}^d} v \, f_t^n(x, v) \, dv$, satisfying

$$\sup_n \int_0^T \int_{\mathbb{R}^d} |J_t^n|^2 \rho_t^n \, dx dt < \infty,$$

we obtain that $t \mapsto \int_{\mathbb{R}^d} \rho_t^n \varphi \, dx$ are equicontinuous for all $\varphi \in C_c^\infty(\mathbb{R}^d)$. The weak* convergence of ρ_t^n to ρ_t in $L^\infty([0, T]; \mathscr{M}_+(\mathbb{R}^d))$ then gives

$$\lim_{n\to\infty} \sup_{t\in[0,T]} \left| \int_{\mathbb{R}^d} \varphi \, d(\rho_t^n - \rho_t) \right| = 0 \qquad \text{for every } \varphi \in C_c^\infty(\mathbb{R}^d).$$

Hence, since the mass of ρ_t^n is uniformly bounded with respect to n and t, it follows from Lemma 8.22 that

$$\int_0^\infty \phi(t) \int_{\mathbb{R}^d} H*\rho_t \, \rho_t \, dx \, dt \le \liminf_{n\to\infty} \int_0^\infty \phi(t) \int_{\mathbb{R}^d} H*\psi_{k_n}*\rho_t^n \, \rho_t^n \, dx \, dt. \tag{8.59}$$

Adding (8.58) and (8.59), by the subadditivity of the lim inf and by the energy bound on approximating solutions (8.49) we find that

$$\begin{aligned}
&\int_0^\infty \phi(t) \Big(\int_{\mathbb{R}^{2d}} |v|^2 f_t \, dx \, dv + \int_{\mathbb{R}^d} H * \rho_t \, \rho_t \, dx \Big) dt \\
&\le \liminf_{n\to\infty} \int_0^\infty \phi(t) \Big(\int_{\mathbb{R}^{2d}} |v|^2 f_t^n \, dx \, dv + \int_{\mathbb{R}^d} H * \psi_{k_n} * \rho_t^n \, \rho_t^n \, dx \Big) dt \\
&= \lim_{n\to\infty} \int_0^\infty \phi(t) \Big(\int_{\mathbb{R}^{2d}} |v|^2 f_0^n \, dx \, dv + \int_{\mathbb{R}^d} H * \psi_{k_n} * \rho_0^n \, \rho_0^n \, dx \Big) dt \\
&= \Big(\int_0^\infty \phi(t) \, dt \Big) \Big(\int_{\mathbb{R}^{2d}} |v|^2 f_0 \, dx \, dv + \int_{\mathbb{R}^d} H * \rho_0 \, \rho_0 \, dx \Big).
\end{aligned}$$

By the arbitrariness of ϕ, we deduce that (8.15) holds for $\mathscr{L}^1$-a.e. $t \in [0, \infty)$.

Step 2: boundedness of the kinetic energy for every time. We show that

$$\sup_{t\in[0,\infty)} \int_{\mathbb{R}^{2d}} |v|^2 f_t \, dx \, dv \leq \int_{\mathbb{R}^{2d}} |v|^2 f_0 \, dx \, dv + \int_{\mathbb{R}^d} H * \rho_0 \, \rho_0 \, dx. \quad (8.60)$$

To this end, let $t \geq 0$ and $t_n \to t$ be a sequence of times such that the energy bound (8.15) holds for every t_n. The strong convergence of f_{t_n} to f_t in L^1_{loc} implies that for every $r > 0$

$$\int_{B_r} |v|^2 f_t \, dx \, dv = \lim_{n\to\infty} \int_{B_r} |v|^2 f_{t_n} \, dx \, dv \leq \liminf_{n\to\infty} \int_{\mathbb{R}^{2d}} |v|^2 f_{t_n} \, dx \, dv.$$

Taking the supremum in r, we deduce that

$$\int_{\mathbb{R}^{2d}} |v|^2 f_t \, dx \, dv \leq \liminf_{n\to\infty} \int_{\mathbb{R}^{2d}} |v|^2 f_{t_n} \, dx \, dv. \quad (8.61)$$

This proves (8.60).

Step 3: strong L^1_{loc}-continuity of the physical density. We prove that ρ_t is strongly L^1_{loc}-continuous, namely that for every sequence of times $t_n \to t$ we have

$$\lim_{n\to\infty} \rho_{t_n} = \rho_t \qquad \text{in } L^1_{\mathrm{loc}}(\mathbb{R}^d).$$

This, in turn, implies that ρ_t is weakly* continuous in time (as measures in $\mathscr{M}(\mathbb{R}^d)$).

Let $r > 0$. For every $R > 0$, noticing that $|f_{t_n} - f_t| \leq |v|^2 R^{-2}(f_{t_n} + f_t)$ when $|v| > R$, we have that

$$\int_{B_r} \int_{\mathbb{R}^d} |f_{t_n} - f_t| \, dv \, dx \leq \int_{B_r} \int_{B_R} |f_{t_n} - f_t| \, dv \, dx + \int_{B_r} \int_{\mathbb{R}^d \setminus B_R} \frac{|v|^2}{R^2} (f_{t_n} + f_t) \, dv \, dx$$

Letting first $n \to \infty$ in the previous equation, by the strong L^1_{loc} continuity of f_t the first term goes to 0. Letting then $R \to \infty$, we deduce that

$$\lim_{n\to\infty} \int_{B_r} \int_{\mathbb{R}^d} |f_{t_n} - f_t| \, dv \, dx = 0.$$

Step 4: strong L^1_{loc}-continuity of the force field. We prove that the force field E_t is strongly L^1_{loc}-continuous with respect to time.

The force field $E_t = K * \rho_t$ is weakly $L^1_{\mathrm{loc}}(\mathbb{R}^d)$ continuous since, by Step 3, ρ_t is weakly* continuous in time (as measures in $\mathscr{M}(\mathbb{R}^d)$).

Since, as observed above, $K = \nabla H \in W^{\alpha,p}_{\mathrm{loc}}(\mathbb{R}^d;\mathbb{R}^d)$ for every $\alpha < 1$ and $p < d/(d-1+\alpha)$, a simple computation shows that, for every $R > 0$ and $t \in (0,\infty)$,

$$\|\rho_t * K\|_{W^{\alpha,p}(B_R;\mathbb{R}^d)} \le \|\rho_t\|_{L^1(\mathbb{R}^d)} \sup_{y\in\mathbb{R}^d} \|K\|_{W^{\alpha,p}(B_R(y);\mathbb{R}^d)} \le C(R).$$

Hence, for every $R > 0$,

$$\sup_{t\in(0,\infty)} \sup_{n\in\mathbb{N}} \|\rho_t * K\|_{W^{\alpha,p}(B_R;\mathbb{R}^d)} < \infty.$$

The strong continuity of the force field $E_t = K * \rho_t$ in $L^1_{\mathrm{loc}}(\mathbb{R}^d)$ follows then from the fractional Rellich theorem, which provides the compact embedding of the fractional space $W^{\alpha,p}(B_R;\mathbb{R}^d)$ in $L^1(B_R;\mathbb{R}^d)$.

Step 5: bound on the total energy for every time. In order to conclude the proof of (8.15), we observe that both the kinetic and the potential energy are lower semicontinuous with respect to strong $L^1_{\mathrm{loc}}(\mathbb{R}^{2d})$-convergence of f and weak convergence (as measures) of ρ, respectively. Indeed, the first has been observed in Step 2 and the second is proved in Lemma 8.22. Then by Step 1 the total energy is bounded by the initial energy for $\mathscr{L}^1$-a.e. time and the lower semicontinuity of the total energy implies that the same property holds for *every* time.

Step 6: global characteristics in dimension 3 **and** 4**.** In order to prove that trajectories do not blow up, we apply Corollary 8.3 and 8.4. The assumptions of these Corollaries are satisfied thanks to the finiteness of energy of Step 5, the fact that $E_t = \nabla(H * \rho_t)$, and Lemma 8.23. We deduce that the trajectories of the Maximal Regular Flow starting at any time t do not blow up for f_t-a.e. $(x,v) \in \mathbb{R}^{2d}$. □

Remark 8.24. In Corollary 8.9 we did not prove that $\|f_t\|_{L^1(\mathbb{R}^d)} = \|f_0\|_{L^1(\mathbb{R}^d)}$ for every $t \in [0,\infty)$. Indeed, although the energy bound prevents mass from escaping in the v variable, one would need more assumptions on the initial datum (for instance, the finiteness of a momentum in the x variable) to prevent a mass loss. The conservation of the L^1-norm of f_t holds along solutions whose flow is globally defined (see Theorem 8.2); in this case, the solution belongs also to $C([0,\infty);L^1(\mathbb{R}^{2d}))$.

Remark 8.25. In the case $\sigma = -1$, Theorem 8.8 allows to show the existence of generalized solutions starting from any finite L^1 datum. The existence of renormalized solutions starting from an L^1 datum of finite (kinetic and potential) energy follows as in Corollary 8.9 provided that on the approximating sequence the kinetic energy is bounded by a fixed constant. This last fact, in turn, cannot be deduced by the boundedness of the energy itself, since in this case the potential energy is not positive any more as in the repulsive case. Hence it can be either assumed on the approximating sequence, or it follows under further integrability assumptions on the initial datum, for instance if $d = 3$ and $f_0 \in L^{9/7}(\mathbb{R}^6)$ (see Remark 8.5).

Remark 8.26. The construction in Theorem 8.8 provides distributional solutions of the Vlasov-Poisson system if further assumptions are assumed on the initial datum such as finiteness of the total energy, as shown in Corollary 8.9. Still, there are examples of infinite energy data such that the generalized solution built in Theorem 8.8 is in fact distributional. For instance, in [110] Perthame considers an initial datum $f_0 \in L^1 \cap L^\infty(\mathbb{R}^6)$ with $(1 + |x|^2) f_0 \in L^1(\mathbb{R}^6)$ and infinite energy, and he shows the existence of a solution $f \in L^\infty([0, \infty); L^1 \cap L^\infty(\mathbb{R}^6))$ of the Vlasov-Poisson system such that the quantities

$$t^{1/2}\|E_t\|_{L^2}, \qquad t^{3/5}\|\rho_t\|_{L^{5/3}}, \qquad \int_{\mathbb{R}^6} \frac{|x - vt|^2}{t} f_t(x, v)\,dx\,dv \tag{8.62}$$

are bounded for all $t \in (0, \infty)$.

It can be easily seen that, under Perthame's assumptions, the construction in the proof of Theorem 8.8 provides a solution of the Vlasov-Poisson equation as the one built in [110]. In particular, thanks to the a priori estimate (8.62) on the approximating sequence, it is easy to see that $\rho^{\text{eff}} = \rho$, therefore providing a Lagrangian (and therefore renormalized and distributional) solution of Vlasov-Poisson.

Similarly, under the assumptions of [126], a similar argument shows that the generalized solutions built in Theorem 8.8 solve the classical Vlasov-Poisson system.

Remark 8.27. A stability result holds for renormalized solutions of the Vlasov-Poisson system (see [33, Theorems 8.2 and 8.3]). For instance, when $\sigma = 1, d = 3$, if f^k is a sequence of renormalized solutions with

$$\sup_{k\in\mathbb{N}} \sup_{t\in[0,\infty)} \|f_t^k\|_{L^1(\mathbb{R}^6)} < \infty,$$

$$\sup_{k\in\mathbb{N}} \sup_{t\in[0,\infty)} \int_{\mathbb{R}^6} |v|^2 f_t^k\,dx\,dv + \int_{\mathbb{R}^3} H * \rho_t^k\,\rho_t^k\,dx < \infty$$

(where $H(x) := \frac{c_d}{d-2}|x|^{2-d}$), then up to subsequences $f^k \rightharpoonup f$ weakly in $L^1([0, T] \times \mathbb{R}^6)$ for any $T > 0$ and f is a renormalized solution of Vlasov-Poisson. Moreover, if $f_0^k \to f_0$ in $L^1(\mathbb{R}^3)$, then $f^k \to f$ strongly in $C([0,T];L^1(\mathbb{R}^6))$, $E^k \to E$ in $C([0,T];L^1_{\mathrm{loc}}(\mathbb{R}^3))$ for any $T>0$ and f is a renormalized solution of the Vlasov-Poisson system starting from f_0. The proof of this fact is an easy consequence of the stability of regular lagrangian flows since by Remark 8.13 the maximal regular flows associated to the solutions f^k do not blow up in finite time. Generalizing this result to any dimension may require some work, because under the assumptions of Corollary 8.9 it is not guaranteed that the maximal regular flows do not blow up in finite time.

Chapter 9
The semigeostrophic system

The semigeostrophic equations are a simple model used in meteorology to describe large scale atmospheric flows. As explained for instance in [31, Section 2.2] and [107, Section 1.1] (see also [68] for a more complete exposition), the semigeostrophic equations can be derived from the 3-d incompressible Euler equations, with Boussinesq and hydrostatic approximations, subject to a strong Coriolis force. Since for large scale atmospheric flows the Coriolis force dominates the advection term, the flow is mostly bi-dimensional. For this reason, the study of the semigeostrophic equations in 2-d or 3-d is pretty similar, and in order to simplify the presentation we focus here on the 2-dimensional periodic case, though the results have been extended to three dimensions and can be found in [8].

The semigeostrophic system on the 2-dimensional torus $\mathbb{T}^2$ is given by

$$\begin{cases} \partial_t \boldsymbol{u}_t^g(x) + \big(\boldsymbol{u}_t(x)\cdot\nabla\big)\boldsymbol{u}_t^g(x) + \nabla p_t(x) & \\ \qquad\qquad = -J\boldsymbol{u}_t(x) & (x,t)\in\mathbb{T}^2\times(0,\infty) \\ \boldsymbol{u}_t^g(x) = J\nabla p_t(x) & (x,t)\in\mathbb{T}^2\times[0,\infty) \\ \nabla\cdot\boldsymbol{u}_t(x) = 0 & (x,t)\in\mathbb{T}^2\times[0,\infty) \\ p_0(x) = p^0(x) & x\in\mathbb{T}^2. \end{cases} \tag{9.1}$$

Here p^0 is the initial datum, J is the rotation matrix

$$J := \begin{pmatrix} 0 & -1 \\ 1 & 0 \end{pmatrix},$$

and the functions $\boldsymbol{u}_t$ and p_t represent respectively the *velocity* and the *pressure*, while $\boldsymbol{u}_t^g$ is the so-called *semi-geostrophic* wind. Clearly the pressure is defined up to a (time-dependent) additive constant. In the sequel we are going to identify functions (and measures) defined on the torus $\mathbb{T}^2$ with $\mathbb{Z}^2$-periodic functions defined on $\mathbb{R}^2$.

Substituting the relation $\boldsymbol{u}_t^g = J\nabla p_t$ into the equation, the system (9.1) can be rewritten as

$$\begin{cases} \partial_t J\nabla p_t + J\nabla^2 p_t \boldsymbol{u}_t + \nabla p_t + J\boldsymbol{u}_t = 0 \\ \nabla \cdot \boldsymbol{u}_t = 0 \\ p_0 = p^0 \end{cases} \tag{9.2}$$

with $\boldsymbol{u}_t$ and p_t periodic.

Energetic considerations (see [68, Section 3.2]) show that it is natural to assume that p_t is (-1)-convex, *i.e.*, the function $P_t(x) := p_t(x) + |x|^2/2$ is convex on $\mathbb{R}^2$. If we denote with $\mathscr{L}_{\mathbb{T}^2}$ the (normalized) Lebesgue measure on the torus, then $\rho_t := (\nabla P_t)_\# \mathscr{L}_{\mathbb{T}^2}$ satisfies the following *dual problem*:

$$\begin{cases} \partial_t \rho_t + \nabla \cdot (\boldsymbol{v}_t \rho_t) = 0 \\ \boldsymbol{v}_t(x) = J(x - \nabla P_t^*(x)) \\ \rho_t = (\nabla P_t)_\# \mathscr{L}_{\mathbb{T}^2} \\ P_0(x) = p^0(x) + |x|^2/2. \end{cases} \tag{9.3}$$

Here P_t^* is the convex conjugate of P_t, namely

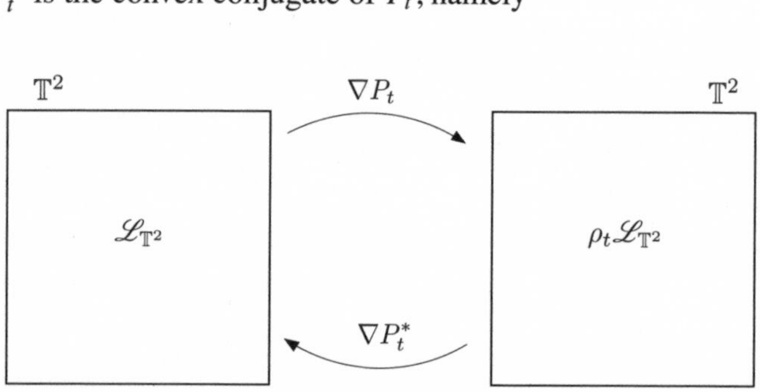

Figure 9.1. The dual change of variables.

$$P_t^*(y) := \sup_{x \in \mathbb{R}^2} (y \cdot x - P_t(x)).$$

Indeed, an easy formal computation allows to obtain (9.3) from (9.2). Taking into account the definition of P_t, the identities $J^2 = -Id$, $\nabla p_t(y) + y = \nabla P_t(y)$, $\nabla^2 p_t(y) + Id = \nabla^2 P_t(y)$ and the fact that $\boldsymbol{u}_t$ is divergence-free, for every test function φ we obtain

$$
\begin{aligned}
\frac{d}{dt}\int_{\mathbb{T}^2}\varphi(x)\,d\rho_t(x) &= \frac{d}{dt}\int_{\mathbb{T}^2}\varphi(\nabla P_t(y))\,dy \\
&= \int_{\mathbb{T}^2}\nabla\varphi(\nabla P_t(y))\cdot\frac{d}{dt}\nabla p_t(y)\,dy \\
&= -\int_{\mathbb{T}^2}\nabla\varphi(\nabla P_t(y))\cdot\Big\{(\nabla^2 p_t(y)+Id)\boldsymbol{u}_t(y) - J\nabla p_t(y)\Big\}\,dy \\
&= -\int_{\mathbb{T}^2}\nabla\big[\varphi(\nabla P_t(y))\big]\cdot\boldsymbol{u}_t(y)\,dy \\
&\quad + \int_{\mathbb{T}^2}\nabla\varphi(\nabla P_t(y))\cdot J(\nabla P_t(y)-y)\,dy \\
&= \int_{\mathbb{T}^2}\nabla\varphi(x)\cdot J(x-\nabla P_t^*(x))\,d\rho_t(x) = \int_{\mathbb{T}^2}\nabla\varphi(x)\cdot\boldsymbol{v}_t(x)\,d\rho_t(x).
\end{aligned}
$$

Notice that this formal derivation holds independently of $\boldsymbol{u}$ (only the divergence-free condition of $\boldsymbol{u}$ is needed), and that $\boldsymbol{u}$ does not appear explicitly in (9.3).

Since $P_t(x) - |x|^2/2$ is periodic, we observe that

$$
\nabla P_t(x+h) = \nabla P_t(x) + h \qquad \forall x\in\mathbb{R}^2,\ h\in\mathbb{Z}^2. \tag{9.4}
$$

Hence ∇P_t can be viewed as a map from $\mathbb{T}^2$ to $\mathbb{T}^2$ and ρ_t is a well defined measure on $\mathbb{T}^2$. One can also verify easily that the inverse map ∇P_t^* satisfies (9.4) as well. Accordingly, we shall understand (9.3) as a PDE on $\mathbb{T}^2$, *i.e.*, using test functions which are $\mathbb{Z}^2$-periodic in space. One may wonder if it is convenient to rewrite the original system (9.2) as an equation for the function P_t instead of p_t, that would look like

$$
\begin{cases}
\partial_t \nabla P_t + (\boldsymbol{u}_t\cdot\nabla)\nabla P_t = J(\nabla P_t - x)\\
\nabla\cdot\boldsymbol{u}_t = 0\\
P_0 = p^0 + |x|^2/2.
\end{cases} \tag{9.5}
$$

Rewriting the system in these terms happens to be a good choice for the corresponding system in 3-space dimensions. However, since we are on the torus and since the map ∇P_t has to be understood with values in $\mathbb{T}^2$, it becomes complicated to give a distributional meaning to (9.5). Indeed, there is no natural notion of duality with test functions for maps with value in a manifold. For this reason, we prefer to deal with p_t and the system (9.2) rather than with P_t and (9.5). Regarding the dual equation, the problem of interpreting ∇P_t (or ∇P_t^*) as a map with values in the torus does not appear. Indeed, the only occurrence of the functions P_t and

P_t^* in the dual equation happens in the formula $\boldsymbol{v}_t(x) = J(x - \nabla P_t^*(x))$; with this definition, $\boldsymbol{v}_t$ is a well defined, $\mathbb{Z}^2$-periodic vector field in $\mathbb{R}^2$.

The dual problem (9.3) is nowadays pretty well understood. In particular, Benamou and Brenier proved in [31] existence of weak solutions to (9.3), see Theorem 9.4 below. On the contrary, much less is known about the original system (9.2).

Formally, given a solution ρ_t of (9.3) and defining P_t^* through the relation $\rho_t = (\nabla P_t)_\# \mathscr{L}_{\mathbb{T}^2}$ (namely the optimal transport map from ρ_t to $\mathscr{L}_{\mathbb{T}^2}$, see Theorem 1.17) the pair $(p_t, \boldsymbol{u}_t)$ given by[1]

$$\begin{cases} p_t(x) := P_t(x) - |x|^2/2 \\ \boldsymbol{u}_t(x) := [\partial_t \nabla P_t^*](\nabla P_t(x)) + [\nabla^2 P_t^*](\nabla P_t(x)) J(\nabla P_t(x) - x) \end{cases} \tag{9.6}$$

solves (9.2). Here, the velocity field $\boldsymbol{u}_t$ has been obtained by substituting the expression for p_t in the first equation of (9.2) and solving for $\boldsymbol{u}_t$. However, being P_t^* just a convex function, *a priori* $\nabla^2 P_t^*$ is just a matrix-valued measure, thus as pointed out in [69] it is not clear the meaning to give to the previous equation.

The formal correspondence between solutions of the dual and of the original equation given by (9.6) appears also when one adopts the Lagrangian point of view. Indeed, we may expect that each particle in the physical space moves along a trajectory and that this trajectory corresponds to a characteristic of the dual velocity $\boldsymbol{v}_t$ when read in the dual variables. Reversing the point of view, given the flow $\boldsymbol{Y}(t, x)$ of $\boldsymbol{v}_t$ in the dual variables, we may look at each characteristic $\boldsymbol{Y}(\cdot, y)$ in the physical variables by performing the change of variables back

$$\boldsymbol{X}(t, x) := \nabla P_t^*(\boldsymbol{Y}(t, \nabla P_0(x)))$$

(see Figure 9.2). With this definition, a simple computation shows that $\boldsymbol{X}(t, x)$ is, formally, the flow of the velocity field $\boldsymbol{u}_t$ defined in (9.6):

$$\begin{aligned} \frac{d}{dt}\mathbf{X}_t &= [\partial_t \nabla P_t^*]\big(\mathbf{Y}_t(\nabla P_0)\big) \\ &\quad + [\nabla^2 P_t^*]\big(\mathbf{Y}_t(\nabla P_0)\big)\frac{d}{dt}\mathbf{Y}_t(\nabla P_0) \\ &= [\partial_t \nabla P_t^*]\big(\mathbf{Y}_t(\nabla P_0)\big) + [\nabla^2 P_t^*]\big(\mathbf{Y}_t(\nabla P_0)\big) J\big[\mathbf{Y}_t(\nabla P_0) \\ &\qquad\qquad - \nabla P_t^*\big(\mathbf{Y}_t(\nabla P_0)\big)\big] \\ &= [\partial_t \nabla P_t^*](\nabla P_t(\mathbf{X}_t)) + [\nabla^2 P_t^*](\nabla P_t(\mathbf{X}_t)) J(\nabla P_t(\mathbf{X}_t) - \mathbf{X}_t). \end{aligned}$$

[1] Because of the many compositions involved in this chapter, we use the notation $[\partial_t f](g)$ (respectively $[\nabla f](g)$) to denote the composition $(\partial_t f) \circ g$ (respectively $(\nabla f) \circ g$), avoiding the ambiguous notation $\partial_t f(g)$ (respectively $\nabla f(g)$)

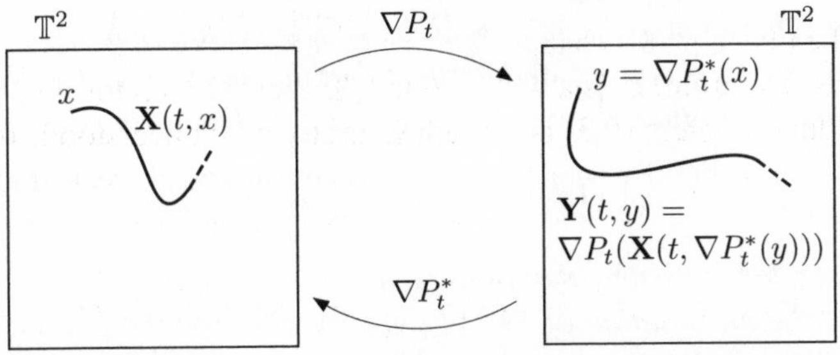

Figure 9.2. The flow of the velocity field in physical and dual variables.

In this chapter we prove that (9.6) is a well defined velocity field, and that the couple $(p_t, \boldsymbol{u}_t)$ is a solution of (9.2) in a distributional sense. In order to carry out our analysis, a fundamental tool is a recent result for solutions of the Monge-Ampère equation, proved by De Philippis and Figalli in [71], showing $L \log^k L$ regularity on $\nabla^2 P_t^*$ (see Theorem 1.18(ii)).

Thanks to this result, we can easily show that the second term appearing in the definition of the velocity $\boldsymbol{u}_t$ in (9.6) is a well defined L^1 function (see the proof of Theorem 9.2). Moreover, following some ideas developed in [106] we can show that the first term is also L^1, thus giving a meaning to $\boldsymbol{u}_t$ (see Proposition 9.6). At this point we can prove that the pair $(p_t, \boldsymbol{u}_t)$ is actually a distributional solution of system (9.2). Let us recall, following [69], the proper definition of *weak Eulerian solution* of (9.2).

Definition 9.1. Let $p : \mathbb{T}^2 \times (0,\infty) \to \mathbb{R}$ and $\boldsymbol{u} : \mathbb{T}^2 \times (0,\infty) \to \mathbb{R}^2$. We say that $(p, \boldsymbol{u})$ is a *weak Eulerian solution* of (9.2) if:

- $|\boldsymbol{u}| \in L^\infty((0,\infty), L^1(\mathbb{T}^2))$, $p \in L^\infty((0,\infty), W^{1,\infty}(\mathbb{T}^2))$, and $p_t(x) + |x|^2/2$ is convex for any $t \geq 0$;
- For every $\phi \in C_c^\infty(\mathbb{T}^2 \times [0,\infty))$, it holds

$$\begin{aligned}\int_0^\infty \int_{\mathbb{T}^2} J\nabla p_t(x) \Big\{\partial_t \phi_t(x) + \boldsymbol{u}_t(x) \cdot \nabla \phi_t(x)\Big\} \\ - \Big\{\nabla p_t(x) + J\boldsymbol{u}_t(x)\Big\} \phi_t(x)\, dx\, dt \\ + \int_{\mathbb{T}^2} J\nabla p_0(x) \phi_0(x)\, dx = 0;\end{aligned} \tag{9.7}$$

- For $\mathscr{L}^1$-a.e. $t \in (0,\infty)$ it holds

$$\int_{\mathbb{T}^2} \nabla \psi(x) \cdot \boldsymbol{u}_t(x)\, dx = 0 \qquad \text{for all } \psi \in C^\infty(\mathbb{T}^2). \tag{9.8}$$

We can now state the main result.

Theorem 9.2. *Let $p_0 : \mathbb{R}^2 \to \mathbb{R}$ be a $\mathbb{Z}^2$-periodic function such that $p_0(x) + |x|^2/2$ is convex, and assume that the measure $(Id + \nabla p_0)_\# \mathscr{L}^2$ is absolutely continuous with respect to $\mathscr{L}^2$ with density ρ_0, namely*

$$(Id + \nabla p_0)_\# \mathscr{L}^2 = \rho_0 \mathscr{L}^2.$$

Moreover, let us assume that both ρ_0 and $1/\rho_0$ belong to $L^\infty(\mathbb{R}^2)$.

Let ρ_t be the solution of (9.3) *(given by Theorem* 9.4 *below), let $P_t : \mathbb{R}^2 \to \mathbb{R}$ be the (unique up to an additive constant) convex function such that $(\nabla P_t)_\# \mathscr{L}^2 = \rho_t \mathscr{L}^2$ and $P_t(x) - |x|^2/2$ is $\mathbb{Z}^2$-periodic, and let $P_t^* : \mathbb{R}^2 \to \mathbb{R}$ its convex conjugate.*

Then the couple $(p_t, \boldsymbol{u}_t)$ defined in (9.6) *is a weak Eulerian solution of* (9.2)*, in the sense of Definition* 9.1.

Although the vector field $\boldsymbol{u}$ provided by the previous theorem is only L^1, as explained in Section 9.3 we can associate to it a measure-preserving Lagrangian flow. In particular we recover (in the case of the 2-dimensional periodic setting) the result of Cullen and Feldman [69] on the existence of Lagrangian solutions to the semigeostrophic equations in physical space.

Many problems regarding the semigeostrophic equation and its dual formulation are nowadays open. Are the distributional solutions of (9.2) and of (9.3) unique? Is the Lagrangian flow associated to $\boldsymbol{u}_t$ unique? (we remark that the Lagrangian flow associated to the dual equation is unique thanks to Theorem 1.5). Does there exist a regular solution for all times if the initial datum is sufficiently smooth? This was proven by Loeper [107] for short times, but any global result is missing.

The chapter is structured as follows: in Section 1.5.1 we recall some preliminary results on optimal transport maps on the torus and their regularity. Then, in Section 9.1 we state the existence result of Benamou and Brenier for solutions to the dual problem (9.3), and we show some important regularity estimates on such solutions, which are used in Section 9.2 to prove Theorem 9.2. In Section 9.3 we prove the existence of a regular Lagrangian flow associated to the vector field $\boldsymbol{u}$ provided by Theorem 9.2.

9.1. The dual problem and the regularity of the velocity field

In this section we recall some properties of solutions of the dual system (9.3), and we show the L^1 integrability of the velocity field $\boldsymbol{u}_t$ defined in (9.6).

Remark 9.3. The dual system (9.3) is made by a continuity equation with an instantaneous coupling between the velocity field and the density through a time independent elliptic PDE (the Monge-Ampère equation). A similar structure was already observed in the Vlasov-Poisson system (8.1), but in this case we needed to consider the equation in the phase space rather than in the physical space. Another equation of this form is the 2-dimensional incompressible Euler equation in the vorticity formulation in $(0,\infty)\times\mathbb{R}^2$

$$\begin{cases} \partial_t\omega_t + \nabla\cdot(\boldsymbol{v}_t\omega_t) = 0 \\ \boldsymbol{v}_t = J\nabla\psi_t \\ \omega_t = \Delta\psi_t. \end{cases} \tag{9.9}$$

Existence and uniqueness (in the class of solutions with bounded vorticities) was proved for this equation by Yudovich [125]. As we will see in Theorem 9.4, despite the nonlinearity of the coupling in the dual semigeostrophic system (which is given by the Monge-Ampère equation and not by the Poisson equation as in (9.9)), we can still prove existence of solutions for the dual semigeostrophic system (9.3). The uniqueness problem, instead, remains open, since the argument of Yudovich cannot be easily adapted. The connection between the 2-dimensional incompressible Euler equation in the vorticity formulation and the dual semigeostrophic system (9.3) is also confirmed by Loeper in [107]: if we "linearize" (9.3) writing $\rho_t = 1+\varepsilon\omega_t + o(\varepsilon)$ and $P_t^* = |x|^2/2 + \varepsilon\psi_t + o(\varepsilon)$ and we rescale the time variable according to $t\to t/\varepsilon$, then, formally, ω and ψ solve (9.9).

We know by Theorem 1.17 that ρ_t uniquely defines P_t (and so also P_t^*) through the relation $(\nabla P_t)_\#\mathscr{L}_{\mathbb{T}^2} = \rho_t$ up to an additive constant. In [31] (see also [69]), the authors prove the existence of distributional solutions to the dual equation by means of an approximation argument, based in turn on the well-posedness and stability of solutions of the transport equation presented in Chapter 1. To be precise, in [31, 69] the proof is given in $\mathbb{R}^3$, but actually it can be rewritten verbatim on the 2-dimensional torus, using the optimal transport maps provided by Theorem 1.17.

Theorem 9.4 (Existence of solutions of (9.3)**).** *Let $P_0:\mathbb{R}^2\to\mathbb{R}$ be a convex function such that $P_0(x)-|x|^2/2$ is $\mathbb{Z}^2$-periodic, $(\nabla P_0)_\#\mathscr{L}_{\mathbb{T}^2} \ll \mathscr{L}_{\mathbb{T}^2}$, and the density ρ_0 satisfies $0<\lambda\le\rho_0\le\Lambda<\infty$. Then there exist convex functions P_t, $P_t^*:\mathbb{R}^2\to\mathbb{R}$, with $P_t(x)-|x|^2/2$ and $P_t^*(y)-|y|^2/2$ periodic, uniquely determined up to time-dependent additive constants, such that $(\nabla P_t)_\#\mathscr{L}_{\mathbb{T}^2} = \rho_t\mathscr{L}_{\mathbb{T}^2}$, $(\nabla P_t^*)_\#\rho_t = \mathscr{L}_{\mathbb{T}^2}$. In addition, setting $\boldsymbol{v}_t(x) = J(x-\nabla P_t^*(x))$, ρ_t is a distributional solution* *to* (9.3), *namely*

$$\int\int_{\mathbb{T}^2} \Big\{\partial_t \varphi_t(x) + \nabla\varphi_t(x) \cdot \boldsymbol{v}_t(x)\Big\} \rho_t(x)\, dx\, dt + \int_{\mathbb{T}^2} \varphi_0(x)\rho_0(x)\, dx = 0 \tag{9.10}$$

for every $\varphi \in C_c^\infty(\mathbb{R}^2 \times [0,\infty))$ $\mathbb{Z}^2$*-periodic in the space variable.*

Finally, the following regularity properties hold:

(i) $\lambda \le \rho_t \le \Lambda$;
(ii) $\rho_t \mathscr{L}^2 \in C([0,\infty), \mathcal{P}_w(\mathbb{T}^2))$;[2]
(iii) $P_t - \fint_{\mathbb{T}^2} P_t$, $P_t^* - \fint_{\mathbb{T}^2} P_t^* \in L^\infty([0,\infty), W^{1,\infty}_{\mathrm{loc}}(\mathbb{R}^2)) \cap C([0,\infty), W^{1,r}_{\mathrm{loc}}(\mathbb{R}^2))$ *for every* $r \in [1,\infty)$;
(iv) $\|\boldsymbol{v}_t\|_\infty \le \sqrt{2}/2$.

Sketch of the proof of Theorem 9.4. We prove the existence of a distributional solution to (9.3) by approximation. We introduce a time discretization with parameter $1/n$, $n \in \mathbb{N}$, and we split $(0,\infty)$ in intervals of the form $((k-1)/n, k/n)$ for $k \in \mathbb{N}$, each of length $1/n$.

We define the approximate solutions $(P_t^n)^*$ and ρ_t^n inductively on k, where $(P_t^n)^*$ are a time-dependent family of convex functions and ρ_t^n are bounded, $\mathbb{Z}^2$-periodic densities with $\lambda \le \rho_t^n \le \Lambda$. In each interval $((k-1)/n, k/n)$ we consider the transport map between the density at the beginning of this interval and the Lebesgue measure; we define $(P_t^n)^*$ to be equal to this map (provided by Theorem 1.17) in the entire time interval

$$\left[\nabla(P^n_{k/n})^*\right]_\# \rho^n_{(k-1)/n} = \mathscr{L}_{\mathbb{T}^2}, \qquad \nabla(P_t^n)^* := \nabla(P^n_{k/n})^*.$$

This transport map induces a velocity field (constant in time in $((k-1)/n, k/n)$, as also $(P_t^n)^*$) through the relation

$$\boldsymbol{v}_t^n := J\big(x - \nabla(P_t^n)^*\big) = J\big(x - \nabla(P^n_{k/n})^*\big).$$

Finally, we let the density at the beginning of our time interval, namely $\rho^n_{(k-1)/n}$, evolve according to this velocity field for time $1/n$. In other words, for $t \in ((k-1)/n, k/n)$ we let ρ_t^n be the solution of

$$\begin{cases} \partial_t \rho + \nabla \cdot (\boldsymbol{v}_t^n \rho) = 0 & \text{in } ((k-1)/n, k/n) \times \mathbb{R}^2 \\ \rho_{(k-1)/n} = \rho^n_{(k-1)/n} & \text{in } \mathbb{R}^2. \end{cases} \tag{9.11}$$

[2] Here $\mathcal{P}_w(\mathbb{T}^2)$ is the space of probability measures on the torus endowed with the *weak* topology induced by the duality with $C(\mathbb{T}^2)$

We remark that the well-posedness of the continuity equation (9.11) is guaranteed the results described in Chapter 1 (and in particular by the main result of [5]) and by the fact that $\boldsymbol{v}_t^n$ is an authonomous, divergence-free, BV vector field in $((k-1)/n, k/n) \times \mathbb{T}^2$. Finally, a solution to (9.3) is obtained by taking the limit as $n \to \infty$ in the discrete scheme presented above; the compactness of the functions $(P_t^n)^* - \fint_{\mathbb{T}^2}(P_t^n)^*$ and ρ_t^n and the equation solved in the limit are studied in [31], where the scheme is performed with a careful regularization of the initial data in order to avoid the use of the results of Chapter 1 on the solutions of the continuity equation with non-smooth vector fields. □

Observe that, by Theorem 9.4(ii), $t \mapsto \rho_t \mathscr{L}_{\mathbb{T}^2}$ is weakly continuous, so ρ_t is a well-defined function *for every* $t \geq 0$. Further regularity properties of ∇P_t and ∇P_t^* with respect to time will be proved in Propositions 9.6 and 9.10.

In the proof of Theorem 9.2 we will need to test with functions which are merely $W^{1,1}$. This is made possible by the following lemma.

Lemma 9.5. *Let ρ_t and P_t be as in Theorem* 9.4. *Then* (9.10) *holds for every $\varphi \in W^{1,1}(\mathbb{T}^2 \times [0,\infty))$ which is compactly supported in time. (Now $\varphi_0(x)$ has to be understood in the sense of traces.)*

Proof. Let $\varphi^n \in C^\infty(\mathbb{T}^2 \times [0,\infty))$ be strongly converging to φ in $W^{1,1}$, so that φ_0^n converges to φ_0 in $L^1(\mathbb{T}^2)$. Taking into account that both ρ_t and $\boldsymbol{v}_t$ are uniformly bounded from above in $\mathbb{T}^2 \times [0,\infty)$, we can apply (9.10) to the test functions φ^n and let $n \to \infty$ to obtain the same formula with φ. □

The following proposition, which provides the Sobolev regularity of $t \mapsto \nabla P_t^*$, is our main technical tool. Notice that, in order to prove Theorem 9.2, only finiteness of the left hand side in (9.12) would be needed, and the proof of this fact involves only a smoothing argument, the regularity estimates of [71] collected in Theorem 1.18(ii), and the argument of [106, Theorem 5.1]. However, the continuity result in [72] allows to show the validity of the natural *a priori* estimate on the left hand side in (9.12).

Proposition 9.6 (Time regularity of optimal maps). *Let ρ_t and P_t be as in Theorem 9.4. Then $\nabla P_t^* \in W^{1,1}_{\mathrm{loc}}(\mathbb{T}^2 \times [0,\infty); \mathbb{R}^2)$, and for every*

$k \in \mathbb{N}$ *there exists a constant* $C(k)$ *such that, for* $\mathscr{L}^1$*-a.e.* $t \geq 0$,

$$\int_{\mathbb{T}^2} \rho_t |\partial_t \nabla P_t^*| \log_+^k(|\partial_t \nabla P_t^*|)\, dx$$
$$\leq C(k)\left(\int_{\mathbb{T}^2} \rho_t |\nabla^2 P_t^*| \log_+^{2k}(|\nabla^2 P_t^*|)\, dx \right. \tag{9.12}$$
$$\left. + \operatorname*{ess\,sup}_{\mathbb{T}^2} \left(\rho_t |\boldsymbol{v}_t|^2\right) \int_{\mathbb{T}^2} |\nabla^2 P_t^*|\, dx \right).$$

Remark 9.7. Under the assumptions of the previous proposition, one could actually prove a slightly stronger statement, showing that the map ∇P_t^* belongs to $W^{1,\gamma_0}_{\mathrm{loc}}(\mathbb{T}^2 \times [0,\infty); \mathbb{R}^2)$ for some $\gamma_0 > 1$. More precisely, there exist constants $C, \gamma_0 > 1$, depending only on λ, such that, for almost every $t \geq 0$,

$$\int_{\mathbb{T}^2} \rho_t |\partial_t \nabla P_t^*|^{\frac{2\gamma_0}{1+\gamma_0}}\, dx$$
$$\leq C\left(\int_{\mathbb{T}^2} \rho_t |\nabla^2 P_t^*|^{\gamma_0}\, dx + \operatorname*{ess\,sup}_{\mathbb{T}^2} \left(\rho_t |\boldsymbol{v}_t|^2\right) \int_{\mathbb{T}^2} |\nabla^2 P_t^*|\, dx \right). \tag{9.13}$$

This estimate, however, is less powerful than (9.12) when dealing with the semigeostrophic system in a non-periodic setting (for instance, see [8], where the semigeostrophic system is studied in $\mathbb{R}^3$), since in the localized version of (1.27) and (9.13) the exponent γ_0 depends also on the set where the estimate is localized, whereas in (9.12) the modulus of integrability of $|\partial_t \nabla P_t^*|$ (namely, the function $t \to t \log_+^k(t)$) does not depend on the set. For this reason we prefer to keep this version of Proposition 9.6.

To prove Proposition 9.6, we need some preliminary results.

Lemma 9.8. *For every* $k \in \mathbb{N}$ *we have*

$$ab \log_+^k(ab)$$
$$\leq 2^{k-1}\left[\left(\frac{k}{e}\right)^k + 1\right] b^2 + 2^{3(k-1)} a^2 \log_+^{2k}(a) \quad \forall\, (a,b) \in \mathbb{R}^+ \times \mathbb{R}^+. \tag{9.14}$$

Proof. From the elementary inequalities

$$\log_+(ts) \leq \log_+(t) + \log_+(s), \;\; (t+s)^k \leq 2^{k-1}(t^k + s^k), \;\; \log_+^k(t) \leq \left(\frac{k}{e}\right)^k t$$

which hold for every $t,\ s > 0$, we infer

$$
\begin{aligned}
ab\log_+^k(ab) &\le ab\left[\log_+\left(\frac{b}{a}\right) + 2\log_+(a)\right]^k \\
&\le 2^{k-1}ab\left[\log_+^k\left(\frac{b}{a}\right) + 2^k\log_+^k(a)\right] \\
&\le 2^{k-1}\left[\left(\frac{k}{e}\right)^k b^2 + 2^k ab\log_+^k(a)\right] \\
&\le 2^{k-1}\left[\left(\frac{k}{e}\right)^k b^2 + b^2 + 2^{2(k-1)}a^2\log_+^{2k}(a)\right],
\end{aligned}
$$

which proves (9.14). □

Lemma 9.9 (Space-time regularity of transport). *Let* $k \in \mathbb{N}\cup\{0\}$, *and let* $\rho \in C^\infty(\mathbb{T}^2 \times [0,\infty))$ *and* $\boldsymbol{v} \in C^\infty(\mathbb{T}^2 \times [0,\infty);\mathbb{R}^2)$ *satisfy*

$$0 < \lambda \le \rho_t(x) \le \Lambda < \infty \qquad \forall\,(x,t) \in \mathbb{T}^2 \times [0,\infty),$$

$$\partial_t\rho_t + \nabla\cdot(\boldsymbol{v}_t\rho_t) = 0 \qquad \textit{in } \mathbb{T}^2 \times [0,\infty),$$

and $\int_{\mathbb{T}^2}\rho_t\,dx = 1$ *for all* $t \ge 0$. *Let us consider convex conjugate maps* P_t *and* P_t^* *such that* $P_t(x) - |x|^2/2$ *and* $P_t^*(y) - |y|^2/2$ *are* $\mathbb{Z}^2$*-periodic,* $(\nabla P_t^*)_\#\rho_t = \mathscr{L}_{\mathbb{T}^2}$, $(\nabla P_t)_\#\mathscr{L}_{\mathbb{T}^2} = \rho_t$. *Then:*

(i) $P_t^* - ⨍_{\mathbb{T}^2} P_t^* \in \mathrm{Lip}_{\mathrm{loc}}([0,\infty); C^k(\mathbb{T}^2))$ *for any* $k \in \mathbb{N}$.
(ii) *The following linearized Monge-Ampère equation holds:*

$$\nabla\cdot\left(\rho_t(\nabla^2P_t^*)^{-1}\partial_t\nabla P_t^*\right) = -\nabla\cdot(\rho_t\boldsymbol{v}_t). \tag{9.15}$$

Proof. Let us fix $T > 0$. From the regularity theory for the Monge-Ampère equation (see Theorem 1.18) we obtain that $P_t \in C^\infty(\mathbb{R}^2)$, uniformly for $t \in [0,T]$, and there exist universal constants $c_1,\ c_2 > 0$ such that

$$c_1 Id \le \nabla^2P_t^*(x) \le c_2 Id \qquad \forall\,(x,t) \in \mathbb{T}^2 \times [0,T]. \tag{9.16}$$

Since ∇P_t^* is the inverse of ∇P_t, by the smoothness of P_t and (9.16) we deduce that $P_t^* \in C^\infty(\mathbb{R}^2)$, uniformly on $[0,T]$.

Now, to prove (i), we need to investigate the time regularity of $P_t^* - ⨍_{\mathbb{T}^2}P_t^*$. Moreover, up to adding a time dependent constant to P_t, we can assume without loss of generality that $\int_{\mathbb{T}^2} P_t^* = 0$ for all t. By the condition $(\nabla P_t^*)_\# \rho_t = \mathscr{L}_{\mathbb{T}^2}$, for any $0 \leq s, t \leq T$ and $x \in \mathbb{R}^2$ it holds

$$\begin{aligned}
\frac{\rho_s(x) - \rho_t(x)}{s-t} &= \frac{\det(\nabla^2 P_s^*(x)) - \det(\nabla^2 P_t^*(x))}{s-t} \\
&= \sum_{i,j=1}^{2} \left(\int_0^1 \frac{\partial \det}{\partial \xi_{ij}} (\tau \nabla^2 P_s^*(x) + (1-\tau)\nabla^2 P_t^*(x))\, d\tau \right) \\
&\quad \times \frac{\partial_{ij} P_s^*(x) - \partial_{ij} P_t^*(x)}{s-t}.
\end{aligned} \tag{9.17}$$

Given a 2×2 matrix $A = (\xi_{ij})_{i,j=1,2}$, we denote by $M(A)$ the cofactor matrix of A. We recall that

$$\frac{\partial \det(A)}{\partial \xi_{ij}} = M_{ij}(A), \tag{9.18}$$

and if A is invertible then $M(A)$ satisfies the identity

$$M(A) = \det(A)\, A^{-1}. \tag{9.19}$$

Moreover, if A is symmetric and satisfies $c_1 Id \leq A \leq c_2 Id$ for some positive constants c_1, c_2, then

$$\frac{c_1^2}{c_2} Id \leq M(A) \leq \frac{c_2^2}{c_1} Id. \tag{9.20}$$

Hence, from (9.17), (9.18), (9.16), and (9.20), for any $0 \leq s, t \leq T$ it follows that

$$\begin{aligned}
&\frac{\rho_s - \rho_t}{s-t} \\
&= \sum_{i,j=1}^{2} \left(\int_0^1 M(\tau \nabla^2 P_s^* + (1-\tau)\nabla^2 P_t^*)\, d\tau \right) \partial_{ij} \left(\frac{P_s^* - P_t^*}{s-t} \right),
\end{aligned} \tag{9.21}$$

with

$$\frac{c_1^2}{c_2} Id \leq \int_0^1 M_{ij}(\tau \nabla^2 P_s^* + (1-\tau)\nabla^2 P_t^*)\, d\tau \leq \frac{c_2^2}{c_1} Id$$

Since $\nabla^2 P_t^*$ is smooth in space, uniformly on $[0, T]$, by classical elliptic regularity theory[3] it follows that for any $k \in \mathbb{N}$ and $\alpha \in (0,1)$ there exists

[3] Note that equation (9.17) is well defined on $\mathbb{T}^2$ since $P_t^* - P_s^*$ is $\mathbb{Z}^2$-periodic. We also observe that $P_t^* - P_s^*$ has average zero on $\mathbb{T}^2$.

a constant $C := C(\|(\rho_s - \rho_t)/(s-t)\|_{C^{k,\alpha}(\mathbb{T}^2\times[0,T])})$ such that

$$\left\| \frac{P_s^*(x) - P_t^*(x)}{s-t} \right\|_{C^{k+2,\alpha}(\mathbb{T}^2)} \leq C.$$

This proves point (i) in the statement. To prove the second part, we let $s \to t$ in (9.21) to obtain

$$\partial_t \rho_t = \sum_{i,j=1}^{2} M_{ij}(\nabla^2 P_t^*(x))\, \partial_t \partial_{ij} P_t^*(x). \tag{9.22}$$

Taking into account the continuity equation and the well-known divergence-free property of the cofactor matrix

$$\sum_i \partial_i M_{ij}(\nabla^2 P_t{}^*(x)) = 0, \qquad j = 1, 2,$$

we can rewrite (9.22) as

$$-\nabla \cdot (\boldsymbol{v}_t \rho_t) = \sum_{i,j=1}^{2} \partial_i \big(M_{ij}(\nabla^2 P_t^*(x))\, \partial_t \partial_j P_t{}^*(x) \big).$$

Hence, using (9.19) and the Monge-Ampère equation $\det(\nabla^2 P_t^*) = \rho_t$, we finally get (9.15). □

Proof of Proposition 9.6. We closely follow the argument of [106, Theorem 5.1], and we split the proof in two parts. In the first step we assume that

$$\rho_t \in C^\infty(\mathbb{T}^2 \times \mathbb{R}), \;\; \boldsymbol{v}_t \in C^\infty(\mathbb{T}^2 \times \mathbb{R}; \mathbb{R}^2)\,, \tag{9.23}$$

$$0 < \lambda \leq \rho_t \leq \Lambda < \infty\,, \tag{9.24}$$

$$\partial_t \rho_t + \nabla \cdot (\boldsymbol{v}_t \rho_t) = 0\,, \tag{9.25}$$

$$(\nabla P_t)_\# \mathscr{L}_{\mathbb{T}^2} = \rho_t \mathscr{L}_{\mathbb{T}^2}, \tag{9.26}$$

and we prove that (9.12) holds for every $t \geq 0$ (in this step, we assume U_t to be given, namely we do not assume any relation between U_t and P_t). In the second step we prove the general case through an approximation argument.

Step 1: The regular case. Let us assume that the regularity assumptions (9.23), (9.24), (9.25), (9.26) hold. Moreover, up to adding a time dependent constant to P_t, we can assume without loss of generality that $\int_{\mathbb{T}^2} P_t^* = 0$ for all $t \geq 0$, so that by Lemma 9.9 we have $\partial_t P_t^* \in C^\infty(\mathbb{T}^2)$.

Fix $t \geq 0$. Multiplying (9.15) by $\partial_t P_t^*$ and integrating by parts, we get

$$\begin{aligned}\int_{\mathbb{T}^2} \rho_t |(\nabla^2 P_t^*)^{-1/2} \partial_t \nabla P_t^*|^2 dx &= \int_{\mathbb{T}^2} \rho_t \partial_t \nabla P_t^* \cdot (\nabla^2 P_t^*)^{-1} \partial_t \nabla P_t^* \, dx \\ &= -\int_{\mathbb{T}^2} \rho_t \partial_t \nabla P_t^* \cdot \boldsymbol{v}_t \, dx.\end{aligned} \tag{9.27}$$

(Since the symmetric matrix $\nabla^2 P_t^*(x)$ is nonnegative, both its square root and the square root of its inverse are well-defined.) From Cauchy-Schwartz inequality it follows that the right-hand side of (9.27) can be rewritten and estimated with

$$\begin{aligned}&-\int_{\mathbb{T}^2} \rho_t \partial_t \nabla P_t^* \cdot (\nabla^2 P_t^*)^{-1/2} (\nabla^2 P_t^*)^{1/2} \boldsymbol{v}_t \, dx \\ &\leq \left(\int_{\mathbb{T}^2} \rho_t |(\nabla^2 P_t^*)^{-1/2} \partial_t \nabla P_t^*|^2 dx \right)^{1/2} \left(\int_{\mathbb{T}^2} \rho_t |(\nabla^2 P_t^*)^{1/2} \boldsymbol{v}_t|^2 dx \right)^{1/2}.\end{aligned} \tag{9.28}$$

Moreover, the second factor in the right-hand side of (9.28) can be estimated with

$$\int_{\mathbb{T}^2} \rho_t \boldsymbol{v}_t \cdot \nabla^2 P_t^* \boldsymbol{v}_t \, dx \leq \max_{\mathbb{T}^2} \left(\rho_t |\boldsymbol{v}_t|^2 \right) \int_{\mathbb{T}^2} |\nabla^2 P_t^*| \, dx. \tag{9.29}$$

Hence, from (9.27), (9.28), and (9.29) it follows that

$$\int_{\mathbb{T}^2} \rho_t |(\nabla^2 P_t^*)^{-1/2} \partial_t \nabla P_t^*|^2 \, dx \leq \max_{\mathbb{T}^2} \left(\rho_t |\boldsymbol{v}_t|^2 \right) \int_{\mathbb{T}^2} |\nabla^2 P_t^*| \, dx. \tag{9.30}$$

We now apply Lemma 9.8 with

$$a = |(\nabla^2 P_t^*)^{1/2}| \text{ and } b = |(\nabla^2 P_t^*)^{-1/2} \partial_t \nabla P_t^*(x)|$$

to deduce the existence of a constant $C(k)$ such that

$$\begin{aligned}&|\partial_t \nabla P_t^*| \log_+^k (|\partial_t \nabla P_t^*|) \\ &\leq C(k) \left(|(\nabla^2 P_t^*)^{1/2}|^2 \log_+^{2k} (|(\nabla^2 P_t^*)^{1/2}|^2) + |(\nabla^2 P_t^*)^{-1/2} \partial_t \nabla P_t^*|^2 \right) \\ &= C(k) \left(|\nabla^2 P_t^*| \log_+^{2k} (|\nabla^2 P_t^*|) + |(\nabla^2 P_t^*)^{-1/2} \partial_t \nabla P_t^*|^2 \right).\end{aligned}$$

Integrating the above inequality over $\mathbb{T}^2$ and using (9.30), we finally obtain

$$\begin{aligned}&\int_{\mathbb{T}^2} \rho_t |\partial_t \nabla P_t^*| \log_+^k (|\partial_t \nabla P_t^*|) \, dx \\ &\leq C(k) \left(\int_{\mathbb{T}^2} \rho_t |\nabla^2 P_t^*| \log_+^{2k} (|\nabla^2 P_t^*|) \, dx + \int_{\mathbb{T}^2} \rho_t |(\nabla^2 P_t^*)^{-1/2} \partial_t \nabla P_t^*|^2 dx \right) \\ &\leq C(k) \left(\int_{\mathbb{T}^2} \rho_t |\nabla^2 P_t^*| \log_+^{2k} (|\nabla^2 P_t^*|) \, dx + \max_{\mathbb{T}^2} \left(\rho_t |\boldsymbol{v}_t|^2 \right) \int_{\mathbb{T}^2} |\nabla^2 P_t^*| dx \right),\end{aligned} \tag{9.31}$$

which proves (9.12).

Step 2: The approximation argument. First of all, we extend the functions ρ_t and $\boldsymbol{v}_t$ for $t \le 0$ by setting $\rho_t = \rho_0$ and $\boldsymbol{v}_t = 0$ for every $t < 0$. We notice that, with this definition, ρ_t solves the continuity equation with velocity $\boldsymbol{v}_t$ on $\mathbb{R}^2 \times \mathbb{R}$.

Fix now $\sigma_1 \in C_c^\infty(\mathbb{R}^2)$, $\sigma_2 \in C_c^\infty(\mathbb{R})$, define the family of mollifiers $(\sigma^n)_{n\in\mathbb{N}}$ as $\sigma^n(x,t) := n^3\sigma_1(nx)\sigma_2(nt)$, and set

$$\rho^n := \rho * \sigma^n, \qquad \boldsymbol{v}^n(x) := \frac{(\rho\boldsymbol{v}) * \sigma^n}{\rho * \sigma^n}.$$

Since $\lambda \le \rho \le \Lambda$ then

$$\lambda \le \rho^n \le \Lambda.$$

Therefore both ρ^n and $\boldsymbol{v}^n$ are well defined and satisfy (9.23), (9.24), (9.25). Moreover for every $t > 0$ the function ρ_t^n is $\mathbb{Z}^2$-periodic and it is a probability density when restricted to $(0,1)^2$ (once again we are identifying periodic functions with functions defined on the torus). Let P_t^n be the only convex function such that $(\nabla P_t^n)_\# \mathscr{L}_{\mathbb{T}^2} = \rho_t^n$ and its convex conjugate P_t^{n*} satisfies $\int_{\mathbb{T}^2} P_t^{n*} = 0$ for all $t \ge 0$. Since $\rho_t^n \to \rho_t$ in $L^1(\mathbb{T}^2)$ for any $t > 0$ (recall that, by Theorem 9.4(ii), ρ_t is weakly continuous in time), from standard stability results for Alexandrov solutions of Monge-Ampère (see for instance [72]) it follows that

$$\nabla P_t^{n*} \to \nabla P_t^* \qquad \text{in } L^1(\mathbb{T}^2) \tag{9.32}$$

for any $t > 0$. Moreover, by Theorems 1.17 and 1.18(ii), for every $k \in \mathbb{N}$ there exists a constant $C := C(\lambda, \Lambda, k)$ such that

$$\int_{\mathbb{T}^2} \rho_t^n |\nabla^2 P_t^{n*}| \log_+^k(|\nabla^2 P_t^{n*}|)\, dx \le C,$$

and by the stability theorem in the Sobolev topology established in [72, Theorem 1.3] it follows that

$$\int_{\mathbb{T}^2} \rho_t^n |\nabla^2 P_t^{n*}| \log_+^k(|\nabla^2 P_t^{n*}|)\, dx \to \int_{\mathbb{T}^2} \rho_t |\nabla^2 P_t^{*}| \log_+^k(|\nabla^2 P_t^{*}|)\, dx, \tag{9.33}$$

$$\int_{\mathbb{T}^2} |\nabla^2 P_t^{n*}|\, dx \to \int_{\mathbb{T}^2} |\nabla^2 P_t^{*}|\, dx. \tag{9.34}$$

Finally, since the function $(w,t) \mapsto F(w,t) = |w|^2/t$ is convex on $\mathbb{R}^2 \times (0,\infty)$, by Jensen inequality we get

$$\|\rho^n |\boldsymbol{v}^n|^2\|_\infty = \|F(\rho^n \boldsymbol{v}^n, \rho^n)\|_\infty \le \|\rho|\boldsymbol{v}|^2\|_\infty. \tag{9.35}$$

Let us fix $T > 0$ and $\phi \in C_c^\infty((0,T))$ nonnegative. From the previous steps and Dunford-Pettis Theorem, it is clear that $\phi(t)\rho_t^n \partial_t \nabla P_t^{n*}$ weakly converge to $\phi(t)\rho_t \partial_t \nabla P_t^*$ in $L^1(\mathbb{T}^2 \times (0,T))$. Moreover, since the function $w \mapsto |w| \log_+^k(|w|/r)$ is convex for every $r \in (0,\infty)$ we can apply Ioffe lower semicontinuity theorem [17, Theorem 5.8] to the functions $\phi(t)\rho_t^n \partial_t \nabla P_t^{n*}$ and $\phi(t)\rho_t^n$ to infer

$$\begin{aligned}
&\int_0^T \phi(t) \int_{\mathbb{T}^2} \rho_t |\partial_t \nabla P_t^*| \log_+^k(|\partial_t \nabla P_t^*|)\, dx\, dt \\
&\leq \liminf_{n\to\infty} \int_0^T \phi(t) \int_{\mathbb{T}^2} \rho_t^n |\partial_t \nabla P_t^{n*}| \log_+^k(|\partial_t \nabla P_t^{n*}|)\, dx\, dt.
\end{aligned} \tag{9.36}$$

By Step 1 we can apply (9.12) to ρ_t^n, $\boldsymbol{v}_t^n$. Taking (9.33), (9.34), (9.35) and (9.36) into account, by Lebesgue dominated convergence theorem we obtain

$$\begin{aligned}
&\int_0^T \phi(t) \int_{\mathbb{T}^2} \rho_t |\partial_t \nabla P_t^*| \log_+^k(|\partial_t \nabla P_t^*|)\, dx\, dt \\
&\leq C(k) \int_0^T \phi(t) \left(\int_{\mathbb{T}^2} \rho_t |\nabla^2 P_t^*| \log_+^{2k}(|\nabla^2 P_t^*|)\, dx \right. \\
&\qquad\qquad \left. + \operatorname*{ess\,sup}_{\mathbb{T}^2} \left(\rho_t |\boldsymbol{v}_t|^2\right) \int_{\mathbb{T}^2} |\nabla^2 P_t^*|\, dx \right) dt.
\end{aligned}$$

Since this holds for every $\phi \in C_c^\infty((0,T))$ nonnegative, we obtain the desired result. □

It is clear from the proof of Proposition 9.6 that the particular coupling between the velocity field $\boldsymbol{v}_t$ and the transport map P_t is not used. Actually, using Theorem 1.18(ii) and [72, Theorem 1.3], and arguing again as in the proof of [106, Theorem 5.1], the following more general statement holds (compare with [106, Theorem 5.1, Equations (27) and (29)]):

Proposition 9.10. *Let ρ_t and v_t be such that $0 < \lambda \leq \rho_t \leq \Lambda < \infty$, $v_t \in L^\infty_{\mathrm{loc}}(\mathbb{T}^2 \times [0,\infty), \mathbb{R}^2)$, and*

$$\partial_t \rho_t + \nabla \cdot (v_t \rho_t) = 0.$$

Assume that $\int_{\mathbb{T}^2} \rho_t\, dx = 1$ for all $t \geq 0$, let P_t be a convex function such that

$$(\nabla P_t)_\# \mathscr{L}_{\mathbb{T}^2} = \rho_t \mathscr{L}_{\mathbb{T}^2},$$

and denote by P_t^* its convex conjugate.

Then ∇P_t *and* ∇P_t^* *belong to* $W^{1,1}_{\mathrm{loc}}(\mathbb{T}^2 \times [0,\infty);\mathbb{R}^2)$. *Moreover, for every* $k \in \mathbb{N}$ *there exists a constant* $C(k)$ *such that, for almost every* $t \geq 0$,

$$\int_{\mathbb{T}^2} \rho_t |\partial_t \nabla P_t^*| \log_+^k(|\partial_t \nabla P_t^*|)\, dx \leq C(k) \left(\int_{\mathbb{T}^2} \rho_t |\nabla^2 P_t^*| \log_+^{2k}(|\nabla^2 P_t^*|)\, dx + \operatorname*{ess\,sup}_{\mathbb{T}^2} \left(\rho_t |v_t|^2\right) \int_{\mathbb{T}^2} |\nabla^2 P_t^*|\, dx \right), \tag{9.37}$$

$$\int_{\mathbb{T}^2} |\partial_t \nabla P_t| \log_+^k(|\partial_t \nabla P_t|)\, dx \leq C(k) \left(\int_{\mathbb{T}^2} |\nabla^2 P_t| \log_+^{2k}(|\nabla^2 P_t|)\, dx + \operatorname*{ess\,sup}_{\mathbb{T}^2} \left(\rho_t |v_t|^2\right) \int_{\mathbb{T}^2} |\nabla^2 P_t^*|\, dx \right). \tag{9.38}$$

Proof. We just give a short sketch of the proof. Equation (9.37) can be proved following the same line of the proof of Proposition 9.6. To prove (9.38) notice that by the approximation argument in the second step of the proof of Proposition 9.6 we can assume that the velocity and the density are smooth and hence, arguing as in Lemma 9.9, we have that $P_t,\ P_t^* \in \mathrm{Lip}_{\mathrm{loc}}([0,\infty), C^\infty(\mathbb{T}^2))$. Now, changing variables in the the left hand side of (9.30) we get

$$\int_{\mathbb{T}^2} \left| \left([\nabla^2 P_t^*](\nabla P_t) \right)^{-1/2} [\partial_t \nabla P_t^*](\nabla P_t) \right|^2 dx \leq \max_{\mathbb{T}^2} \left(\rho_t |v_t|^2\right) \int_{\mathbb{T}^2} |\nabla^2 P_t^*|\, dx. \tag{9.39}$$

Taking into account the identities

$$[\nabla^2 P_t^*](\nabla P_t) = \left(\nabla^2 P_t\right)^{-1} \text{ and } [\partial_t \nabla P_t^*](\nabla P_t) + [\nabla^2 P_t^*](\nabla P_t)\partial_t \nabla P_t = 0$$

which follow differentiating with respect to time and space $\nabla P_t^* \circ \nabla P_t = Id$, we can rewrite (9.39) as

$$\int_{\mathbb{T}^2} |(\nabla^2 P_t)^{-1/2} \partial_t \nabla P_t|^2\, dx \leq \max_{\mathbb{T}^2} \left(\rho_t |v_t|^2\right) \int_{\mathbb{T}^2} |\nabla^2 P_t^*|\, dx.$$

At this point the proof of (9.38) is obtained arguing as in Proposition 9.6. □

9.2. Existence of an Eulerian solution

In this section we prove Theorem 9.2.

Proof of Theorem 9.2. First of all notice that, thanks to Theorem 1.18(i) and Proposition 9.6, it holds $|\nabla^2 P_t^*|$, $|\partial_t \nabla P_t^*| \in L^\infty_{loc}([0,\infty), L^1(\mathbb{T}^2))$. Moreover, since $(\nabla P_t)_\# \mathscr{L}_{\mathbb{T}^2} = \rho_t \mathscr{L}_{\mathbb{T}^2}$, it is immediate to check that the function $\boldsymbol{u}$ in (9.6) is well-defined[4] and $|\boldsymbol{u}|$ belongs to $L^\infty_{loc}([0,\infty), L^1(\mathbb{T}^2))$.

Let $\phi \in C_c^\infty(\mathbb{R}^2 \times [0,\infty))$ be a $\mathbb{Z}^2$-periodic function in space and let us consider the function $\varphi : \mathbb{R}^2 \times [0,\infty) \to \mathbb{R}^2$ given by

$$\varphi_t(y) := J(y - \nabla P_t^*(y))\phi_t(\nabla P_t^*(y)). \tag{9.40}$$

By Theorem 1.17 and the periodicity of ϕ, $\varphi_t(y)$ is $\mathbb{Z}^2$-periodic in the space variable. Moreover φ_t is compactly supported in time, and Proposition 9.6 implies that $\varphi \in W^{1,1}(\mathbb{R}^2 \times [0,\infty))$. So, by Lemma 9.5, each component of the function $\varphi_t(y)$ is an admissible test function for (9.10). For later use, we write down explicitly the derivatives of φ:

$$\begin{aligned} \partial_t \varphi_t(y) = {} & -J[\partial_t \nabla P_t^*](y)\phi_t(\nabla P_t^*(y)) \\ & + J(y - \nabla P_t^*(y))[\partial_t \phi_t](\nabla P_t^*(y)) \\ & + J(y - \nabla P_t^*(y))\big([\nabla \phi_t](\nabla P_t^*(y)) \cdot \partial_t \nabla P_t^*(y)\big), \end{aligned} \tag{9.41}$$

$$\begin{aligned} \nabla \varphi_t(y) = {} & J(Id - \nabla^2 P_t^*(y))\phi_t(\nabla P_t^*(y)) \\ & + J(y - \nabla P_t^*(y)) \otimes \big([\nabla^T \phi_t](\nabla P_t^*(y))\nabla^2 P_t^*(y)\big). \end{aligned}$$

Taking into account that $(\nabla P_t)_\# \mathscr{L}_{\mathbb{T}^2} = \rho_t \mathscr{L}_{\mathbb{T}^2}$ and that $[\nabla P_t^*](\nabla P_t(x)) = x$ almost everywhere, we can rewrite the boundary term in (9.10) as

$$\begin{aligned} \int_{\mathbb{T}^2} \varphi_0(y)\rho_0(y)\,dy &= \int_{\mathbb{T}^2} J(\nabla P_0(x) - x)\phi_0(x)\,dx \\ &= \int_{R^2} J\nabla p_0(x)\phi_0(x)\,dx. \end{aligned} \tag{9.42}$$

[4] Note that the composition of $\nabla^2 P_t^*$ with ∇P_t makes sense. Indeed, by the conditions $(\nabla P_t)_\# \mathscr{L}_{\mathbb{T}^2} = \rho_t \mathscr{L}_{\mathbb{T}^2} \ll \mathscr{L}_{\mathbb{T}^2}$, if we change the value of $\nabla^2 P_t^*$ in a set of measure zero, also $[\nabla^2 P_t^*](\nabla P_t)$ will change only on a set of measure zero.

In the same way, since $\boldsymbol{v}_t(y) = J(y - \nabla P_t^*(y))$, we can use (9.41) to rewrite the other term as

$$\begin{aligned}
&\int_0^\infty \int_{\mathbb{T}^2} \Big\{\partial_t \varphi_t(y) + \nabla\varphi_t(y)\cdot \boldsymbol{v}_t(y)\Big\}\rho_t(y)\,dy\,dt \\
&= \int_0^\infty \int_{\mathbb{T}^2} \Big\{ - J[\partial_t \nabla P_t^*](\nabla P_t(x))\phi_t(x) + J(\nabla P_t(x) - x)\partial_t\phi_t(x) \\
&\quad + J(\nabla P_t(x) - x)\big(\nabla\phi_t(x)\cdot[\partial_t \nabla P_t^*](\nabla P_t(x))\big) \\
&\quad + J(Id - \nabla^2 P_t^*(\nabla P_t(x)))\phi_t(x) J(\nabla P_t(x) - x) \\
&\quad + J(\nabla P_t(x) - x)\otimes\big(\nabla^T\phi_t(x)\nabla^2 P_t^*(\nabla P_t(x))\big)J(\nabla P_t(x) - x)\Big\}dx\,dt
\end{aligned} \tag{9.43}$$

which, taking into account the formula (9.6) for $\boldsymbol{u}$, after rearranging the terms turns out to be equal to

$$\begin{aligned}
\int_0^\infty \int_{\mathbb{T}^2} &\{J\nabla p_t(x)\big(\partial_t\phi_t(x) + \boldsymbol{u}_t(x)\cdot\nabla\phi_t(x)\big) \\
&+\big(-\nabla p_t(x) - J\boldsymbol{u}_t(x)\big)\phi_t(x)\}\;dx\,dt.
\end{aligned} \tag{9.44}$$

Hence, combining (9.42), (9.43), (9.44), and (9.10), we obtain the validity of (9.7).

Now we prove (9.8). Given $\phi \in C_c^\infty(0,\infty)$ and a $\mathbb{Z}^2$-periodic function $\psi \in C^\infty(\mathbb{R}^2)$, let us consider the function $\varphi : \mathbb{R}^2 \times [0,\infty) \to \mathbb{R}$ defined by

$$\varphi_t(y) := \phi(t)\psi(\nabla P_t{}^*(y)). \tag{9.45}$$

As in the previous case, we have that φ is $\mathbb{Z}^2$-periodic in the space variable and $\varphi \in W^{1,1}(\mathbb{T}^2 \times [0,\infty))$, so we can use φ as a test function in (9.8). Then, identities analogous to (9.41) yield

$$\begin{aligned}
0 &= \int_0^\infty \int_{\mathbb{T}^2} \{\partial_t\varphi_t(y) + \nabla\varphi_t(y)\cdot\boldsymbol{v}_t(y)\}\,\rho_t(y)\,dy\,dt \\
&= \int_0^\infty \phi'(t)\int_{\mathbb{T}^2}\psi(x)\,dx\,dt + \int_0^\infty \phi(t)\int_{\mathbb{T}^2}\Big\{\nabla\psi(x)\cdot\partial_t\nabla P_t{}^*(\nabla P_t(x)) \\
&\qquad + \nabla^T\psi(x)\nabla^2 P_t^*(\nabla P_t(x))J(\nabla P_t(x) - x)\Big\}\,dx\,dt \\
&= \int_0^\infty \phi(t)\int_{\mathbb{T}^2}\nabla\psi(x)\cdot\boldsymbol{u}_t(x)\,dx\,dt.
\end{aligned}$$

Since ϕ is arbitrary we obtain

$$\int_{\mathbb{T}^2}\nabla\psi(x)\cdot\boldsymbol{u}_t(x)\,dx = 0 \qquad \text{for } \mathscr{L}^1\text{-a.e. } t > 0.$$

By a standard density argument it follows that the above equation holds outside a negligible set of times independent of the test function ψ, thus proving (9.8). □

9.3. Existence of a regular Lagrangian flow for the semigeostrophic velocity

We recall the notion of regular Lagrangian flow of a Borel vector field on the 2-dimensional torus, introduced in Definition 1.4 in $\mathbb{R}^d$; as observed in Section 1.1, this definition does not require any regularity of $\boldsymbol{b}$ and, by Fubini's theorem, it does not depend on the choice of the representative of $\boldsymbol{b}$ in the Lebesgue equivalence class.

Definition 9.11. Given a Borel, locally integrable vector field $\boldsymbol{b} : \mathbb{T}^2 \times (0, \infty) \to \mathbb{R}^2$, we say that a Borel function $\boldsymbol{X} : \mathbb{T}^2 \times [0, \infty) \to \mathbb{T}^2$ is a *regular Lagrangian flow associated to* $\boldsymbol{b}$ if the following two conditions are satisfied.

(i) For almost every $x \in \mathbb{T}^2$ the map $t \mapsto \boldsymbol{X}(\cdot, x)$ is locally absolutely continuous in $[0, \infty)$ and

$$\boldsymbol{X}(t, x) = x + \int_0^t \boldsymbol{b}_s(\boldsymbol{X}(s, x))dx \qquad \forall t \in [0, \infty). \tag{9.46}$$

(ii) For every $t \in [0, \infty)$ it holds $\boldsymbol{X}(t, \cdot)_\# \mathscr{L}_{\mathbb{T}^2} \le C \mathscr{L}_{\mathbb{T}^2}$, with $C \in [0, \infty)$ independent of t.

A particular class of regular Lagrangian flows is the collection of the measure-preserving ones, where (ii) is strengthened to

$$\boldsymbol{X}(t, \cdot)_\# \mathscr{L}_{\mathbb{T}^2} = \mathscr{L}_{\mathbb{T}^2} \qquad \forall t \ge 0.$$

We show existence of a measure-preserving regular Lagrangian flow associated to the vector field $\boldsymbol{u}$ defined by

$$\boldsymbol{u}_t(x) = [\partial_t \nabla P_t^*](\nabla P_t(x)) + [\nabla^2 P_t^*](\nabla P_t(x)) J(\nabla P_t(x) - x), \tag{9.47}$$

where P_t and P_t^* are as in Theorem 9.2. Recall also that, under these assumptions, $|\boldsymbol{u}| \in L^\infty_{\text{loc}}([0, \infty), L^1(\mathbb{T}^2))$.

Existence for a weaker notion of Lagrangian flow of the semigeostrophic equations was proved by Cullen and Feldman, see [69, Definition 2.4], but since at that time the results of [71] were not available the velocity could not be defined, not even as a function. Hence, they had to adopt a more indirect definition. We shall prove indeed that their flow is a flow according to Definition 9.11. We discuss the uniqueness issue in the last section.

Theorem 9.12. *Let us assume that the hypotheses of Theorem 9.2 are satisfied, and let P_t and P_t^* be the convex functions such that*

$$(\nabla P_t)_{\#}\mathscr{L}_{\mathbb{T}^2} = \rho_t \mathscr{L}_{\mathbb{T}^2}, \qquad (\nabla P_t^*)_{\#}\rho_t \mathscr{L}_{\mathbb{T}^2} = \mathscr{L}_{\mathbb{T}^2}.$$

Then, for $\boldsymbol{u}_t$ given by (9.47) there exists a measure-preserving regular Lagrangian flow $\boldsymbol{X}$ associated to $\boldsymbol{u}_t$. Moreover $\boldsymbol{X}$ is invertible in the sense that for all $t \geq 0$ there exist Borel maps $\boldsymbol{X}^{-1}(t,\cdot)$ such that $\boldsymbol{X}^{-1}(t, \boldsymbol{X}(t,x)) = x$ and $\boldsymbol{X}(t, \boldsymbol{X}^{-1}(t,x)) = x$ for $\mathscr{L}^2$-a.e. $x \in \mathbb{T}^2$.

Proof. Let us consider the velocity field in the dual variables $\boldsymbol{v}_t(x) = J(x - \nabla P_t^*(x))$. Since P_t^* is convex, $\boldsymbol{v}_t \in BV(\mathbb{T}^2;\mathbb{R}^2)$ uniformly in time (actually, by Theorem 1.18(ii) $\boldsymbol{v}_t \in W^{1,1}(\mathbb{T}^2;\mathbb{R}^2)$). Moreover $\boldsymbol{v}_t$ is divergence-free. Hence, by the theory of regular Lagrangian flows associated to BV vector fields of Theorem 1.5 (notice that, since we are on the torus, no growth conditions are required and trajectories cannot blow up), there exists a unique measure-preserving regular Lagrangian flow $\boldsymbol{Y} : \mathbb{T}^2 \times [0,\infty) \to \mathbb{T}^2$ associated to $\boldsymbol{v}$.

We now define (see also Figure 9.1)[5]

$$\boldsymbol{X}(t,x) := \nabla P_t^*(\boldsymbol{Y}(t, \nabla P_0(x))). \tag{9.48}$$

The validity of property (b) in Definition 9.11 and the invertibility of $\boldsymbol{X}$ follow from the same arguments of [69, Propositions 2.14 and 2.17]. Hence we only have to show that property (a) in Definition 9.11 holds.

Let us define $Q^n := B * \sigma^n$, where B is a Sobolev and uniformly continuous extension of ∇P^* to $\mathbb{T}^2 \times \mathbb{R}$, and σ^n is a standard family of mollifiers in $\mathbb{T}^2 \times \mathbb{R}$. It is well known that $Q^n \to \nabla P^*$ locally uniformly and in the strong topology of $W^{1,1}_{\mathrm{loc}}(\mathbb{T}^2 \times [0,\infty))$. Thus, using the measure-preserving property of $\boldsymbol{Y}(t,\cdot)$, for all $T > 0$ we get

$$\begin{aligned}
0 &= \lim_{n\to\infty} \int_{\mathbb{T}^2}\int_0^T \Big\{ |Q_t^n - \nabla P_t^*| + |\partial_t Q_t^n - \partial_t \nabla P_t^*| + |\nabla Q_t^n - \nabla^2 P_t^*| \Big\}\, dy\, dt \\
&= \lim_{n\to\infty} \int_{\mathbb{T}^2}\int_0^T \Big\{ |Q_t^n(\boldsymbol{Y}(t,\cdot) - \nabla P_t^*(\boldsymbol{Y}(t,\cdot))| + |[\partial_t Q_t^n](\boldsymbol{Y}(t,\cdot)) \\
&\qquad - [\partial_t \nabla P_t^*](\boldsymbol{Y}(t,\cdot))| \\
&\qquad + |[\nabla Q_t^n](\boldsymbol{Y}(t,\cdot)) - [\nabla^2 P_t^*](\boldsymbol{Y}(t,\cdot))| \Big\}\, dx\, dt.
\end{aligned}$$

[5] Observe that the definition of $\boldsymbol{X}$ makes sense. Indeed, by Theorem 1.18(i), both maps ∇P_0 and ∇P_t^* are Hölder continuous in space. Morever, by the weak continuity in time of $t \mapsto \rho_t$ (Theorem 9.4(ii)) and the stability results for Alexandrov solutions of Monge-Ampère, ∇P^* is continuous both in space and time. Finally, since $(\nabla P_0)_{\#}\mathscr{L}_{\mathbb{T}^2} \ll \mathscr{L}_{\mathbb{T}^2}$, if we change the value of $\boldsymbol{Y}$ in a set of measure zero, also $\boldsymbol{X}$ will change only on a set of measure zero.

Up to a (not re-labeled) subsequence the previous convergence is pointwise in space, namely, for almost every $x \in \mathbb{T}^2$,

$$\lim_{n\to\infty} \int_0^T \Big\{ |Q_t^n(\boldsymbol{Y}(t,x)) - \nabla P_t^*(\boldsymbol{Y}(t,x))| + |[\partial_t Q_t^n](\boldsymbol{Y}(t,x)) - [\partial_t \nabla P_t^*](\boldsymbol{Y}(t,x))| + |[\nabla Q_t^n](\boldsymbol{Y}(t,x)) - [\nabla^2 P_t^*](\boldsymbol{Y}(t,x))| \Big\}\, dt = 0. \tag{9.49}$$

Hence, since $\boldsymbol{Y}$ is a regular Lagrangian flow and by assumption

$$(\nabla P_0)\mathscr{L}_{\mathbb{T}^2} \ll \mathscr{L}_{\mathbb{T}^2},$$

for almost every y we have that (9.49) holds at $x = \nabla P_0(y)$, and the function $t \mapsto \boldsymbol{Y}(t,x)$ is absolutely continuous on $[0,T]$, with derivative given by

$$\frac{d}{dt}\boldsymbol{Y}(t,x) = \boldsymbol{v}_t(\boldsymbol{Y}(t,x)) = J(\boldsymbol{Y}(t,x) - \nabla P_t^*(\boldsymbol{Y}(t,x))) \text{ for } \mathscr{L}^1\text{-a.e. } t \in [0,T].$$

Let us fix such an y. Since Q^n is smooth, the function $Q_t^n(\boldsymbol{Y}(t,x))$ is absolutely continuous in $[0,T]$ and its time derivative is given by

$$\frac{d}{dt}\big(Q_t^n(\boldsymbol{Y}(t,x))\big) = [\partial_t Q_t^n](\boldsymbol{Y}(t,x)) + [\nabla Q_t^n](\boldsymbol{Y}(t,x))J(\boldsymbol{Y}(t,x) - \nabla P_t^*(\boldsymbol{Y}(t,x))).$$

Hence, since $J(\boldsymbol{Y}(t,x) - \nabla P_t^*(\boldsymbol{Y}(t,x))) = \boldsymbol{v}_t(\boldsymbol{Y}(t,x))$ is uniformly bounded, from (9.49) we get

$$\lim_{n\to\infty} \frac{d}{dt}\big(Q_t^n(\boldsymbol{Y}(t,x))\big) = [\partial_t \nabla P_t^*](\boldsymbol{Y}(t,x)) + [\nabla^2 P_t^*](\boldsymbol{Y}(t,x))J(\boldsymbol{Y}(t,x) - \nabla P_t^*(\boldsymbol{Y}(t,x))) := \boldsymbol{w}_t(y) \qquad \text{in } L^1(0,T). \tag{9.50}$$

Recalling that

$$\lim_{n\to\infty} Q_t^n(\boldsymbol{Y}(t,x)) = \nabla P_t^*(\boldsymbol{Y}(t,x)) = \boldsymbol{X}(t,y) \qquad \forall t \in [0,T],$$

we infer that $\boldsymbol{X}(t,y)$ is absolutely continuous in $[0,T]$ (being the limit in $W^{1,1}(0,T)$ of absolutely continuous maps). Moreover, by taking the limit as $n \to \infty$ in the identity

$$Q_t^n(\boldsymbol{Y}(t,x)) = Q_0^n(\boldsymbol{Y}(0,x)) + \int_0^t \frac{d}{ds}\big(Q_s^n(\boldsymbol{Y}(s,x))\big)\, ds,$$

thanks to (9.50) we get

$$\boldsymbol{X}(t, y) = \boldsymbol{X}(0, y) + \int_0^t \boldsymbol{w}_s(y)\, ds. \tag{9.51}$$

To obtain (9.46) we only need to show that $\boldsymbol{w}_t(y) = \boldsymbol{u}_t(\boldsymbol{X}(t, y))$, which follows at once from (9.47), (9.48), and (9.50). □

Acknowledgements

I am deeply grateful to my advisors, Luigi Ambrosio and Alessio Figalli. They introduced me to their vision of the subject and, at the same time, encouraged me to search and develop my own point of view. As soon as they were not in meetings, the doors of their offices were open and they were available to discuss and spend time with my questions. If I have the possibility to teach mathematics to somebody in the future, I wish to be able to be a teacher as they have been for me.

During my PhD, I had the opportunity to spend several months at the University of Texas at Austin. I want to thank in particular, besides Alessio Figalli who invited me, Luis Caffarelli, Francesco Maggi, Filippo Cagnetti, Giuseppe Mingione (who visited the UT math department while I was there), and the analysis group. I also visited for shorter periods the Mathematical Sciences Research Institute in Berkeley, the University of Basel, invited by Gianluca Crippa, and the ETH in Zürich, upon invitation of Tristan Rivière. Each of these visits was an opportunity to learn, discover, ask and dialogue about new unexpected aspects of interesting problems.

In the last three years, while working at the Scuola Normale Superiore, I was lucky to share many classes, days of study, thoughts and ideas with other students. My gratitude goes in particular to Guido De Philippis, for many cigarettes that we smoke together (of course I don't smoke, but we took this occasion to talk of beautiful theorems and ideas!) and his patience and help when I approached for the first time the optimal transport and regularity theory. I also thank the other (present and former) PhD students at the Scuola Normale Superiore, among whom there are Paolo Baroni, Simone Di Marino, and Federico Stra, and the PhD students at UT Austin, in particular Javier Morales, Robin Neumayer, Cornelia Mihaila, and Yash Jhavieri. Finally, I wish to thank all the colleagues, professors, and friends that taught and shared with me so much about mathematics. Thanks!

References

[1] A. ABBONDANDOLO and A. FIGALLI, *High action orbits for Tonelli Lagrangians and superlinear Hamiltonians on compact configuration spaces*, J. Differential Equations **234** (2007), 626–653.

[2] E. ACERBI and N. FUSCO, *A regularity theorem for minimizers of quasiconvex integrals*, Arch. Rational Mech. Anal. **99** (1987), 261–281.

[3] E. ACERBI and N. FUSCO, *Local regularity for minimizers of non-convex integrals*, Ann. Scuola Norm. Sup. Pisa Cl. Sci. **16** (1989), 603–636.

[4] G. ALBERTI, S. BIANCHINI and G. CRIPPA, *A uniqueness result for the continuity equation in two dimensions*, J. Eur. Math. Soc. **16** (2014), 201–234.

[5] L. AMBROSIO, *Transport equation and Cauchy problem for BV vector fields*, Invent. Math. **158** (2004), 227–260.

[6] L. AMBROSIO, *Transport equation and Cauchy problem for non-smooth vector fields*, In: "Calculus of Variations and Non-Linear Partial Differential Equations" (CIME Series, Cetraro, 2005), B. Dacorogna, P. Marcellini (eds.), Lecture Notes in Mathematics, Vol. 1927, 2008, 2–41.

[7] L. AMBROSIO, M. COLOMBO, G. DE PHILIPPIS and A. FIGALLI, *Existence of Eulerian solutions to the semigeostrophic equations in physical space: the 2-dimensional periodic case*, Comm. Partial Differential Equations **37** (2012), 2209–2227.

[8] L. AMBROSIO, M. COLOMBO, G. DE PHILIPPIS and A. FIGALLI, *A global existence result for the semigeostrophic equations in three dimensional convex domains*, special issue on Optimal Transport, Discr. Cont. Dyn. Sys. **34** (2014), 1251–1268.

[9] L. AMBROSIO, M. COLOMBO and S. DI MARINO, *Sobolev spaces in metric measure spaces: reflexivity and lower semicon-*

tinuity of slope, Advanced Studies in Pure Mathematics **67** (2015), Variational Methods for Evolving Objects, 1–58.

[10] L. AMBROSIO, M. COLOMBO and A. FIGALLI, *Existence and uniqueness of Maximal Regular Flows for non-smooth vector fields*, Arch. Ration. Mech. Anal. **218** (2015), 1043–1081.

[11] L. AMBROSIO, M. COLOMBO and A. FIGALLI, *On the Lagrangian structure of transport equations: the Vlasov-Poisson system*, preprint (2014).

[12] L. AMBROSIO and G. CRIPPA, Existence, uniqueness, stability and differentiability properties of the flow associated to weakly differentiable vector fields, Lecture Notes of the Unione Matematica Italiana **5** (2008), 3–54.

[13] L. AMBROSIO and G. CRIPPA, Continuity equations and ODE flows with non-smooth velocity, Lecture Notes of a course given at Heriott-Watt University, Edinburgh. Proceeding of the Royal Society of Edinburgh, Section A: Mathematics, **144** (2014), 1191–1244.

[14] L. AMBROSIO, G. CRIPPA and S. MANIGLIA, *Traces and fine properties of a BD class of vector fields and applications*, Ann. Sci. Toulouse **14** (2005), 527–561.

[15] L. AMBROSIO, C. DE LELLIS and J. MALÝ, *On the chain rule for the divergence of BV like vector fields: applications, partial results, open problems*, In: "Perspectives in Nonlinear Partial Differential Equations: in honour of Haim Brezis", Contemp. Math. **446** (2007), 31–67.

[16] L. AMBROSIO and A. FIGALLI, *On flows associated to Sobolev vector fields in Wiener spaces: an approach à la DiPerna-Lions*, J. Funct. Anal. **256** (2009), 179–214.

[17] L. AMBROSIO, N. FUSCO and D. PALLARA, "Functions of Bounded Variation and Free Discontinuity Problems", Oxford Mathematical Monographs. The Clarendon Press, Oxford University Press (2000).

[18] L. AMBROSIO and N. GIGLI, *A user's guide to optimal transport*, In: "Modelling and Optimisation of Flows on Networks", Lecture Notes in Math., Springer, Heidelberg **2061** (2013), 1–155.

[19] L. AMBROSIO, N. GIGLI and G. SAVARÉ, "Gradient Flows in Metric Spaces and in the Wasserstein Space of Probability Measures", Lectures in Mathematics, ETH Zurich, Birkhäuser, 2005, second edition in 2008.

[20] L. AMBROSIO, N. GIGLI and G. SAVARÉ, *Calculus and heat flow in metric measure spaces and applications to spaces with Ricci bounds from below*, Invent. Math. **195** (2014), 289–391.

[21] L. AMBROSIO, N. GIGLI and G. SAVARÉ, *Density of Lipschitz functions and equivalence of weak gradients in metric measure spaces*, Rev. Mat. Iberoam. **29** (2013), 969–996.

[22] L. AMBROSIO, M. LECUMBERRY and S. MANIGLIA, *Lipschitz regularity and approximate differentiability of the DiPerna-Lions flow*, Rend. Sem. Mat. Univ. Padova **114** (2005), 29–50.

[23] L. AMBROSIO and J. MALÝ, *Very weak notions of differentiability*, Proc. Roy. Soc. Edinburgh Sect. A **137** (2007), 447–455.

[24] G. ANZELLOTTI and M. GIAQUINTA, *Convex functionals and partial regularity*, Arch. Rational Mech. Anal. **102** (1988), 243–272.

[25] A. A. ARSEN'EV, *Existence in the large of a weak solution of Vlasov's system of equations*, Ž Vyčisl. Mat. i Mat. Fiz. **15** (1975), 136–147, 276.

[26] C. BARDOS and P. DEGOND, *Global existence for the Vlasov-Poisson equation in 3 space variables with small initial data*, Ann. Inst. H. Poincaré Anal. Non Linéaire **2** (1985), 101–118.

[27] P. BARONI, M. COLOMBO and G. MINGIONE, *Non-autonomous functionals, borderline cases and related function classes*, St. Petersburg Math. J. **27** (2016), 347–379.

[28] P. BARONI, M. COLOMBO and G. MINGIONE, *Harnack inequalities for double phase functionals*, Nonlin. Anal. **121** (2015), special issue in honor of Enzo Mitidieri for his 60th birthday, 206–222.

[29] J. BATT, *Global symmetric solutions of the initial value problem of stellar dynamics*, J. Differential Equations **25** (1977), 342–364.

[30] M. BECKMANN, *A continuous model of transportation*, Econometrica **20** (1952), 643–660.

[31] J.-D. BENAMOU and Y. BRENIER, *Weak existence for the semigeostrophic equation formulated as a coupled Monge-Ampère/transport problem*, SIAM J. Appl. Math. **58** (1998), 1450–1461.

[32] A. BOHUN, F. BOUCHUT and G. CRIPPA, *Lagrangian flows for vector fields with anisotropic regularity*, Ann. Inst. H. Poincaré Anal. Non Linéaire **33** (2016), 1409–1429.

[33] A. BOHUN, F. BOUCHUT and G. CRIPPA, *Lagrangian solutions to the Vasov-Poisson system with L^1 density*, J. Differential Equations **260** (2016), 3576–3597.

[34] F. BOUCHUT, *Renormalized solutions to the Vlasov equation with coefficients of bounded variation*, Arch. Rational Mech. Anal. **157** (2001), 75–90.

[35] F. BOUCHUT and G. CRIPPA, *Équations de transport à coefficient dont le gradient est donné par une intégrale singulière, (French)*

[Transport equations with a coefficient whose gradient is given by a singular integral], Séminaire: Équations aux Dérivées Partielles. 2007–2008, Exp. No. I, Sémin. Équ. Dériv. Partielles, École Polytech., Palaiseau (2009).

[36] F. BOUCHUT and G. CRIPPA, *Lagrangian flows for vector fields with gradient given by a singular integral*, J. Hyperbolic Differ. Equ. **10** (2013), 235–282.

[37] L. BRASCO, *Global L^∞ gradient estimates for solutions to a certain degenerate elliptic equation*, Nonlinear Anal. **74** (2011), 516–531.

[38] L. BRASCO, G. CARLIER and F. SANTAMBROGIO, *Congested traffic dynamics, weak flows and very degenerate elliptic equations*, J. Math. Pures Appl. **93** (2010), 652–671.

[39] J. E. BROTHERS and W. P. ZIEMER, *Minimal rearrangements of Sobolev functions*, J. Reine Angew. Math. **384** (1988), 153–179.

[40] G. BUTTAZZO, L. DE PASCALE and P. GORI-GIORGI, *Optimal transport formulation of electronic density-functional theory*, Phys. Rev. A **85** (2012), 062502.

[41] L. CAFFARELLI, *A localization property of viscosity solutions to the Monge-Ampère equation and their strict convexity*, Ann. of Math. **131** (1990), 129–134.

[42] L. CAFFARELLI, *Interior $W^{2,p}$ estimates for solutions of the Monge-Ampère equation*, Ann. of Math. **131** (1990), 135–150.

[43] L. CAFFARELLI, *Some regularity properties of solutions to Monge-Ampére equations*, Comm. Pure Appl. Math. **44** (1991), 965–969.

[44] L. CAFFARELLI and X. CABRÉ, *Fully nonlinear elliptic equations*, American Mathematical Society Colloquium Publications **43**, American Mathematical Society (1995).

[45] L. CAFFARELLI, L. NIRENBERG and J. SPRUCK, *The Dirichlet problem for the degenerate Monge-Ampère equation*, Rev. Mat. Iberoamericana **2** (1986), 19–27.

[46] E. CAGLIOTI, S. CAPRINO, C. MARCHIORO and M. PULVIRENTI, *The Vlasov equation with infinite mass*, Arch. Ration. Mech. Anal. **159** (2001), 85–108.

[47] F. CAGNETTI, M. COLOMBO, G. DE PHILIPPIS and F. MAGGI, *Essential connectedness and the rigidity problem for Gaussian symmetrization*, J. Eur. Math. Soc. (JEMS) **19** (2017), 395–439.

[48] F. CAGNETTI, M. COLOMBO, G. DE PHILIPPIS and F. MAGGI, *Rigidity of equality cases in Steiner's perimeter inequality*, Anal. PDE **7** (2015), 1535–1593.

[49] G. CARLIER, C. JIMENEZ and F. SANTAMBROGIO, *Optimal transportation with traffic congestion and Wardrop equilibria*, SIAM J. Control Optim. **47** (2008), 1330–1350.

[50] M. CHLEBIK, A. CIANCHI, and N. FUSCO, *The perimeter inequality under Steiner symmetrization: cases of equality*, Ann. of Math. **162** (2005), 525–555.

[51] F. COLOMBINI and N. LERNER, *Uniqueness of continuous solutions for BV vector fields*, Duke Math. J. **111** (2002), 357–384.

[52] F. COLOMBINI and N. LERNER, *Uniqueness of L^∞ solutions for a class of conormal BV vector fields*, Contemp. Math. **368** (2005), 133–156.

[53] M. COLOMBO and S. DI MARINO, *Equality between Monge and Kantorovich multimarginal problems with Coulomb cost*, Ann. Mat. Pura Appl. **194** (2015), 307–320.

[54] M. COLOMBO, S. DI MARINO and L. DE PASCALE, *Multimarginal optimal transport maps for 1-dimensional repulsive costs*, Canad. J. Math. **67** (2015), 350–368.

[55] M. COLOMBO, A. FIGALLI and Y. JHAVERI, *Lipschitz Changes of Variables between Perturbations of Log-concave Measures*, submitted paper, 2015.

[56] M. COLOMBO, G. CRIPPA and S. SPIRITO, *Renormalized solutions to the continuity equation with an integrable damping term*, Calc. Var. Partial Differential Equations **54** (2015), 1831–1845.

[57] M. COLOMBO, G. CRIPPA and S. SPIRITO, *Logarithmic estimates for continuity equations*, Networks and Heterogeneous Media **16** (2016), 301–311.

[58] M. COLOMBO and A. FIGALLI, *Regularity results for very degenerate elliptic equations*, J. Math. Pures Appl. **101** (2014), 94–117.

[59] M. COLOMBO and A. FIGALLI, *An excess-decay result for a class of degenerate elliptic equations*, Discr. Cont. Dyn. Sys. Ser. S **7** (2014), 631–652.

[60] M. COLOMBO and E. INDREI, *Obstructions to regularity in the classical Monge problem*, Math. Res. Lett. **21** (2014), 697–712.

[61] M. COLOMBO and G. MINGIONE, *Regularity for double phase variational problems*, Arch. Ration. Mech. Anal. **215** (2015), 443–496.

[62] M. COLOMBO and G. MINGIONE, *Bounded minimisers of double phase variational integrals*, Arch. Ration. Mech. Anal. **218** (2015), 219–273.

[63] M. COLOMBO and G. MINGIONE, *Calderón-Zygmund estimates and non-uniformly elliptic operators*, J. Funct. Anal. **270** (2015), 1416–1478.

[64] M. COLOMBO and F. STRA, *Counterexamples in multimarginal optimal transport with Coulomb cost and radially symmetric data*, Math. Models Methods Appl. Sci. **26** (2016), 1025–1049.

[65] D. CORDERO ERAUSQUIN, *Sur le transport de mesures périodiques*, C. R. Acad. Sci. Paris Sér. I Math. **329** (1999), 199–202.

[66] C. COTAR, G. FRIESECKE and C. KLÜPPELBERG, *Density functional theory and optimal transportation with Coulomb cost*, Comm. Pure Appl. Math. **66** (2013), 548–599.

[67] G. CRIPPA and C. DE LELLIS, *Estimates for transport equations and regularity of the DiPerna-Lions flow*, J. Reine Angew. Math. **616** (2008), 15–46.

[68] M. CULLEN, "A Mathematical Theory of Large-scale Atmosphere/ocean Flow", Imperial College Press, 2006.

[69] M. CULLEN and M. FELDMAN, *Lagrangian solutions of semigeostrophic equations in physical space*, SIAM J. Math. Anal. **37** (2006), 1371–1395.

[70] N. DEPAUW, *Non unicité des solutions bornées pour un champ de vecteurs BV en dehors d'un hyperplan*, C.R. Math. Sci. Acad. Paris **337** (2003), 249–252.

[71] G. DE PHILIPPIS and A. FIGALLI, *$W^{2,1}$ regularity for solutions of the Monge-Ampère equation*, Invent. Math. **192** (2013), 55–69.

[72] G. DE PHILIPPIS and A.FIGALLI, *Second order stability for the Monge-Ampère equation and strong Sobolev convergence of optimal transport maps*, Anal. PDE **6** (2013), 993–1000.

[73] G. DE PHILIPPIS and A. FIGALLI, *Optimal regularity of the convex envelope*, Trans. Amer. Math. Soc. **367** (2015), 4407–4422.

[74] D. DE SILVA and O. SAVIN, *Minimizers of convex functionals arising in random surfaces*, Duke Math. J. **151** (2010), 487–532.

[75] E. DIBENEDETTO, *$C^{1+\alpha}$ local regularity of weak solutions of degenerate elliptic equations*, Nonlinear Anal. **7** (1983), 827–850.

[76] E. DIBENEDETTO and N.S. TRUDINGER, *Harnack inequalities for quasiminima of variational integrals*, Ann. Inst. H. Poincaré Anal. Non Linéaire **1** (1984), 295–308.

[77] E. DIBENEDETTO and V. VESPRI, *On the singular equation $\beta(u)_t = \Delta u$*, Arch. Rational Mech. Anal. **132** (1995), 247–309.

[78] R. J. DIPERNA and P.-L. LIONS, *Solutions globales d'équations du type Vlasov-Poisson* (French), [Global solutions of Vlasov-Poisson type equations] C. R. Acad. Sci. Paris Sér. I Math. **307** (1988), 655–658.

[79] R. J. DIPERNA and P.-L. LIONS, *Global weak solutions of kinetic equations*, Rend. Sem. Mat. Univ. Politec. Torino **46** (1988), 259–288.

[80] R. J. DiPerna and P.-L. Lions, *Global weak solutions of Vlasov-Maxwell systems*, Comm. Pure Appl. Math. **42** (1989), 729–757.

[81] R. J. DiPerna and P.-L. Lions, *Ordinary differential equations, transport theory and Sobolev spaces*, Invent. Math. **98** (1989), 511–547.

[82] R. L. Dobrushin, *Vlasov Equations*, Funktsional. Anal. i Prilozhen. **13** (1979), 48–58.

[83] L. Esposito, F. Leonetti and G. Mingione, *Sharp regularity for functionals with (p, q) growth*, J. Differential Equations **204** (2004), 5–55.

[84] L. Esposito, G. Mingione and C. Trombetti, *On the Lipschitz regularity for certain elliptic problems*, Forum Math. **18** (2006), 263–292.

[85] L. C. Evans, *A new proof of local $C^{1,\alpha}$ regularity for solutions of certain degenerate elliptic p.d.e*, J. Differential Equations **45** (1982), 356–373.

[86] H. Federer, "Geometric Measure Theory", Die Grundlehren der mathematischen Wissenschaften, **153**, Springer-Verlag New York Inc., New York, 1969.

[87] I. Fonseca, N. Fusco and P. Marcellini, *An existence result for a nonconvex variational problem via regularity*, ESAIM Control Optim. Calc. Var. **7** (2002), 69–95.

[88] P. Gérard, *Moyennes de solutions d'equations aux derivees partielles*, French, [Means of solutions of partial differential equations] Séminaire sur les équations aux dérivées partielles 1986–1987, Exp. No. XI (1987), École Polytech., Palaiseau.

[89] M. Giaquinta, "Multiple Integrals in the Calculus of Variations and Nonlinear Elliptic Systems", Princeton Univ. Press, Princeton, 1983.

[90] M. Giaquinta and G. Modica, *Partial regularity of minimizers of quasiconvex integrals*, Ann. Inst. H. Poincaré Analyse non linéaire **3** (1986), 185–208.

[91] D. Gilbarg and N. S. Trudinger, *Elliptic partial differential equations of second order*, Reprint of the 1998 edition. Classics in Mathematics. Springer-Verlag, Berlin (2001).

[92] F. Golse, P.-L. Lions, B. Perthame and R. Sentis, *Regularity of the moments of the solution of a transport equation*, J. Funct. Anal. **76** (1988), 110–125.

[93] F. Golse, B. Perthame and R. Sentis, *Un résultat de compacité pour les Équations de transport et application au calcul de la limite de la valeur propre principale d'un opèrateur de trans-*

port, French, [A compactness result for transport equations and application to the calculation of the limit of the principal eigenvalue of a transport operator] C. R. Acad. Sci. Paris Sér. I Math. **301** (1985), 341–344.

[94] C. GUTIERREZ, *The Monge-Ampére equation*, Progress in Nonlinear Differential Equations and their Applications, **44**, Birkhäuser Boston, MA, 2001.

[95] E. HORST, *On the classical solutions of the initial value problem for the unmodified nonlinear Vlasov equation. I. General theory. II: Special cases*, Math. Methods Appl. Sci. **3** (1981), 229–248, **4** (1982), 19–32.

[96] E. HORST and R. HUNZE, *Weak solutions of the initial value problem for the unmodified nonlinear Vlasov equation*, Math. Methods Appl. Sci. **6** (1984), 262–279.

[97] R. ILLNER and H. NEUNZERT, *An existence theorem for the unmodified Vlasov equation*, Math. Methods Appl. Sci. **1** (1979), 530–544.

[98] C. IMBERT and L. SILVESTRE, *Estimates on elliptic equations that hold only where the gradient is large*, J. Eur. Math. Soc. (JEMS) **18** (2016), 1321–1338.

[99] S. V. IORDANSKII, *The Cauchy problem for the kinetic equation of plasma*, Trudy Mat. Inst. Steklov. **60** (1961), 181–194.

[100] J. KRISTENSEN and G. MINGIONE, *The singular set of minima of integral functionals*, Arch. Ration. Mech. Anal. **180** (2006), 331–398.

[101] C. LE BRIS and P.-L. LIONS, *Renormalized solutions of some transport equations with partially $W^{1,1}$ velocities and applications*, Ann. Mat. Pura Appl. **183** (2003), 97–130.

[102] N. LERNER, *Transport equations with partially BV velocities*, Ann. Scuola Norm. Sup. Pisa Cl. Sci. **3** (2004), 681–703.

[103] J. L. LEWIS, *Regularity of the derivatives of solutions to certain degenerate elliptic equations*, Indiana Univ. Math. J. **32** (1983), 849–858.

[104] G. M. LIEBERMAN, *The natural generalization of the natural conditions of Ladyzhenskaya and Ural'tseva for elliptic equations*, Comm. Partial Differential Equations **16** (1991), 311–361.

[105] P.-L. LIONS and B. PERTHAME, *Propagation of moments and regularity for the 3-dimensional Vlasov-Poisson system*, Invent. Math. **105** (1991), 415–430.

[106] G. LOEPER, *On the regularity of the polar factorization for time dependent maps*, Calc. Var. Partial Differential Equations **22** (2005), 343–374.

[107] G. LOEPER, *A fully non-linear version of the incompressible Euler equations: The semi-geostrophic system*, SIAM J. Math. Anal. **38** (2006), 795–823.

[108] G. LOEPER, *Uniqueness of the solution to the Vlasov-Poisson system with bounded density*, J. Math. Pures Appl. **86** (2006), 68–79.

[109] R. J. MC CANN, *Polar factorization of maps on manifolds*, Geom. Funct. Anal. **11** (2001), 589–608.

[110] B. PERTHAME, *Time decay, propagation of low moments and dispersive effects for kinetic equations*, Comm. Partial Differential Equations **21** (1996), 659–686.

[111] K. PFAFFELMOSER, *Global classical solutions of the Vlasov-Poisson system in three dimensions for general initial data*, J. Differential Equations **95** (1992), 281–303.

[112] G. REIN, *Collisionless kinetic equations from astrophysics: the Vlasov-Poisson system*, In: "Handbook of Differential Equations: Evolutionary Equations", Handb. Differ. Equ., Elsevier/North-Holland, Amsterdam **3** (2007), 383–476.

[113] F. SANTAMBROGIO, *Models and applications of Optimal Transport in Economics, Traffic and Urban Planning*, Optimal Transportation, theory and applications, London Math. Soc. (2014).

[114] F. SANTAMBROGIO and V. VESPRI, *Continuity in two dimensions for a very degenerate elliptic equation*, Nonlinear Anal. **73** (2010), 3832–3841.

[115] O. SAVIN: *Small perturbation solutions for elliptic equations*. Comm. Partial Differential Equations **32** (2007), 557–578.

[116] J. SCHAEFFER, *Global existence for the Poisson-Vlasov system with nearly symmetric data*, J. Differential Equations **69** (1987), 111–148.

[117] E. M. STEIN, "Harmonic Analysis, Real-variable Methods, Orthogonality, and Oscillatory Integrals", With the assistance of Timothy S. Murphy. Princeton Mathematical Series, Vol. 43, Monographs in Harmonic Analysis, III. Princeton University Press, Princeton, 1993.

[118] P. TOLKSDORFF, *Regularity for a more general class of quasilinear elliptic equations*, J. Differential Equations **51** (1984), 126–150.

[119] K. UHLENBECK, *Regularity for a class of non-linear elliptic systems*, Acta Math. **138** (1977), 219–240.

[120] S. UKAI and T. OKABE, *On classical solutions in the large in time of two-dimensional Vlasov's equation*, Osaka J. Math. **15** (1978), 245–261.

[121] N. N. URALTSEVA, *Degenerate quasilinear elliptic systems*, Zap. Naučn. Sem. Leningrad. Otdel. Mat. Inst. Steklov. (LOMI) **7** (1968), 184–222.

[122] L. WANG, *Compactness methods for certain degenerate elliptic equations*, J. Differential Equations **107** (1994), 341–350.

[123] S. WOLLMAN: *Global-in-time solutions of the two-dimensional Vlasov-Poisson system*. Comm. Pure Appl. Math. **33** (1980), 173–197.

[124] C. VILLANI, "Optimal Transport, Old and New", Grundlehren des mathematischen Wissenschaften [Fundamental Principles of Mathematical Sciences], Vol. 338, Springer-Verlag, Berlin-New York, 2009.

[125] V. I. YUDOVICH, *Nonstationary flow of an ideal incompressible liquid*, Zhurn. Vych. Mat. **3** (1963), 1032–1066.

[126] X. ZHANG and J. WEI, *The Vlasov-Poisson system with infinite kinetic energy and initial data in* $L^p(\mathbb{R}^6)$. J. Math. Anal. Appl. **341** (2008), 548–558.

THESES

This series gathers a selection of outstanding Ph.D. theses defended at the Scuola Normale Superiore since 1992.

Published volumes

1. F. COSTANTINO, *Shadows and Branched Shadows of* 3 *and* 4*-Manifolds*, 2005. ISBN 88-7642-154-8
2. S. FRANCAVIGLIA, *Hyperbolicity Equations for Cusped* 3*-Manifolds and Volume-Rigidity of Representations*, 2005. ISBN 88-7642-167-x
3. E. SINIBALDI, *Implicit Preconditioned Numerical Schemes for the Simulation of Three-Dimensional Barotropic Flows*, 2007. ISBN 978-88-7642-310-9
4. F. SANTAMBROGIO, *Variational Problems in Transport Theory with Mass Concentration*, 2007. ISBN 978-88-7642-312-3
5. M. R. BAKHTIARI, *Quantum Gases in Quasi-One-Dimensional Arrays*, 2007. ISBN 978-88-7642-319-2
6. T. SERVI, *On the First-Order Theory of Real Exponentiation*, 2008. ISBN 978-88-7642-325-3
7. D. VITTONE, *Submanifolds in Carnot Groups*, 2008. ISBN 978-88-7642-327-7
8. A. FIGALLI, *Optimal Transportation and Action-Minimizing Measures*, 2008. ISBN 978-88-7642-330-7
9. A. SARACCO, *Extension Problems in Complex and CR-Geometry*, 2008. ISBN 978-88-7642-338-3
10. L. MANCA, *Kolmogorov Operators in Spaces of Continuous Functions and Equations for Measures*, 2008. ISBN 978-88-7642-336-9

11. M. LELLI, *Solution Structure and Solution Dynamics in Chiral Ytterbium(III) Complexes*, 2009. ISBN 978-88-7642-349-9
12. G. CRIPPA, *The Flow Associated to Weakly Differentiable Vector Fields*, 2009. ISBN 978-88-7642-340-6
13. F. CALLEGARO, *Cohomology of Finite and Affine Type Artin Groups over Abelian Representations*, 2009. ISBN 978-88-7642-345-1
14. G. DELLA SALA, *Geometric Properties of Non-compact CR Manifolds*, 2009. ISBN 978-88-7642-348-2
15. P. BOITO, *Structured Matrix Based Methods for Approximate Polynomial GCD*, 2011. ISBN: 978-88-7642-380-2; e-ISBN: 978-88-7642-381-9
16. F. POLONI, *Algorithms for Quadratic Matrix and Vector Equations*, 2011. ISBN: 978-88-7642-383-3; e-ISBN: 978-88-7642-384-0
17. G. DE PHILIPPIS, *Regularity of Optimal Transport Maps and Applications*, 2013. ISBN: 978-88-7642-456-4; e-ISBN: 978-88-7642-458-8
18. G. PETRUCCIANI, *The Search for the Higgs Boson at CMS*, 2013. ISBN: 978-88-7642-481-6; e-ISBN: 978-88-7642-482-3
19. B. VELICHKOV, *Existence and Regularity Results for Some Shape Optimization Problems*, 2015. ISBN: 978-88-7642-526-4; e-ISBN: 978-88-7642-527-1
20. M. RUGGIERO, *Rigid Germs, the Valuative Tree, and Applications to Kato Varieties*, 2015. ISBN: 978-88-7642-558-5 e-ISBN: 978-88-7642-559-2
21. A. BEVILACQUA, *Doubly Stochastic Models for Volcanic Hazard Assessment at Campi Flegrei Caldera*, 2016. ISBN: 978-88-7642-556-1 e-ISBN: 978-88-7642-577-6
22. M. COLOMBO, *Flows of Non-smooth Vector Fields and Degenerate Elliptic Equations with Applications to the Vlasov-Poisson and Semi-geostrophic Systems*, 2017. ISBN: 978-88-7642-606-3 e-ISBN: 978-88-7642-607-0

Volumes published earlier

H. Y. FUJITA, *Equations de Navier-Stokes stochastiques non homogènes et applications*, 1992.

G. GAMBERINI, *The minimal supersymmetric standard model and its phenomenological implications*, 1993. ISBN 978-88-7642-274-4

C. DE FABRITIIS, *Actions of Holomorphic Maps on Spaces of Holomorphic Functions*, 1994. ISBN 978-88-7642-275-1

C. PETRONIO, *Standard Spines and 3-Manifolds*, 1995. ISBN 978-88-7642-256-0

I. DAMIANI, *Untwisted Affine Quantum Algebras: the Highest Coefficient of* det H_η *and the Center at Odd Roots of 1*, 1996. ISBN 978-88-7642-285-0

M. MANETTI, *Degenerations of Algebraic Surfaces and Applications to Moduli Problems*, 1996. ISBN 978-88-7642-277-5

F. CEI, *Search for Neutrinos from Stellar Gravitational Collapse with the MACRO Experiment at Gran Sasso*, 1996. ISBN 978-88-7642-284-3

A. SHLAPUNOV, *Green's Integrals and Their Applications to Elliptic Systems*, 1996. ISBN 978-88-7642-270-6

R. TAURASO, *Periodic Points for Expanding Maps and for Their Extensions*, 1996. ISBN 978-88-7642-271-3

Y. BOZZI, *A study on the activity-dependent expression of neurotrophic factors in the rat visual system*, 1997. ISBN 978-88-7642-272-0

M. L. CHIOFALO, *Screening effects in bipolaron theory and high-temperature superconductivity*, 1997. ISBN 978-88-7642-279-9

D. M. CARLUCCI, *On Spin Glass Theory Beyond Mean Field*, 1998. ISBN 978-88-7642-276-8

G. LENZI, *The MU-calculus and the Hierarchy Problem*, 1998. ISBN 978-88-7642-283-6

R. SCOGNAMILLO, *Principal G-bundles and abelian varieties: the Hitchin system*, 1998. ISBN 978-88-7642-281-2

G. ASCOLI, *Biochemical and spectroscopic characterization of CP20, a protein involved in synaptic plasticity mechanism*, 1998. ISBN 978-88-7642-273-7

F. PISTOLESI, *Evolution from BCS Superconductivity to Bose-Einstein Condensation and Infrared Behavior of the Bosonic Limit*, 1998. ISBN 978-88-7642-282-9

L. PILO, *Chern-Simons Field Theory and Invariants of 3-Manifolds*,1999. ISBN 978-88-7642-278-2

P. ASCHIERI, *On the Geometry of Inhomogeneous Quantum Groups*, 1999. ISBN 978-88-7642-261-4

S. CONTI, *Ground state properties and excitation spectrum of correlated electron systems*, 1999. ISBN 978-88-7642-269-0

G. GAIFFI, *De Concini-Procesi models of arrangements and symmetric group actions*, 1999. ISBN 978-88-7642-289-8

N. DONATO, *Search for neutrino oscillations in a long baseline experiment at the Chooz nuclear reactors*, 1999. ISBN 978-88-7642-288-1

R. CHIRIVÌ, *LS algebras and Schubert varieties*, 2003. ISBN 978-88-7642-287-4

V. MAGNANI, *Elements of Geometric Measure Theory on Sub-Riemannian Groups*, 2003. ISBN 88-7642-152-1

F. M. ROSSI, *A Study on Nerve Growth Factor (NGF) Receptor Expression in the Rat Visual Cortex: Possible Sites and Mechanisms of NGF Action in Cortical Plasticity*, 2004. ISBN 978-88-7642-280-5

G. PINTACUDA, *NMR and NIR-CD of Lanthanide Complexes*, 2004. ISBN 88-7642-143-2

Fotocomposizione "CompoMat", Loc. Braccone, 02040 Configni (RI) Italia
Finito di stampare nel mese di aprile 2017
presso le Industrie Grafiche della Pacini Editore S.r.l.
Via A. Gherardesca, 56121 Ospedaletto, Pisa, Italia